Introduction to

Fuzzy Sets, Fuzzy Logic, and Fuzzy Control Systems

Introduction to

Fuzzy Sets, Fuzzy Logic, and Fuzzy Control Systems

Guanrong Chen
University of Houston
Houston, Texas

Trung Tat Pham
University of Houston, Clear Lake
Houston, Texas

CRC Press is an imprint of the
Taylor & Francis Group, an **informa** business

CRC Press
Taylor & Francis Group
6000 Broken Sound Parkway NW, Suite 300
Boca Raton, FL 33487-2742

First issued in paperback 2019

ISBN-13: 978-0-8493-1658-6 (hbk)
ISBN-13: 978-0-367-39788-3 (pbk)

Library of Congress Card Number 00-045431

Library of Congress Cataloging-in-Publication Data

Chen, G. (Guanrong)
Introduction to fuzzy sets, fuzzy logic, and fuzzy control systems / Guanrong Chen, Trung Tat Pham.
p. cm.
Includes bibliographical references and index.
ISBN 0-8493-1658-8 (alk. paper)
1. Soft computing. 2. Fuzzy systems. I. Pham, Trung Tat. II. Title.

QA76.9.S63 C48 2000
0063—dc21 00-045431

Visit the Taylor & Francis Web site at
http://www.taylorandfrancis.com

and the CRC Press Web site at
http://www.crcpress.com

Preface

This textbook is an enlarged and enhanced version of the authors' lecture notes used for a graduate course in fuzzy sets, fuzzy logic, fuzzy systems, and fuzzy control theories. This course has been taught for seven years at the University of Houston, with emphasis on fuzzy systems and fuzzy control, regarding both basic mathematical theories and their potential engineering applications.

The word "fuzzy" is perhaps no longer fuzzy to many engineers today. Introduced in the earlier 1970s, fuzzy systems and fuzzy control theories as an emerging technology targeting industrial applications have added a promising new dimension to the existing domain of conventional control systems engineering. It is now a common belief that when a complex physical system does not provide a set of differential or difference equations as a precise or reasonably accurate mathematical model, particularly when the system description requires certain human experience in linguistic terms, fuzzy systems and fuzzy control theories have some salient features and distinguishing merits over many other approaches.

Fuzzy control methods and algorithms, including many specialized software and hardware available on the market today, may be classified as one type of intelligent control. This is because fuzzy systems modeling, analysis, and control incorporate a certain amount of human knowledge into its components (fuzzy sets, fuzzy logic, and fuzzy rule base). Using human expertise in system modeling and controller design is not only advantageous but often necessary. Classical controller design has already incorporated human skills and knowledge: for instance, what type of controller to use and how to determine the controller structure and parameters largely depend on the decision and preference of the designer, especially when multiple choices are possible. The relatively new fuzzy control technology provides one more choice for this consideration; it has the intention to be an alternative, rather than a simple replacement, of the existing control techniques such as classical control and other intelligent control methods (e.g., neural networks, expert systems, etc.). Together, they supply systems and control engineers with a more complete toolbox to deal with the complex, dynamic, and uncertain real world. Fuzzy control technology is one of the many tools in this toolbox that is developed not only for elegant mathematical theories but, more importantly, for many practical problems with various technical challenges.

Compared with conventional approaches, fuzzy control utilizes more information from domain experts and relies less on mathematical modeling about a physical system.

On the one hand, fuzzy control theory can be quite heuristic and somewhat ad hoc. This sometimes is preferable or even desirable, particularly when low-cost and easy operations are required where mathematical rigor is not the main concern. There are many examples of this kind in industrial applications, for which fuzzy sets and fuzzy logic are easy to use. Within this context, determining a fuzzy set or a fuzzy rule base seems to be somewhat subjective, where human knowledge about the underlying physical system comes into play. However, this may not be any more subjective than selecting a suitable mathematical model in the deterministic control approach ("linear or nonlinear?" "if linear, what's the order or dimension and, yet, if nonlinear, what kind of nonlinearity?" "what kind of optimality criterion to use?" "what kind of norm for robustness measure?" etc.). It is also not much more subjective than choosing a suitable distribution function in the stochastic control approach ("Gaussian or non-Gaussian noise?" "white noise or just unknown but bounded uncertainty?" and the like). Although some of these questions can be answered on the basis of statistical analysis of available empirical data in classical control systems, the same is true for establishing an initial fuzzy rule base in fuzzy control systems.

On the other hand, fuzzy control theory can be rigorous and fuzzy controllers can have precise and analytic structures with guaranteed closed-loop system stability and some performance specifications, if such characteristics are intended. In this direction, the ultimate objective of the current fuzzy systems and fuzzy control research is appealing: the fuzzy control system technology is moving toward a solid foundation as part of the modern control theory. The trend of a rigorous approach to fuzzy control, starting from the mid-1980s, has produced many exciting and promising results. For instance, some analytic structures of fuzzy controllers, particularly fuzzy PID controllers, and their relationship with corresponding conventional controllers are much better understood today. Numerous analysis and design methods have been developed, which have turned the earlier "art" of building a working fuzzy controller to the "science" of systematic design. As a consequence, the existing analytical control theory has made the fuzzy control systems practice safer, more efficient, and more cost-effective.

This textbook represents a continuing effort in the pursuit of analytic theory and rigorous design for fuzzy control systems. More specifically, the basic notion of fuzzy mathematics (Zadeh fuzzy set theory, fuzzy membership functions, interval and fuzzy number arithmetic operations) is first studied in this text. Consequently, in a comparison with the classical two-valued logic, the fundamental concept of fuzzy logic is introduced. The ultimate goal of this course is to develop an elementary practical theory for automatic control of uncertain or imperfectly modeled systems encountered in engineering applications using fuzzy mathematics and fuzzy logic, thereby offering an alternative approach to control systems design and analysis under irregular conditions, for which conventional control systems theory may not be able to manage or well perform. Therefore, this part of the text on fuzzy mathematics and fuzzy logic is followed by the basic fuzzy systems theory (Mamdani and Takagi-Sugeno modeling, along with parameter estimation and system identification) and fuzzy control theory. Here, fuzzy control theory is introduced, first based on the developed fuzzy system modeling, along with the concepts of controllability, observability, and stability, and then based on the well-known classical Proportional-Integral-Derivative (PID) controllers theory and design methods. In particular, fuzzy PID controllers are studied in greater detail. These controllers have precise analytic structures, with rigorous analysis and guaranteed closed-loop system stability; they are comparable, and also compatible, with the classical PID controllers. To that end, fuzzy adaptive and optimal control issues are also discussed, albeit only briefly, followed by some potential industrial application examples.

The primary purpose of this course is to provide some rather systematic training for systems and control majors, both senior undergraduate and first-year graduate students, and to familiarize them with some fundamental mathematical theory and design methodology in fuzzy control systems. We have tried to make this book self-contained, so that no preliminary knowledge of fuzzy mathematics and fuzzy control systems theory is needed to understand the material presented in this textbook. Although we assume that the students are aware of the classical set theory, two-valued logic, and elementary classical control systems theory, the fundamentals of these subjects are briefly reviewed throughout for their convenience.

Some familiar terminology in the field of fuzzy control systems has become quite standard today. Therefore, as a textbook written in a classical style, we have taken the liberty to omit some personal and specialized names such as "TS fuzzy model" and "t-norm." One reason is that too many names have to be given to too many items in doing so. Nevertheless, closely related references are given at the end of each chapter for crediting and for the reader's further reading. Also, we have indicated by * in the Table of Contents those relatively advanced materials that are beyond the basic scope of the present text; they are used for reader's further studies of the subject.

It is our hope that students will benefit from this textbook in obtaining some relatively comprehensive knowledge about fuzzy control systems theory which, together with their mathematical foundations, can in a way better prepare them for the rapidly developing applied control technologies in modern industry.

The Authors

Table of Contents

1. Fuzzy Set Theory 1

I. Classical Set Theory 1
A. Fundamental Concepts 1
B. Elementary Measure Theory of Sets 3

II. Fuzzy Set Theory 5

III. Interval Arithmetic 9
A. Fundamental Concepts 9
B. Interval Arithmetic 12
C. Algebraic Properties of Interval Arithmetic 13
D. Measure Theory of Intervals 17
E. Properties of the Width of an Interval 20
F. Interval Evaluation 22
G. Interval Matrix Operations 25
H. Interval Matrix Equations and Interval Matrix Inversion 30

IV: Operations on Fuzzy Sets 37
A. Fuzzy Subsets 37
B. Fuzzy Numbers and Their Arithmetic 42

Problems 54

References 56

2. Fuzzy Logic Theory 57

I. Classical Logic Theory 58
A. Fundamental Concepts 58
B. Logical Functions of the Two-Valued Logic 61

II. The Boolean Algebra 62
A. Basic Operations of the Boolean Algebra 62
B. Basic Properties of the Boolean Algebra 63

III. Multi-Valued Logic 65
A. Three-Valued Logic 65
B. n-Valued Logic 65

IV. Fuzzy Logic and Approximate Reasoning 66

V. Fuzzy Relations 69

VI. Fuzzy Logic Rule Base 75
A. Fuzzy IF-THEN Rules 75
B. Fuzzy Logic Rule Base 77
C. Interpretation of Fuzzy IF-THEN Rules 80

D. Evaluation of Fuzzy IF-THEN Rules 82

Problems 84

References 87

3. Fuzzy System Modeling 89

I. Modeling of Static Fuzzy Systems 90
A. Fuzzy Logic Description of Input-Output Relations 90
B. Parameters Identification in Static Fuzzy Modeling 96

II. Discrete-Time Dynamic Fuzzy Systems and Their Stability Analysis 102
A. Dynamic Fuzzy Systems without Control 103
B. Dynamic Fuzzy Systems with Control 109

III. Modeling of Continuous-Time Dynamic Fuzzy Control Systems 114
A. Fuzzy Interval Partitioning 115
B. Dynamic Fuzzy System Modeling 119

IV. Stability Analysis of Continuous-Time Dynamic Fuzzy Systems 124

V. Controllability Analysis of Continuous-Time Dynamic Fuzzy Systems 129

VI. Analysis of Nonlinear Continuous-Time Dynamic Fuzzy Systems 133

Problems 136

References 138

4. Fuzzy Control Systems 139

I. Classical Programmable Logic Control 140

II. Fuzzy Logic Control (I): A General Model-Free Approach 145
A. A Closed-Loop Set-Point Tracking System 147
B. Design Principle of Fuzzy Logic Controllers 150
C. Examples of Model-Free Fuzzy Controller Design 160

III. Fuzzy Logic Control (II): A General Model-Based Approach 170

Problems 178

References 182

5. Fuzzy PID Controllers 183

I. Conventional PID Controllers: Design 183

II. Fuzzy PID Controllers Design 192
A. Fuzzy PD Controller 193
B. Fuzzy PI Controller 207
C. Fuzzy PI+D Controller 209

III. Fuzzy PID Controllers: Stability Analysis 223
A. BIBO Stability and the Small Gain Theorem 223
B. BIBO Stability of Fuzzy PD Control Systems 226
C. BIBO Stability of Fuzzy PI Control Systems 229
D. BIBO Stability of Fuzzy PI+D Control Systems 231
E. Graphical Stability Analysis of Fuzzy PID Control Systems 232

Problems 236

References 237

6. Adaptive Fuzzy Control 239

I. Fundamental Adaptive Fuzzy Control Concept 240
A. Operational Concepts 240
B. System Parameterization 242
C. Adjusting Mechanism 243
D. Guidelines for Selecting an Adaptive Fuzzy Controller 245

II. Gain Scheduling 246

III. Fuzzy Self-Tuning Regulator 252

IV. Model Reference Adaptive Fuzzy Systems 255

V. Dual Control 257

VI. Sub-Optimal Fuzzy Control 258
A. SISO Control Systems 259
B. MIMO Control Systems 260

Problems 266

References 269

7. Some Applications of Fuzzy Control 271

I. Health Monitoring Fuzzy Diagnostic Systems 271
A. Fuzzy Rule-Based Health Monitoring Expert Systems 272
B. Computer Simulations 276
C. Numerical Results 277

II. Fuzzy Control of Image Sharpness for Autofocus Cameras 281
A. Basic Image Processing Techniques 282
B. Fuzzy Control Model 283
C. Computer Simulation Results 286

III. Fuzzy Control for Servo Mechanic Systems 291
A. Fuzzy Modeling of a Servo Mechanic System 292
B. Fuzzy Controller of a Servo Mechanic System...................... 294
C. Computer Simulations and Numerical Results 295

IV. Fuzzy PID Controllers for Servo Mechanic Systems................. 300
A. Fuzzy PID Controller of a Servo Mechanic System 300
B. Adaptive Fuzzy Controller of a Servo Mechanic System 301

V. Fuzzy Controller for Robotic Manipulator................................ 302
A. Fuzzy Modeling of a 2-Link Planar Manipulator.................. 304
B. Fuzzy Controller of a 2-Link Planar Manipulator................. 307
C. Numerical Simulations... 307

Problems... 311

References 313

Index... 315

CHAPTER 1

Fuzzy Set Theory

The classical set theory is built on the fundamental concept of "set" of which an individual is either a member or not a member. A sharp, crisp, and unambiguous distinction exists between a member and a nonmember for any well-defined "set" of entities in this theory, and there is a very precise and clear boundary to indicate if an entity belongs to the set. In other words, when one asks the question "Is this entity a member of that set?" The answer is either "yes" or "no." This is true for both the deterministic and the stochastic cases. In probability and statistics, one may ask a question like "What is the probability of this entity being a member of that set?" In this case, although an answer could be like "The probability for this entity to be a member of that set is 90%," the final outcome (i.e., conclusion) is still either "it is" or "it is not" a member of the set. The chance for one to make a correct prediction as "it is a member of the set" is 90%, which does not mean that it has 90% membership in the set and in the meantime it possesses 10% non-membership. Namely, in the classical set theory, it is not allowed that an element is in a set and not in the set at the same time. Thus, many real-world application problems cannot be described and handled by the classical set theory, including all those involving elements with only partial membership of a set. On the contrary, fuzzy set theory accepts partial memberships, and, therefore, in a sense generalizes the classical set theory to some extent.

In order to introduce the concept of fuzzy sets, we first review the elementary set theory of classical mathematics. It will be seen that the fuzzy set theory is a very natural extension of the classical set theory, and is also a rigorous mathematical notion.

I. CLASSICAL SET THEORY

A. Fundamental Concepts

Let **S** be a nonempty set, called the *universe set* below, consisting of all the possible elements of concern in a particular context. Each of these elements is called a *member*, or an *element*, of **S**. A union of several (finite or infinite) members of **S** is called a *subset* of **S**. To indicate that a member s of **S** belongs to a subset S of **S**, we write

$$s \in S.$$

If s is not a member of S, we write

$$s \notin S.$$

To indicate that S is a subset of **S**, we write

$$S \subset \mathbf{S}.$$

Usually, this notation implies that S is a strictly proper subset of **S** in the sense that there is at least one member $x \in \mathbf{S}$ but $x \notin S$. If it can be either $S \subset \mathbf{S}$ or $S = \mathbf{S}$, we write

$$S \subseteq \mathbf{S}.$$

An empty subset is denoted by $\varnothing$. A subset of certain members that have properties $P_1, \ldots, P_n$ will be denoted by a capital letter, say A, as

$$A = \{\, a \mid a \text{ has properties } P_1, \ldots, P_n \,\}.$$

An important and frequently used universe set is the n-dimensional Euclidean space R^n. A subset $A \subseteq \mathrm{R}^n$ that is said to be *convex* if

$$\boldsymbol{x} = \begin{bmatrix} x_1 \\ \vdots \\ x_n \end{bmatrix} \in A \qquad \text{and} \qquad \boldsymbol{y} = \begin{bmatrix} y_1 \\ \vdots \\ y_n \end{bmatrix} \in A$$

implies

$$\lambda \boldsymbol{x} + (1-\lambda)\boldsymbol{y} \in A \qquad \text{for any } \lambda \in [0,1].$$

Let A and B be two subsets. If every member of A is also a member of B, i.e., if $a \in A$ implies $a \in B$, then A is said to be a *subset* of B. We write $A \subset B$. If both $A \subset B$ and $B \subset A$ are true, then they are *equal*, for which we write $A = B$. If it can be either $A \subset B$ or $A = B$, then we write $A \subseteq B$. Therefore, $A \subset B$ is equivalent to both $A \subseteq B$ and $A \neq B$.

The *difference* of two subsets A and B is defined by

$$A - B = \{\, c \mid c \in A \text{ and } c \notin B \,\}.$$

In particular, if $A = \mathbf{S}$ is the universe set, then $\mathbf{S} - B$ is called the *complement* of B, and is denoted by $\overline{B}$, i.e.,

$$\overline{B} = \mathbf{S} - B.$$

Obviously,

$$\overline{\overline{B}} = B, \qquad \overline{\mathbf{S}} = \varnothing, \qquad \text{and} \qquad \overline{\varnothing} = \mathbf{S}.$$

Let $r \in \mathrm{R}$ be a real number and A be a subset of R. Then the *multiplication* of r and A is defined to be

$$rA = \{\, ra \mid a \in A \,\}.$$

The *union* of two subsets A and B is defined by

$$A \cup B = B \cup A = \{\, c \mid c \in A \text{ or } c \in B \,\}.$$

Thus, we always have

$$A \cup \mathbf{S} = \mathbf{S}, \qquad A \cup \varnothing = A, \qquad \text{and} \qquad A \cup \overline{A} = \mathbf{S}.$$

The *intersection* of two subsets A and B is defined by

$$A \cap B = B \cap A = \{\, c \mid c \in A \text{ and } c \in B \,\}.$$

Obviously,

$$A \cap \mathbf{S} = A, \qquad A \cap \varnothing = \varnothing, \qquad \text{and} \qquad A \cap \overline{A} = \varnothing.$$

Two subsets A and B are said to be *disjoint* if

$$A \cap B = \varnothing.$$

Basic properties of the classical set theory are summarized in Table 1.1, where $A \subseteq \mathbf{S}$ and $B \subseteq \mathbf{S}$.

Table 1.1 Properties of Classical Set Operations

Involutive law	$\overline{\overline{A}} = A$
Commutative law	$A \cup B = B \cup A$
	$A \cap B = B \cap A$
Associative law	$(A \cup B) \cup C = A \cup (B \cup C)$
	$(A \cap B) \cap C = A \cap (B \cap C)$
Distributive law	$A \cap (B \cup C) = (A \cap B) \cup (A \cap C)$
	$A \cup (B \cap C) = (A \cup B) \cap (A \cup C)$
	$A \cup A = A$
	$A \cap A = A$
	$A \cup (A \cap B) = A$
	$A \cap (A \cup B) = A$
	$A \cup (\overline{A} \cap B) = A \cup B$
	$A \cap (\overline{A} \cup B) = A \cap B$
	$A \cup \mathbf{S} = \mathbf{S}$
	$A \cap \varnothing = \varnothing$
	$A \cup \varnothing = A$
	$A \cap \mathbf{S} = A$
	$A \cap \overline{A} = \varnothing$
	$A \cup \overline{A} = \mathbf{S}$
DeMorgan's law	$\overline{A \cap B} = \overline{A} \cup \overline{B}$
	$\overline{A \cup B} = \overline{A} \cap \overline{B}$

In order to simplify the notation throughout the rest of the book, if the universe set **S** has been specified or is not of concern, we simply call any of its subsets a *set*. Thus, we can consider two sets A and B in **S**, and if $A \subset B$ then A is called a *subset* of B.

For any set A, the *characteristic function* of A is defined by

$$\mathbf{X}_A(x) = \begin{cases} 1 & \text{if } x \in A, \\ 0 & \text{if } x \notin A. \end{cases}$$

It is easy to verify that for any two sets A and B in the universe set **S** and for any element $x \in \mathbf{S}$, we have

$$\mathbf{X}_{A \cup B}(x) = \max\{ \mathbf{X}_A(x), \mathbf{X}_B(x) \},$$
$$\mathbf{X}_{A \cap B}(x) = \min\{ \mathbf{X}_A(x), \mathbf{X}_B(x) \},$$
$$\mathbf{X}_{\overline{A}}(x) = 1 - \mathbf{X}_A(x).$$

B*. Elementary Measure Theory of Sets

In this subsection, we briefly review the basic notion of *measure* in the classical set theory which, although may not be needed throughout this book, will be useful in further studies of some advanced fuzzy mathematics.

Let **S** be the universe set and **A** a nonempty family of subsets of **S**. Let, moreover,

$$\mu: \quad \mathbf{A} \to [0,\infty]$$

be a nonnegative real-valued function defined on (subsets of) **A**, which may assume the value ∞.

A set B in **A**, denoted as an element of **A** by $B \in \mathbf{A}$, is called a *null set* with respect to μ if $\mu(B) = 0$, where

$$\mu(B) = \{ \mu(b) \mid b \in B \}.$$

μ is said to be *additive* if

$$\mu(\bigcup_{i=1}^{n} A_i) = \bigcup_{i=1}^{n} \mu(A_i)$$

for any finite collection $\{A_1,\ldots,A_n\}$ of sets in **A** satisfying both $\bigcup_{i=1}^{n} A_i \in \mathbf{A}$ and $A_i \cap A_j = \varnothing$, $i \neq j$, $i,j=1,\ldots,n$. μ is said to be *countably additive* if $n = \infty$ in the above. Moreover, μ is said to be *subtractive* if

$$A \in \mathbf{A}, \quad B \in \mathbf{A}, \quad A \subseteq B, \quad B - A \in \mathbf{A}, \quad \text{and} \quad \mu(B) < \infty$$

together imply

$$\mu(B - A) = \mu(B) - \mu(A).$$

It can be verified, however, that if μ is additive then it is also subtractive.

Now, μ is called a *measure* on **A** if it is countably additive and there is a nonempty set $C \in \mathbf{A}$ such that $\mu(C) < \infty$.

For example, if we define a function μ by $\mu(A) = 0$ for all $A \in \mathbf{A}$, then μ is a measure on **A**, which is called the *trivial measure*. As the second example, suppose that **A** contains at least one finite set and define μ by $\mu(A)$ = the number of elements belonging to **A**. Then μ is a measure on **A**, which is called the *natural measure*.

A measure μ on **A** has the following two simple properties: (i) $\mu(\varnothing) = 0$, and (ii) μ is finitely additive.

Let μ be a measure on **A**. Then a set $A \in \mathbf{A}$ is said to have a *finite measure* if $\mu(A) < \infty$, and have a *σ-finite measure* if there is a sequence $\{A_i\}$ of sets in **A** such that

$$A \subseteq \bigcup_{i=1}^{\infty} A_i \qquad \text{and} \qquad \mu(A_i) < \infty \qquad \text{for all } i = 1,2,\cdots.$$

μ is *finite* (resp., *σ-finite*) on **A** if every set in **A** has a finite (resp., σ-finite) measure.

A measure μ on **A** is said to be *complete* if

$$B \in \mathbf{A}, \qquad A \subseteq B, \qquad \text{and} \qquad \mu(B) = 0$$

together imply $\mu(A) = 0$. μ is said to be *monotone* if

$$A \in \mathbf{A}, \qquad B \in \mathbf{A}, \qquad \text{and} \qquad A \subseteq B$$

together imply

$$\mu(A) \leq \mu(B).$$

μ is said to be *subadditive* if

$$\mu(A) \leq \mu(A_1) + \mu(A_2)$$

for any $A, A_1, A_2 \in \mathbf{A}$ with $A = A_1 \cup A_2$. μ is said to be *finitely subadditive* if

$$\mu(A) \le \sum_{i=1}^{n} \mu(A_i)$$

for any finite collection $\{A, A_1, \ldots, A_n\}$ of subsets in $\mathbf{A}$ satisfying $A = \bigcup_{i=1}^{n} A_i$, and μ is said to be *countably subadditive* if $n = \infty$ in the above.

It can be shown that if μ is countably subadditive and $\mu(\varnothing) = 0$, then it is also finitely subadditive.

Let $A \in \mathbf{A}$. A measure μ on $\mathbf{A}$ is said to be *continuous from below* at A if

$$\{A_i\} \subset \mathbf{A}, \qquad A_1 \subseteq A_2 \subseteq \ldots, \qquad \text{and} \qquad \lim_{i\to\infty} A_i = A$$

together imply

$$\lim_{i\to\infty} \mu(A_i) = \mu(A),$$

and μ is said to be *continuous from above* at A if

$$\{A_i\} \subset \mathbf{A}, \qquad A_1 \supseteq A_2 \supseteq \ldots, \qquad \mu(A_1) < \infty, \qquad \text{and} \qquad \lim_{i\to\infty} A_i = A$$

together imply

$$\lim_{i\to\infty} \mu(A_i) = \mu(A).$$

μ is continuous from below (resp., above) on $\mathbf{A}$ if and only if it is continuous from below (resp., above) at every set $A \in \mathbf{A}$, and μ is said to be *continuous* if it is continuous both from below and from above (at A, or on $\mathbf{A}$).

Let $\mathbf{A}_1$ and $\mathbf{A}_2$ be families of subsets of $\mathbf{A}$ such that $\mathbf{A}_1 \subseteq \mathbf{A}_2$, and let μ_1 and μ_2 be measures on $\mathbf{A}_1$ and $\mathbf{A}_2$, respectively. μ_2 is said to be an *extension* of μ_1 if $\mu_1(A) = \mu_2(A)$ for every $A \in \mathbf{A}_1$.

For example, let $\mathbf{A} = (-\infty,\infty)$, $\mathbf{A}_1 = \{\ [a,b) \mid -\infty < a < b < \infty\ \}$, $\mathbf{A}_2$ = family of all finite, disjoint unions of bounded intervals of the form $[c,d)$, and a measure μ_1 be defined on $\mathbf{A}_1$ by

$$\mu_1([a,b)) = b - a.$$

Then μ_1 is countably additive and so is a finite measure on $\mathbf{A}_1$. This μ_1 can be extended to a finite measure μ_2 on $\mathbf{A}_2$ by defining

$$\mu_2([a,b)) = \mu_1([a,b)) \qquad \text{for all } [a,b) \in \mathbf{A}_1.$$

More generally, if f is a finite, nondecreasing, and left-continuous real-valued function of a real variable, then

$$\mu_f([a,b)) := f(b) - f(a) \qquad \text{for all } [a,b) \in \mathbf{A}_1,$$

defines a finite measure on $\mathbf{A}_1$, and it can be extended to be a finite measure μ_2 on $\mathbf{A}_2$.

II. FUZZY SET THEORY

In Section I.A, we have defined the characteristic function $\mathbf{X}_A$ of a set A by

$$\mathbf{X}_A(x) = \begin{cases} 1 & \text{if } x \in A, \\ 0 & \text{if } x \notin A, \end{cases}$$

which is an indicator of members and nonmembers of the crisp set A. In the case that an element has only partial membership of the set, we need to generalize this characteristic function to describe the membership grade of this element in the set: larger values denote higher degrees of the membership.

To give more motivation for this concept of partial membership, let us consider the following examples.

Example 1.1. Let **S** be the set of all human beings, used as the universe set, and let

$$S_f = \{ s \in \mathbf{S} \mid s \text{ is old} \}.$$

Then S_f is a "fuzzy subset" of **S** because the property "old" is not well defined and cannot be precisely measured: given a person who is 40 year old, it is not clear if this person belongs to the set S_f. Thus, to make the subset S_f well-defined, we have to quantify the concept "old," so as to characterize the subset S_f in a precise and rigorous way.

For the time being, let us say, we would like to describe the concept "old" by the curve shown in Figure 1.1(a) using common sense, where the only people who are considered to be "absolutely old" are those 120 years old or older, and the only people who are considered to be "absolutely young" are those newborns. Meanwhile, all the other people are old as well as young, depending on their actual ages. For example, a person 40 years old is considered to be "old" with "degree 0.5" and at the same time also "young" with "degree 0.5" according to the measuring curve that we used. We cannot exclude this person from the set S_f described above, nor include him completely. Thus, the curve that we introduce in Figure 1.1(a) establishes a mathematical measure for the "oldness" of a human being, and hence can be used to define the partial membership of any person relative to the subset S_f described above. The curve shown in Figure 1.1(a), which is indeed a generalization of the classical characteristic function $\mathbf{X}_{S_f}$ (it can be used to conclude a person who either "is" or "is not" a member of the subset S_f), is called a *membership function* associated with the subset S_f.

Of course, one may also use the piecewise linear membership function shown in Figure 1.1(b) to describe the same concept of oldness for the same subset S_f, depending on whichever is more meaningful and more convenient in one's concern, where both are reasonable and acceptable in common sense. The reader may suggest many more good candidates for such a membership function for the subset S_f described above. There is yet no fixed, unique, and universal rule or criterion for selecting a membership function for a particular "fuzzy subset" in general: a correct and good membership function is determined by the user based on his scientific knowledge, working experience, and actual need for the particular application in question. This selection is more or less subjective, but the situation is just like in the classical probability theory and statistics where if one says "we assume that the noise is Gaussian and white," what he uses to start with all the rigorous mathematics is a subjective hypothesis that may not be very true, simply because the noise in question may not be exactly Gaussian and may not be perfectly white. Using the same approach, we can say, "we assume that the membership function that

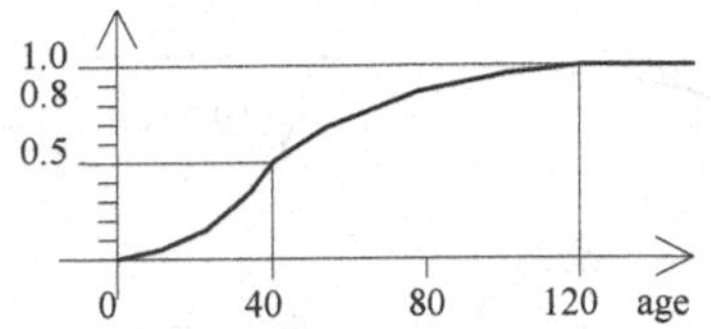

Figure 1.1(a) An example of membership functions.

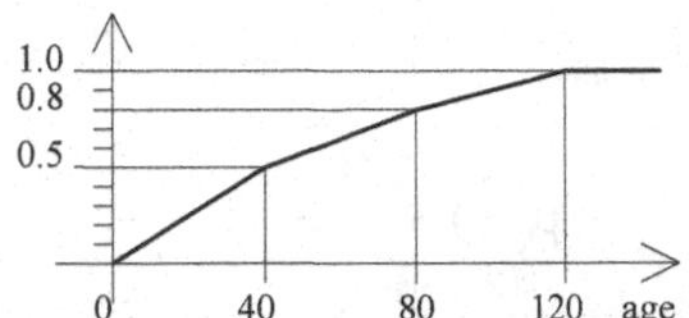

Figure 1.1(b) Another example of membership functions.

describes the oldness is the one given in Figure 1.1(a)," to start with all the rigorous mathematics in the rest of the investigation.

The fuzzy set theory is taking the same logical approach as what people have been doing with the classical set theory: in the classical set theory, as soon as the two-valued characteristic function has been defined and adopted, rigorous mathematics follows; in the fuzzy set case, as soon as a multi-valued characteristic function (the membership function) has been chosen and fixed, a rigorous mathematical theory can be fully developed.

Now, we return to the subset S_f introduced above. Suppose that the membership function associated with it, say the one shown in Figure 1.1(a), has been chosen and fixed. Then, this subset S_f along with the membership function used, which we will denote by $\mu_{S_f}(s)$ with $s \in S_f$, is called a *fuzzy subset* of the universe set **S**. A fuzzy subset thus consists of two components: a subset and a membership function associated with it. This is different from the classical set theory, where all sets and subsets share the same (and the unique) membership function: the two-valued characteristic function mentioned above.

Throughout this book, if no confusion would arise, we will simply call a fuzzy subset a *fuzzy set*, keeping in mind that it has to be a subset of some universe set and has to have a pre-described membership function associated with it.

To familiarize this new concept, let us now discuss one more example.

Example 1.2. Let **S** be the (universe) set of all real numbers, and let

$$S_f = \{ s \in \mathbf{S} \mid s \text{ is positive and large} \}.$$

This subset, S_f, is not well-defined in the classical set theory because, although the statement "s is positive" is precise, the statement "s is large" is vague. However, if we introduce a membership function that is reasonable and meaningful for a particular application for the characterization or measure of the property "large," say the one shown in Figure 1.2 quantified by the function

$$\mu_{S_f}(s) = \begin{cases} 0 & \text{if } s \le 0, \\ 1 - e^{-s} & \text{if } s > 0, \end{cases}$$

then the fuzzy subset S_f, associated with this membership function $\mu_{S_f}(s)$, is well defined.

Similarly, a membership function for the subset

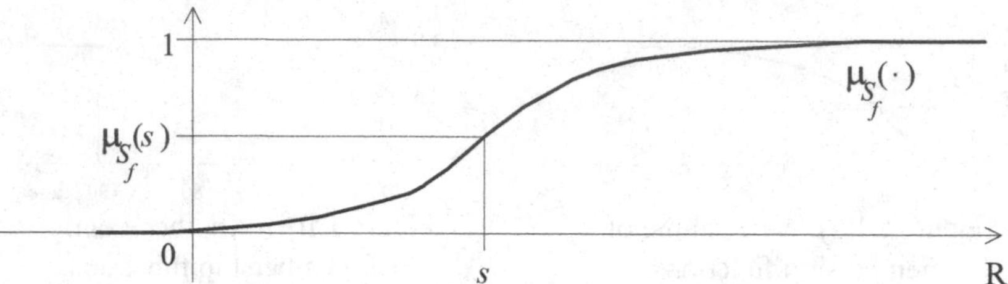

Figure 1.2 A membership function for a positive and large real number.

$$\widetilde{S}_f = \{ s \in \mathbf{S} \mid |s| \text{ is small} \}$$

may be chosen to be the one shown in Figure 1.3, where the cutting edge E is determined by the user according to his concern in the application.

Other commonly used membership functions for fuzzy sets that are convenient in various applications are shown in Figure 1.4, where we have normalized their maximum value to be 1, as usual, since 1 = 100% describes a full membership and is convenient to use.

Obviously, a membership function is a nonnegative-valued function, which differs from the probability density functions in that the area under the curve of a membership function need not be equal to unity (in fact, it can be any value between 0 and ∞, including 0 and ∞). Another distinction between the fuzzy set theory and the classical one (actually, the entire theory of classical mathematics) is that a member of a fuzzy set may assume two or more (even conflicting) membership values. For example, if we use the two membership functions shown in Figure 1.5 to measure "positive and large" and "negative and small," respectively, then a member $s = 0.1$ has the first membership value 0.095 and the second 0.08: they do not sum up to 1.0 nor cancel out to be 0. Moreover, the two concepts are conflicting: s is positive and in the meantime negative, a situation just like someone is old and also is young, which classical mathematics cannot accept. Such a vague and conflicting description of a fuzzy set is acceptable by the fuzzy mathematics, however, which turns out to be very useful in many real-world applications. More importantly, the use of conflicting membership functions like this will not cause any logical or mathematical problems in the consequence, provided that a correct approach is taken in the sequel.

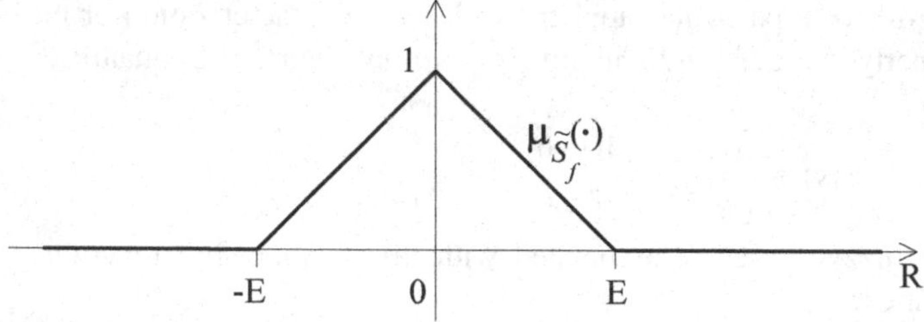

Figure 1.3 A membership function for a real number of small magnitude.

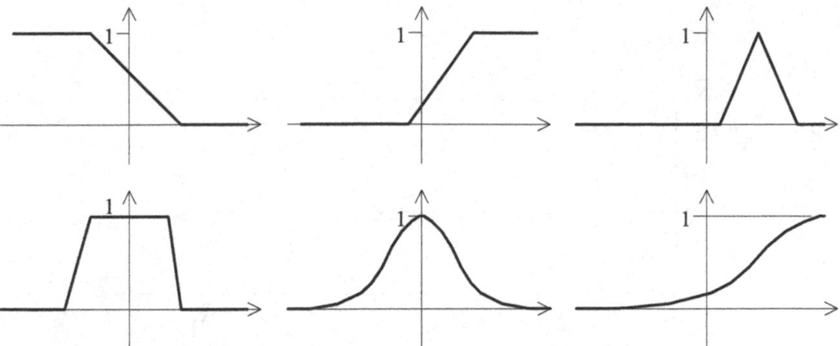

Figure 1.4 Various shapes of commonly used membership functions.

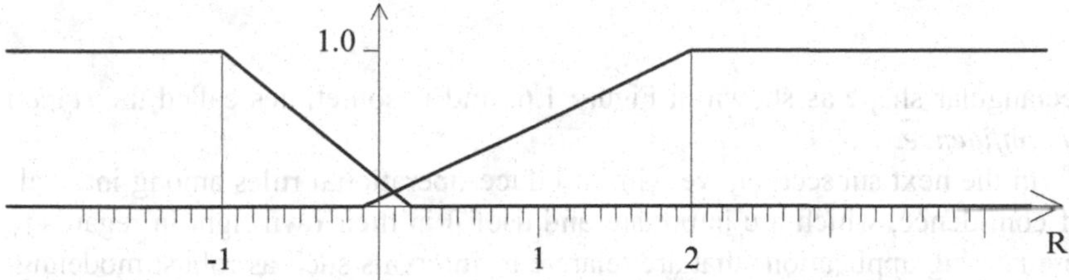

Figure 1.5 The real number $s = 0.1$ is both "positive large" and "negative small" at the same time.

III. INTERVAL ARITHMETIC

In the last section, we introduced the concept of *fuzzy (sub)sets*, which consists of two parts: a (sub)set defined in the classical sense and a membership function defined on the (sub)set that is also defined in the ordinary sense. In this section, we first study some fundamental properties and operation rules pertaining to a special yet important kind of sets – intervals – and then in the next section, we will study properties and operations of membership functions defined on intervals.

A. Fundamental Concepts

Our concern here is the situation where the value of a member s of a set is uncertain. We assume, however, that the information on the uncertain value of s provides an acceptable range:

$$\underline{s} \leq s \leq \overline{s},$$

where $[\,\underline{s}\,,\ \overline{s}\,] \subset R$ is called the *interval of confidence* about the values of s. As a special case, when $\underline{s} = \overline{s}$, we have the certainty of confidence $[\,\underline{s}\,,\ \overline{s}\,] = [s\,,\ s\,] = s$. We mainly study closed intervals in this book; so an interval will always mean a closed and bounded interval throughout, unless otherwise indicated. In the two-dimensional case, an interval of confidence has a

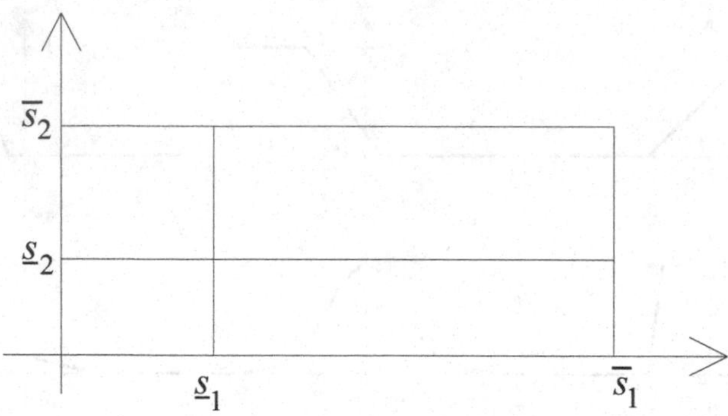

Figure 1.6 An interval of confidence in the two-dimensional case.

rectangular shape as shown in Figure 1.6, and is sometimes called the *region of confidence.*

In the next subsection, we will introduce operational rules among intervals of confidence, which are important and useful in their own right in regards to engineering applications that are relative to intervals such as robust modeling, robust stability, and robust control. To prepare for that, we first give the following definitions.

Definition 1.1.

(a) *Equality*: Two intervals $[\underline{s}_1, \overline{s}_1]$ and $[\underline{s}_2, \overline{s}_2]$ are said to be equal:

$$[\underline{s}_1, \overline{s}_1] = [\underline{s}_2, \overline{s}_2]$$

if and only if $\underline{s}_1 = \underline{s}_2$ and $\overline{s}_1 = \overline{s}_2$.

(b) *Intersection*: The *intersection* of two intervals $[\underline{s}_1, \overline{s}_1]$ and $[\underline{s}_2, \overline{s}_2]$ is defined to be

$$[\underline{s}_1, \overline{s}_1] \cap [\underline{s}_2, \overline{s}_2] = [\max\{\underline{s}_1, \underline{s}_2\}, \min\{\overline{s}_1, \overline{s}_2\}]$$

and

$$[\underline{s}_1, \overline{s}_1] \cap [\underline{s}_2, \overline{s}_2] = \varnothing$$

if and only if $\underline{s}_1 > \overline{s}_2$ or $\underline{s}_2 > \overline{s}_1$.

(c) *Union*: The *union* of two intervals $[\underline{s}_1, \overline{s}_1]$ and $[\underline{s}_2, \overline{s}_2]$ is defined to be

$$[\underline{s}_1, \overline{s}_1] \cup [\underline{s}_2, \overline{s}_2] = [\min\{\underline{s}_1, \underline{s}_2\}, \max\{\overline{s}_1, \overline{s}_2\}],$$

provided that $[\underline{s}_1, \overline{s}_1] \cap [\underline{s}_2, \overline{s}_2] \neq \varnothing$. Otherwise, it is undefined (since the result is not an interval).

(d) *Inequality*: Interval $[\underline{s}_1, \overline{s}_1]$ is said to be *less than* (resp., *greater than*) interval $[\underline{s}_2, \overline{s}_2]$, denoted by

$$[\underline{s}_1, \overline{s}_1] < [\underline{s}_2, \overline{s}_2] \qquad (\text{resp., } [\underline{s}_1, \overline{s}_1] > [\underline{s}_2, \overline{s}_2])$$

if and only if $\overline{s}_1 < \underline{s}_2$ (resp., $\underline{s}_1 > \overline{s}_2$). Otherwise, they cannot be compared. Note that the relations $\leq$ and $\geq$ are not defined for intervals.

(e) *Inclusion*: The interval $[\underline{s}_1, \overline{s}_1]$ is said to be included in $[\underline{s}_2, \overline{s}_2]$, denoted by

$$[\underline{s}_1, \overline{s}_1] \subseteq [\underline{s}_2, \overline{s}_2]$$

if and only if $\underline{s}_2 \leq \underline{s}_1$ and $\overline{s}_1 \leq \overline{s}_2$. This is equivalent to saying that the interval $[\underline{s}_1, \overline{s}_1]$ is a *subset* or *subinterval* of $[\underline{s}_2, \overline{s}_2]$.

(f) *Width*: The *width* of an interval $[\underline{s}, \overline{s}]$ is defined to be

$$\mathrm{w}\{[\underline{s}, \overline{s}]\} = \overline{s} - \underline{s}.$$

Hence, a singleton $s = [s,s]$ has a width zero: $\mathrm{w}\{s\} = \mathrm{w}\{[s,s]\} = 0$, for all $s \in R$.

(g) *Absolute Value*: The *absolute value* of an interval $[\underline{s}, \overline{s}]$ is defined to be

$$|[\underline{s}, \overline{s}]| = \max\{|\underline{s}|, |\overline{s}|\}.$$

Thus, the absolute value of a singleton $s = [s,s]$ is its usual absolute value: $|[s,s]| = |s|$ for all $s \subset R$.

(h) *Midpoint* (*mean*): The *midpoint* (or *mean*) of an interval $[\underline{s}, \overline{s}]$ is defined to be

$$\mathrm{m}\{[\underline{s}, \overline{s}]\} = \frac{1}{2}(\underline{s} + \overline{s}).$$

(i) *Symmetry*: Interval $[\underline{s}, \overline{s}]$ is said to be *symmetric* if and only if

$$\underline{s} = -\overline{s} \qquad \text{or} \qquad \mathrm{m}\{[\underline{s}, \overline{s}]\} = 0.$$

Example 1.3. For three given intervals, $S_1 = [-1,0]$, $S_2 = [-1,2]$, and $S_3 = [2,10]$, we have

$$
\begin{aligned}
&S_1 \cap S_2 = [-1,0] \cap [-1,2] = [-1,0],\\
&S_1 \cap S_3 = [-1,0] \cap [2,10] = \varnothing,\\
&S_2 \cap S_3 = [-1,2] \cap [2,10] = [2,2] = 2,\\
&S_1 \cup S_2 = [-1,0] \cup [-1,2] = [-1,2],\\
&S_1 \cup S_3 = [-1,0] \cup [2,10] = \text{undefined},\\
&S_2 \cup S_3 = [-1,2] \cup [2,10] = [-1,10],\\
&S_1 = [-1,0] < [2,10] = S_3,\\
&S_1 = [-1,0] \subset [-1,2] = S_2,\\
&\mathrm{w}\{S_1\} = \mathrm{w}\{[-1,0]\} = 0 - (-1) = 1,\\
&\mathrm{w}\{S_2\} = \mathrm{w}\{[-1,2]\} = 2 - (-1) = 3,\\
&\mathrm{w}\{S_3\} = \mathrm{w}\{[2,10]\} = 10 - 2 = 8,\\
&|S_1| = |[-1,0]| = \max\{|-1|, |0|\} = 1,\\
&|S_2| = |[-1,2]| = \max\{|-1|, |2|\} = 2,\\
&|S_3| = |[2,10]| = \max\{|2|, |10|\} = 10,\\
&\mathrm{m}\{S_1\} = \mathrm{m}\{[-1,0]\} = \frac{1}{2}(-1+0) = -\frac{1}{2},
\end{aligned}
$$

$$m\{S_2\} = m\{\,[-1,2]\,\} = \frac{1}{2}(-1+2) = \frac{1}{2},$$

$$m\{S_3\} = m\{\,[2,10]\,\} = \frac{1}{2}(2+10) = 6.$$

B. Interval Arithmetic

Let $[\,\underline{s}\,,\,\overline{s}\,]$, $[\,\underline{s}_1, \overline{s}_1\,]$, and $[\,\underline{s}_2, \overline{s}_2\,]$ be intervals. The basic arithmetic of intervals is defined as follows.

Definition 1.2.

(1) *Addition.*

$$[\,\underline{s}_1, \overline{s}_1\,] + [\,\underline{s}_2, \overline{s}_2\,] = [\,\underline{s}_1 + \underline{s}_2,\ \overline{s}_1 + \overline{s}_2\,].$$

(2) *Subtraction.*

$$[\,\underline{s}_1, \overline{s}_1\,] - [\,\underline{s}_2, \overline{s}_2\,] = [\,\underline{s}_1 - \overline{s}_2\,,\ \overline{s}_1 - \underline{s}_2\,].$$

(3) *Reciprocal.*

If $0 \notin [\,\underline{s}\,,\,\overline{s}\,]$ then $[\,\underline{s}\,,\,\overline{s}\,]^{-1} = [\,1/\overline{s}\,,\,1/\underline{s}\,]$;
if $0 \in [\,\underline{s}\,,\,\overline{s}\,]$ then $[\,\underline{s}\,,\,\overline{s}\,]^{-1}$ is undefined.

(4) *Multiplication.*

$$[\,\underline{s}_1, \overline{s}_1\,] \cdot [\,\underline{s}_2, \overline{s}_2\,] = [\,\underline{p},\ \overline{p}\,].$$

Here

$$\underline{p} = \min\{\,\underline{s}_1\underline{s}_2,\ \underline{s}_1\overline{s}_2\,,\ \overline{s}_1\underline{s}_2,\ \overline{s}_1\,\overline{s}_2\,\},$$

$$\overline{p} = \max\{\,\underline{s}_1\underline{s}_2,\ \underline{s}_1\overline{s}_2\,,\ \overline{s}_1\underline{s}_2,\ \overline{s}_1\,\overline{s}_2\,\}.$$

(5) *Division.*

$$[\,\underline{s}_1, \overline{s}_1\,] \,/\, [\,\underline{s}_2, \overline{s}_2\,] = [\,\underline{s}_1, \overline{s}_1\,] \cdot [\,\underline{s}_2, \overline{s}_2\,]^{-1},$$

provided that $0 \notin [\,\underline{s}_2, \overline{s}_2\,]$.

Here, it is very important to note that interval arithmetic intends to obtain an interval as the result of an operation such that the resulting interval contains all possible solutions. Therefore, all these operational rules are defined in a conservative way in the sense that it intends to make the resulting interval as large as necessary to avoid loosing any true solution. For example, $[1,2]-[0,1]=[0,2]$ means for any $a \in [1,2]$ and any $b \in [0,1]$, it is guaranteed that $a-b \in [0,2]$.

It is also very important to point out that the conservatism may cause some unusual results that could seem to be inconsistent with the ordinary numerical solutions. For instance, according to the subtraction rule (2), we have $[1,2] - [1,2] = [-1,1] \neq [0,0] = 0$. The result $[-1,1]$ here contains 0, but not only 0. The reason is that there can be other possible solutions: if we take 1.5 from the first interval and 1.0 from the second, then the result is 0.5 rather than 0; and 0.5 is indeed contained in $[-1,1]$. Thus, an interval subtract itself is equal to zero (a point) only if this interval is itself a point (a trivial interval). In general, we have the following:

For any interval Z,

$$Z - Z = 0; \qquad \text{or} \qquad Z / Z = I \ (0 \notin Z)$$

only if $w\{Z\} = 0$, i.e., $Z = [z,z]$ is a point, where $I = [1,1]$.

Associated with this is the following:

For any intervals X, Y, and Z,

$$X + Z = Y + Z \Rightarrow X = Y.$$

Moreover, we have the following:

For any interval Z, with $0 \in Z$,

$$Z^2 = Z \cdot Z = [\underline{z}, \overline{z}] \cdot [\underline{z}, \overline{z}] = [\underline{p}, \overline{p}],$$

where

$$\underline{p} = \min\{\underline{z}^2, \underline{z}\overline{z}, \overline{z}^2\} = \underline{z}\overline{z},$$
$$\overline{p} = \max\{\underline{z}^2, \underline{z}\overline{z}, \overline{z}^2\} = \max\{\underline{z}^2, \overline{z}^2\}.$$

It should be noted that this is consistent with the definition of interval multiplication, Definition 1.2 (4). However, if one changed it to the following:

$$Z^2 = ([\underline{z}, \overline{z}])^2 = \{z^2 \mid z \in Z\} = [0, \max\{\underline{z}^2, \overline{z}^2\}],$$

then it would be more natural in the sense that a square is always nonnegative, but it is not consistent with the interval multiplication definition. Observe, moreover, that if we take a negative number from the first interval and a positive one from the second to multiply, a negative number does result. Therefore, we will use the first square rule shown above, although, oftentimes, it gives a more conservative result in an interval operation involving interval squares.

For three intervals, $X = [\underline{x}, \overline{x}]$, $Y = [\underline{y}, \overline{y}]$, and $Z = [\underline{z}, \overline{z}]$, if we consider the interval operations of addition (+), subtraction (–), multiplication (×), and division (/) to be (set-variable and set-valued) *functions*, namely,

$$Z = f(X,Y) = X * Y, \qquad * \in \{+, -, \cdot, /\},$$

then it can be verified that all these four functions are continuous on compact sets such as intervals (see Section III-D below). Each function, $f(X,Y)$, therefore, assumes a maximum and a minimum value, as well as all values in between, on any (closed and bounded) interval. Thus $X * Y$ is, again, a (closed and bounded) interval. The set of intervals is therefore closed under the four operations $\{+, -, \cdot, /\}$ defined above.

It is also clear that the real numbers $x, y, z, \ldots$. are isomorphic to intervals of the form $[x,x]$, $[y,y]$, $[z,z]$, ... For this reason, we will simplify the notation $[x,x] * Y$ of a point-interval operation to $x * Y$. On the other hand, the multiplication symbol "·" will often be dropped, and the division symbol "/" may sometimes be replaced by "÷" for convenience.

We collect together the most important and useful interval operation rules in the next subsection.

C. Algebraic Properties of Interval Arithmetic

Let X, Y, and Z be intervals. We first have the following simple but important rules.

Theorem 1.1. The addition and multiplication operations of intervals are commutatitve and associative but not distributive. More precisely:

(1) $X+Y=Y+X$;

(2) $Z+(X+Y)=(Z+X)+Y$;

(3) $Z(XY)=(ZX)Y$;

(4) $XY=YX$;

(5) $Z+\mathbf{0}=\mathbf{0}+Z=Z$ and $Z\mathbf{0}=\mathbf{0}Z=\mathbf{0}$, where $\mathbf{0}=[0,0]$;

(6) $ZI=IZ=Z$, where $I=[1,1]$;

(7) $Z(X+Y)\neq ZX+ZY$, except when

(a) $Z=[z,z]$ is a point; or

(b) $X=Y=\mathbf{0}$; or

(c) $xy\geq 0$ for all $x\in X$ and $y\in Y$.

In general, we only have the subdistributive law:

$$Z(X+Y)\subseteq ZX+ZY.$$

Proof. For (1) and (4), let $*\in\{+,\cdot\}$. Then

$$\begin{aligned} X*Y &= \{x*y\,|\,x\in X, y\in Y\}\\ &= \{y*x\,|\,y\in Y, x\in X\}\\ &= Y*X. \end{aligned}$$

For (2) and (3), let $*\in\{+,\cdot\}$. Then

$$\begin{aligned} Z*(X*Y) &= \{z*a\,|\,z\in Z, a\in X*Y\}\\ &= \{z*(x*y)\,|\,z\in Z, x\in X, y\in Y\}\\ &= \{(z*x)*y\,|\,z\in Z, x\in X, y\in Y\}\\ &= \{b*y\,|\,b\in Z*X, y\in Y\}\\ &= (Z*X)*Y. \end{aligned}$$

For (5) and (6), let $*\in\{+,\cdot\}$. Then

$$\begin{aligned} Z*\mathbf{0} &= \{z*0\,|\,z\in Z, 0\in\mathbf{0}\}\\ &= \{0*z\,|\,0\in\mathbf{0}, z\in Z\}\\ &= \mathbf{0}*Z.\\ Z\cdot I &= \{z\cdot 1\,|\,z\in Z, 1\in I\}\\ &= \{1\cdot z\,|\,1\in I, z\in Z\}\\ &= \{z\,|\,z\in Z\}\\ &= Z. \end{aligned}$$

For (7):

(a)
$$\begin{aligned} z(X+Y) &= \{z\cdot a\,|\,a\in X+Y\}\\ &= \{z\cdot(x+y)\,|\,x\in X, y\in Y\}\\ &= \{z\cdot x+z\cdot y\,|\,x\in X, y\in Y\}\\ &= zX+zY. \end{aligned}$$

(b)
$$\begin{aligned} Z(\mathbf{0}+\mathbf{0}) &= Z\mathbf{0}\\ &= \mathbf{0}\\ &= Z\mathbf{0}+Z\mathbf{0} \quad \text{by (5)} \end{aligned}$$

(c) Without loss of generality, we only consider the case where $\underline{x} \geq 0$ and $\underline{y} \geq 0$ in $X = [\,\underline{x},\ \overline{x}\,]$ and $Y = [\,\underline{y},\ \overline{y}\,]$. If $\underline{z} \geq 0$, then we have

$$Z(X+Y) = [\,\underline{z}(\underline{x}+\underline{y}),\ \overline{z}\,(\,\overline{x}+\overline{y}\,)]$$

and

$$\begin{aligned} ZX+ZY &= [\,\underline{z}\,\underline{x},\ \overline{z}\,\overline{x}\,]+[\,\underline{z}\,\underline{y},\ \overline{z}\,\overline{y}\,] \\ &= [\,\underline{z}(\underline{x}+\underline{y}),\ \overline{z}(\,\overline{x}+\overline{y})\,], \end{aligned}$$

i.e., the equality holds. If $\overline{z} \leq 0$ then by considering $-Z$ we have the same situation and result. Now, if $\underline{z}\,\overline{z} < 0$, then we have

$$Z(X+Y) = [\,\underline{z}(\,\overline{x}+\overline{y}\,),\ \overline{z}\,(\,\overline{x}+\overline{y}\,)]$$

and

$$\begin{aligned} ZX+ZY &= [\,\underline{z}\,\overline{x},\ \overline{z}\,\overline{x}\,]+[\,\underline{z}\,\overline{y},\ \overline{z}\,\overline{y}\,] \\ &= [\,\underline{z}(\,\overline{x}+\overline{y}),\ \overline{z}(\,\overline{x}+\overline{y})\,], \end{aligned}$$

which proves the final case.

We note that in case (7), we do not even have the distributive law $Z(x+y) = Zx + Zy$ for points x and y. We also note that more conditions under which this equality holds can be found in Theorem 1.3. A counterexample for the distributive law is the following.

Example 1.4. Let $Z = [1,2]$, $X = I = [1,1]$, and $Y = -I = [-1,-1]$. Then we have

$$\begin{aligned} Z(X+Y) &= [1,2](I-I) = [1,2]\cdot\mathbf{0} = \mathbf{0}; \\ ZX+ZY &= [1,2]\cdot[1,1]+[1,2]\cdot[-1,-1] = [-1,1] \neq \mathbf{0}. \end{aligned}$$

A more general rule for interval arithmetic operations is the following fundamental *monotonic inclusion law*.

Theorem 1.2. Let X_1, X_2, Y_1, and Y_2 be intervals such that

$$X_1 \subseteq Y_1 \qquad \text{and} \qquad X_2 \subseteq Y_2.$$

Then for the operations $* \in \{+, -, \cdot, /\}$, we have

$$X_1 * X_2 \subseteq Y_1 * Y_2.$$

Proof. Since $X_1 \subseteq Y_1$ and $X_2 \subseteq Y_2$, it follows that

$$\begin{aligned} X_1 * X_2 &= \{\, x_1 * x_2 \mid x_1 \in X_1, x_2 \in X_2 \,\} \\ &\subseteq \{\, y_1 * y_2 \mid y_1 \in Y_1, y_2 \in Y_2 \,\} \\ &= Y_1 * Y_2. \end{aligned}$$

In particular, for any interval Z we have

$$0 \in Z - Z \qquad \text{and} \qquad 1 \in Z/Z \qquad (0 \notin Z),$$

as discussed before.

A particularly useful special case of the theorem is the following.

Corollary 1.1. Let X and Y be intervals with $x \in X$ and $y \in Y$. Then

$$x * y \in X * Y$$

for any $* \in \{+, -, \cdot, /\}$.

The following properties can be easily verified by following the method of proof of Theorem 1.1.

Theorem 1.3. Let X, Y, and Z be symmetric intervals in the sense that the means $m\{X\} = 0$, $m\{Y\} = 0$, and $m\{Z\} = 0$. Then

(1) $|Z| = \frac{1}{2} w\{Z\}$;

(2) $Z = |Z|[-1,1]$;

(3) $X + Y = X - Y = (|X| + |Y|)[-1,1]$;

(4) $XY = |X||Y|[-1,1]$;

(5) $Z(X \pm Y) = ZX \pm ZY = |Z|(|X| + |Y|)[-1,1]$;

(6) $Z = m\{Z\} + \frac{1}{2} w\{Z\}[-1,1]$;

(7) if X and Y are symmetric but Z is arbitrary, then
  (a) $ZX = |Z|X$; and
  (b) $Z(X + Y) = ZX + ZY$.

We note that the equality $X + Y = X - Y$ in (3) seems to be very strange, but it is true for symmetric intervals X and Y, which can be verified as follows:

$$\begin{aligned} X - Y &= [\underline{x}, \overline{x}] - [\underline{y}, \overline{y}] = [\min\{\underline{x}-\underline{y}, \underline{x}-\overline{y}\}, \max\{\overline{x}-\underline{y}, \overline{x}-\overline{y}\}] \\ &= [\min\{\underline{x}+|y|, \underline{x}-|y|\}, \max\{\overline{x}+|y|, \overline{x}-|y|\}] \\ &= [\underline{x}+\underline{y}, \overline{x}+\overline{y}] \\ &= X + Y, \end{aligned}$$

where one should recall that $Y = [-|y|, |y|]$ because it is symmetric.

We close this subsection by discussing the problem of solving the *interval equation*

$$AX = B,$$

where A and B are both given intervals with $0 \notin A$, and X is to be determined.

Theorem 1.4. Let X be a solution of the interval equation

$$AX = B, \qquad 0 \notin A.$$

Then $X \subseteq B / A$.

Proof. For any $x \in X$, there exist $a \in A$ and $b \in B$ such that $ax = b$. Hence, $x = b/a \in B / A$ since $0 \notin A$.

Note, however, that even if B/A is undefined, the interval equation $AX = B$ may still have a solution. For example, the equation

$$[-1/3,1]\, X = [-1,2]$$

has $0 \in [-1/3,1]$, so that $[-1,2]/[-1/3,1]$ is undefined. Yet $X = [-1,2]$ is its unique solution. This can be verified by direct multiplication $[-1/3,1] \times [-1,2]$, which gives $[-1,2]$. The uniqueness is verified by examining all possible solutions for $X = [\underline{x}, \overline{x}]$ that satisfy

$$[-1/3,1] \times [\underline{x}, \overline{x}] = [-1,2]$$

via multiplication operations.

Note also that the above interval equation may not have a solution at all even if $0 \notin A$. One example is $[1,2]X = [2,3]$.

The matrix equation $AX = B$ and its solvability will be further discussed in Section III.H.

D. Measure Theory of Intervals

In this subsection, we only introduce the notion of *distance* in the concern of measure theory for a family of intervals.

Definition 1.3. Let $X = [\,\underline{x},\ \overline{x}\,]$ and $Y = [\,\underline{y},\ \overline{y}\,]$ be intervals. The *distance* between X and Y is defined by

$$d(X,Y) = \max\{\, |\underline{x} - \underline{y}|,\ |\overline{x} - \overline{y}|\,\}$$

It can be verified that the set-variable function $d(\cdot,\cdot)$ satisfies the following properties:

(1) $d(X,Y) \geq 0$, and $d(X,Y) = 0$ if and only if $X = Y$;

(2) $d(X,Y) \leq d(X,Z) + d(Z,Y)$ for any interval Z (the *triangular inequality*).

The triangular inequality can be verified as follows:

$$\begin{aligned} & d(X,Z) + d(Z,Y) \\ &\quad = \max\{\, |\underline{x} - \underline{z}|,\ |\overline{x} - \overline{z}|\,\} + \max\{\, |\underline{z} - \underline{y}|,\ |\overline{z} - \overline{y}|\,\} \\ &\quad \geq \max\{\, |\underline{x} - \underline{z}| + |\underline{z} - \underline{y}|,\ |\overline{x} - \overline{z}| + |\overline{z} - \overline{y}|\,\} \\ &\quad \geq \max\{\, |\underline{x} - \underline{y}|,\ |\overline{x} - \overline{y}|\,\} \\ &\quad = d(X,Y). \end{aligned}$$

For real numbers x and y, this distance reduces to the standard one:

$$d(\,[x,x],\ [y,y]\,) = |x - y|.$$

It can also be verified that the interval distance function $d(\cdot,\cdot)$ defined here induces a *metric*, a *Hausdorff metric*, which is a generalization of the distance between two singleton points in a usual metric space. In fact, for any two nonempty compact sets X and Y of real numbers, including intervals, the *Hausdorff distance* is defined by

$$h(X,Y) = \max\{\ \sup_{y\in Y} \inf_{x\in X} d(x,y),\ \sup_{x\in X} \inf_{y\in Y} d(x,y)\ \}.$$

The introduction of a metric into the family of intervals, denoted **I**, makes it a metric space. Thus, the concepts of convergence and continuity can be defined and used in the standard way.

Definition 1.4. Let $\{X_n\}_{n=1}^{\infty}$ be a sequence of intervals in **I**. This sequence is said to be convergent to an interval $X \in \mathbf{I}$ if the sequences of the upper and lower bounds of the individual members of the sequence of intervals converge to the corresponding bounds of $X = [\,\underline{x},\ \overline{x}\,]$, namely,

$$\lim_{n\to\infty} \underline{x}_n = \underline{x} \quad \text{and} \quad \lim_{n\to\infty} \overline{x}_n = \overline{x},$$

where $X_n = [\underline{x}_n, \overline{x}_n\,]$. In this case, we write

$$\lim_{n\to\infty} X_n = X.$$

Then, it can be verified by a standard argument that every Cauchy sequence of intervals converges to an interval in **I**, that is, we have the following result.

Theorem 1.5. The topological space **I**, when equipped with the metric defined by d(×,×), is a complete metric space (**I**,d).

The next result characterizes the convergence behavior of an important class of interval sequences.

Theorem 1.6. Let $\{X_n\}_{n=1}^{\infty}$ be a sequence of intervals such that

$$X_1 \supseteq X_2 \supseteq X_3 \supseteq \ldots$$

Then $\lim_{n\to\infty} X_n = X$, where

$$X = \bigcap_{n=1}^{\infty} X_n .$$

Proof. Consider the sequence of bounds

$$\underline{x}_1 \le \underline{x}_2 \le \underline{x}_3 \le \ldots \le \overline{x}_3 \le \overline{x}_2 \le \overline{x}_1 .$$

The sequence of the lower bounds of $\{X_n\}_{n=1}^{\infty}$ is a monotonic nondecreasing sequence of real numbers with an upper bound, say $\overline{x}_1 < \infty$. Thus, it converges to a real number, $\underline{x}$. Similarly, the monotonic nonincreasing sequence of real numbers $\{\overline{x}_n\}_{n=1}^{\infty}$ converges to a real number, $\overline{x}$, for which $\underline{x} \le \overline{x}$. Hence, it follows that

$$\lim_{n\to\infty} X_n = [\underline{x}, \overline{x}] = X = \bigcap_{n=1}^{\infty} X_n .$$

Corollary 1.2. If

$$X_1 \supseteq X_2 \supseteq X_3 \supseteq \ldots \supseteq Y,$$

then

$$\lim_{n\to\infty} X_n = X \quad \text{with} \quad X \supseteq Y.$$

The next important property is fundamental, as mentioned in the last subsection.

Theorem 1.7. The interval operations { +, –, ·, / }, introduced in Section III-B, are continuous functions of intervals.

Proof. We only show the addition operation; the others are similar.

Let $\{X_n\}_{n=1}^{\infty}$ and $\{Y_n\}_{n=1}^{\infty}$ be two sequences of intervals in **I**, with

$$\lim_{n\to\infty} X_n = X \quad \text{and} \quad \lim_{n\to\infty} Y_n = Y.$$

The sequence of interval sums $\{X_n + Y_n\}_{n=1}^{\infty}$ then satisfies

$$\begin{aligned} \lim_{n\to\infty} (X_n+Y_n) &= \lim_{n\to\infty} [\underline{x}_n+\underline{y}_n, \overline{x}_n + \overline{y}_n] \\ &= [\lim_{n\to\infty} (\underline{x}_n+\underline{y}_n), \lim_{n\to\infty} (\overline{x}_n + \overline{y}_n)] \\ &= [\underline{x}+\underline{y}, \overline{x} + \overline{y}] \\ &= X + Y. \end{aligned}$$

Corollary 1.3. Let f be an ordinary continuous function and X be an interval. Let

$$f_I(X) = [\min_{x\in X} f(x), \max_{x\in X} f(x)].$$

Then $f_{\mathbf{I}}(X)$ is a continuous interval-variable and interval-valued function.

This corollary can be easily verified by using the continuity of the real function f, which guarantees the continuity of all important interval-variable and interval-valued functions like X^n, e^X, $sin(X)$, $\sqrt{|X|}$, etc.

Theorem 1.8. Let $X = [\underline{x}, \overline{x}]$, $Y = [\underline{y}, \overline{y}]$, $Z = [\underline{z}, \overline{z}]$, and $S = [\underline{s}, \overline{s}]$ be intervals in **I**. Then

(1) $d(X+Y, X+Z) = d(Y,Z)$;
(2) $d(X+Y, Z+S) \le d(X,Z) + d(Y,S)$;
(3) $d(\lambda X, \lambda Y) = |\lambda|\, d(X,Y), \quad \lambda \in R$;
(4) $d(XY, XZ) \le |X|\, d(Y,Z)$.

Proof. For (1), it follows from the definition of $d(\cdot,\cdot)$ that

$$\begin{aligned} d(X+Y,X+Z) &= \max\{\,|(\underline{x}+\underline{y})-(\underline{x}+\underline{z})|, |(\overline{x}+\overline{y})-(\overline{x}+\overline{z})|\,\} \\ &= \max\{\,|\underline{y}-\underline{z}|, |\overline{y}-\overline{z}|\,\} \\ &= d(Y,Z). \end{aligned}$$

For (2), using the triangular inequality, part (1) above, and the symmetry of $d(\cdot,\cdot)$, we have

$$\begin{aligned} d(X+Y,Z+S) &\le d(X+Y,Y+Z) + d(Z+S,Y+Z) \\ &\le d(X,Z) + d(Y,S). \end{aligned}$$

For (3), for any real number $\lambda \in R$, we have

$$\begin{aligned} d(\lambda X,\lambda Y) &= \max\{\,|\lambda\underline{x}-\lambda\underline{y}|, |\lambda\overline{x}-\lambda\overline{y}|\,\} \\ &= |\lambda| \max\{\,|\underline{x}-\underline{y}|, |\overline{x}-\overline{y}|\,\} \\ &= |\lambda|\, d(X,Y). \end{aligned}$$

For (4), for an interval $A = [\underline{a}, \overline{a}]$, we will use $l(A) = \underline{a}$ and $u(A) = \overline{a}$ for convenience in this proof. Then, what we need to show is

$$max\{\,|\,l(XY)-l(XZ)\,|, |\,u(XY)-u(XZ)\,|\,\} \le |X|\, d(Y,Z).$$

We only show that

$$|\,l(XY)-l(XZ)\,| \le |X|\, d(Y,Z)$$

and the inequality $|\,u(XY) - u(XZ)\,| \le |X|\, d(Y,Z)$ can be verified in the same manner. Without loss of generality, assume that

$$l(XY) \ge l(XZ);$$

the case of $l(XY) < l(XZ)$ can be similarly analyzed. Then, since

$$XZ = \{\, xz \mid x \in X, z \in Z \,\},$$

there exists an $x \in X$ such that

$$l(XZ) = l(xZ).$$

On the other hand, we have

$$xY \subseteq XY,$$

which implies that

$$l(xY) \ge l(XY).$$

Hence, we have

$$l(xY) - l(xZ) \ge l(XY) - l(XZ) \ge 0,$$

so that

$$|\,l(XY) - l(XZ)\,| \;=\; l(XY) - l(XZ)$$

$$\leq \; l(xY) - l(xZ)$$
$$= \; | \, l(xY) - l(xZ) \, |$$
$$\leq \; |x| \, d(Y,Z)$$
$$\leq \; |X| \, d(Y,Z).$$

E. Properties of the Width of an Interval

In this subsection, we summarize some interesting and useful properties of the width of an interval, which is defined in Definition 1.1 (f), as follows. For an interval $S = [\underline{s}, \bar{s}\,]$, the width of S is $w\{S\} = \underline{s} - \bar{s}$, which is equivalent to

$$w\{S\} = \max_{s_1, s_2 \in S} | s_1 - s_2 |.$$

In addition to the properties listed in Problem *P1.2*, we have the following.

Theorem 1.9. Let X and Y be intervals. Then

(1) $w\{XY\} \leq w\{X\} \, |Y| + |X| \, w\{Y\}$;

(2) $w\{XY\} \geq max\{ \, |X| \, w\{Y\} \, , |Y| \, w\{X\} \, \}$;

(3) $w\{X^n\} \leq n \, |X|^{n-1} \; w\{X\}, \qquad n=1,2,\cdots$;

(4) $w\{ \, (X-x)^n \, \} \leq 2 \, (\, w\{X\} \,)^n, \quad x \in X, n=1,2,\cdots$.

Proof. For (1), using the equivalent definition $w\{S\} = \max_{s_1, s_2 \in S} |s_1 - s_2|$, we have

$$\begin{aligned}
w\{XY\} &= \max_{x_1,x_2 \in X; y_1,y_2 \in Y} | x_1 y_1 - x_2 y_2 | \\
&= \max_{x_1,x_2 \in X; y_1,y_2 \in Y} | x_1 y_1 - x_1 y_2 + x_1 y_2 - x_2 y_2 | \\
&\leq \max_{x_1,x_2 \in X; y_1,y_2 \in Y} \{ | x_1 (y_1 - y_2) | + | y_2 (x_1 - x_2) | \} \\
&\leq \max_{x_1,x_2 \in X; y_1,y_2 \in Y} |x_1| \, |(y_1 - y_2)| + \max_{x_1,x_2 \in X; y_1,y_2 \in Y} |y_2| \, |(x_1 - x_2)| \\
&= (\max_{x_1 \in X} |x_1|) (\max_{y_1,y_2 \in Y} |(y_1 - y_2)|) \\
&\qquad + (\max_{y_2 \in Y} |y_2|) (\max_{x_1,x_2 \in X} |(x_1 - x_2)|) \\
&= |X| \, w\{Y\} + w\{X\} \, |Y|.
\end{aligned}$$

For (2), we first have

$$\begin{aligned}
w\{XY\} &= \max_{x_1,x_2 \in X; y_1,y_2 \in Y} | x_1 y_1 - x_2 y_2 | \\
&\geq \max_{x_1,x_2 \in X; y_1,y_2 \in Y} | x_1 y_1 - x_1 y_2 | \\
&= \max_{x_1,x_2 \in X; y_1,y_2 \in Y} | x_1 | \, | y_1 - y_2 | \\
&= |X| \, w\{Y\}.
\end{aligned}$$

Similarly, we can show that

$$w\{XY\} \geq |Y| \, w\{X\}.$$

Hence, the inequality (2) follows.

For (3), we use mathematical induction. First, for $n = 1$, the inequality holds. If the inequality is true for $n \geq 1$, then it follows from part (1) above that

$$
\begin{aligned}
w\{X^{n+1}\} &= w\{X^n X\} \\
&\le n\,|X|^{n-1}\, w\{X\}\,|X| + |X|^n\, w\{X\} \\
&= (n+1)\,|X|^n\, w\{X\}.
\end{aligned}
$$

This completes the induction.

For (4), since $x \in X$, we have $w\{x\} \le w\{X\}$ (see Problem *P1.2*), so that

$$
\begin{aligned}
w\{(X-x)^n\} &\le w\{(X-X)^n\} \\
&= w\{\,[-w\{X\}, w\{X\}]^n\,\} \\
&= w\{\,[-(\,w\{X\}\,)^n, (\,w\{X\}\,)^n\,]\,\} \\
&= 2\,(\,w\{X\}\,)^n,
\end{aligned}
$$

where we recall that $X - X$ is not zero in general; but if $X - X = \mathbf{0}$ the above inequality holds as well.

Theorem 1.10. Let X and Y be intervals. Then

(1) $X \subseteq Y$ implies that

$$\frac{1}{2}(\,w\{Y\} - w\{X\}\,) \le d(X,Y) \le w\{Y\} + w\{X\};$$

(2) if X is symmetric, then

$$XY = |Y|\,X \qquad \text{and} \qquad w\{XY\} = |Y|\,w\{X\},$$

where the second equality is also valid for the case of $0 \in X$, if either $\underline{y} \ge 0$ or $\overline{y} \le 0$.

Proof. For (1), since $X \subseteq Y$, we have $\underline{y} \le \underline{x} \le \overline{x} \le \overline{y}$, so that

$$
\begin{aligned}
d(X,Y) &= \max\{\,|\,\underline{x} - \underline{y}\,|, |\,\overline{x} - \overline{y}\,|\,\} \\
&= \max\{\,\underline{x} - \underline{y},\ \overline{y} - \overline{x}\,\} \\
&\le \overline{y} - \overline{x} + \underline{x} - \underline{y} \\
&= (\,\overline{y} - \underline{y}\,) - (\,\overline{x} - \underline{x}\,) \\
&= w\{Y\} - w\{X\}.
\end{aligned}
$$

On the other hand, we have

$$
\begin{aligned}
d(X,Y) &= \max\{\,\underline{x} - \underline{y},\ \overline{y} - \overline{x}\,\} \\
&\ge \frac{1}{2}(\,\underline{x} - \underline{y} + \overline{y} - \overline{x}\,) \\
&= \frac{1}{2}(\,w\{Y\} - w\{X\}\,).
\end{aligned}
$$

For (2), X is symmetric means that $X = -X$, or equivalently, $\underline{x} = -\overline{x} = x$. Thus, we have

$$
\begin{aligned}
XY &= [\,\min\{x\underline{y}, x\overline{y}, -x\underline{y}, -x\overline{y}\,\}, \max\{x\underline{y}, x\overline{y}, -x\underline{y}, -x\overline{y}\,\}\,] \\
&= [\,x\min\{\underline{y}, -\underline{y}, \overline{y}, -\overline{y}\,\}, x\max\{\underline{y}, -\underline{y}, \overline{y}, -\overline{y}\,\}\,] \\
&= [\,x\,(-|Y|\,), x\,|Y|\,] \\
&= [-x, x]\,|Y| \\
&= |Y|\,X.
\end{aligned}
$$

It then follows from $w\{\lambda X\} = |\lambda|\,w\{X\}$ (see Problem *P1.2*) that

$$w\{XY\} = |Y|\,w\{X\}.$$

The other cases can be similarly verified.

F. Interval Evaluation

In this subsection, we extend the ordinary real-variable and real-valued functions f:R→R to interval-variable and interval-valued functions f:**I**→**I**, where **I** is the family of intervals. This will further generalize the function

$$Z = f(X,Y) = X * Y, \qquad * \in \{+, -, \times, /\},$$

where $X, Y, Z \in$ **I**, which was discussed in Section III.C.

We first recall from Corollary 1.3 of Section III.C that for any ordinary continuous function f:R→R and any interval $X \in$ **I**, the interval-variable and interval-valued function

$$f_I(X) = [\ \min_{x \in X} f(x),\ \max_{x \in X} f(x)\ ]$$

is also continuous in the interval metric. We further extend this function, $f_I(\cdot)$, as follows. Let A_1, ..., A_m be intervals in **I**, and suppose that each constant $a_k \in A_k$, $k = 1, 2, ..., m$, assumes only one value in any evaluation of a continuous function f:R→R in the following discussions. For any interval $X \in$ **I**, consider a function $f(x;a_1,...,a_m)$ defined on X, depending on m parameters $a_k \in A_k$, $k = 1, 2, ..., m$. Define an *interval expression* of f by

$$\begin{aligned} f_I(X;A_1,...,A_m) &= \{ f(x;a_1,...,a_m) \mid x \in X,\ a_k \in A_k,\ 1 \le k \le m \} \\ &= [\ \min_{\substack{x \in X \\ a_k \in A_k, 1 \le k \le m}} f(x;a_1,...,a_m),\ \max_{\substack{x \in X \\ a_k \in A_k, 1 \le k \le m}} f(x;a_1,...,a_m)\ ]. \end{aligned}$$

In the following, we always assume that an interval expression can be computed in finitely many steps. In this case, the expression is called a *finite interval expression*. Moreover, we only consider those finite interval expressions that consist of *rational operations*, namely, addition, subtraction, multiplication, and division. An interval expression of this kind is called a *rational interval expression*, which is similarly defined for multi-valuable case.

Example 1.5. Consider the real-variable and real-valued function

$$f(x;a) = \frac{ax}{1-x}, \qquad x \neq 1, \qquad x \neq 0.$$

If $X = [2,3]$ and $A = [0,2]$ are intervals with $x \in X$ and $a \in A$, then we have the rational interval expression of f given by

$$\begin{aligned} f_I(X;A) &= \left\{ \frac{ax}{1-x} \,\middle|\, 2 \le x \le 3, 0 \le a \le 2 \right\} \\ &= [\ \min_{\substack{2 \le x \le 3, \\ 0 \le a \le 2}} \frac{ax}{1-x},\ \max_{\substack{2 \le x \le 3, \\ 0 \le a \le 2}} \frac{ax}{1-x}\ ] \\ &= [-4,0]. \end{aligned}$$

The following theorem is very important and useful, which states that all rational interval expressions have the *inclusion monotonic property*.

Theorem 1.11. Let f:R^{n+m}→R be a real-variable and real-valued function that has a rational interval expression $f_I(X_1,...,X_n,A_1,...,A_m)$. Then for all

$$X_k \subseteq Y_k, \qquad k = 1,2,\ldots,n,$$
$$A_l \subseteq B_l, \qquad l = 1,2\ldots,m,$$

we have

$$f_{\mathrm{I}}(X_1,\ldots,X_n,A_1,\ldots,A_m) \subseteq f_{\mathrm{I}}(Y_1,\ldots,Y_n,B_1,\ldots,B_m).$$

Proof. It is a consequence of Theorem 1.2 and Corollary 1.1, established in Section III.C, and the definition of rational interval expressions.

Example 1.6. Let $X = [0.2,0.4]$ and $Y = [0.1,0.5]$. Then $X \subset Y$.

(a) $X^{-1} = \dfrac{1}{[0.2,0.4]} = [2.5,5.0],$

$Y^{-1} = \dfrac{1}{[0.1,0.5]} = [2.0,10.0],$

$X^{-1} \subset Y^{-1}.$

(b) $1 - X = [1.0,1.0] - [0.2,0.4] = [0.6,0.8],$

$1 - Y = [1.0,1.0] - [0.1,0.5] = [0.5,0.9],$

$1 - X \subset 1 - Y.$

(c) $\dfrac{1}{1-X} = \dfrac{1}{[0.6,0.8]} = [5/4,5/3],$

$\dfrac{1}{1-Y} = \dfrac{1}{[0.5,0.9]} = [10/9,2.0],$

$\dfrac{1}{1-X} \subset \dfrac{1}{1-Y}.$

Next, we should point out a very important issue in the evaluation of an interval expression: the number of intervals involved in a rational interval expression should be reduced (whenever possible) before evaluating (i.e., computing) the expression, in order to obtain less-conservative lower and upper bounds of the resulting interval expression.

Example 1.7. Consider, again, the function f defined in Example 1.5, namely,

$$f(x;a) = \frac{ax}{1-x}, \qquad x \neq 1, \qquad x \neq 0.$$

We now rewrite it as

$$\tilde{f}(x;a) = \frac{a}{\dfrac{1}{x} - 1}, \qquad x \neq 1, \qquad x \neq 0.$$

As a real-variable and real-valued function, $\tilde{f} \equiv f$. Actually, for $x \in X = [2,3]$ and $a \in A = [0,2]$, we have

$$f_{\mathrm{I}}(X;A) = [-4,0]$$

from Example 1.5, and also have

$$\tilde{f}_{\mathrm{I}}(X;A) = \left\{ \frac{a}{1/x-1} \;\middle|\; 2 \leq x \leq 3,\, 0 \leq a \leq 2 \right\}$$

$$= [\min_{\substack{2\le x\le 3\\0\le a\le 2}} \frac{a}{1/x-1}, \max_{\substack{2\le x\le 3\\0\le a\le 2}} \frac{a}{1/x-1}]$$
$$= [-4,0]$$
$$= f_I(X;A),$$

as expected. However, if we formally perform interval arithmetic (rather than numerical minimization and maximization as we just did), then we have

$$f_I(X;A) = \frac{[0,2].[2,3]}{1-[2,3]} = \frac{[0,6]}{[-2,-1]} = [0,6] . [-1,-1/2] = [-6,0]$$

and

$$\widetilde{f}_I(X;A) = \frac{[0,2]}{\frac{1}{[2,3]}-1} = \frac{[0,2]}{[\frac{1}{3},\frac{1}{2}]-1} = \frac{[0,2]}{[-\frac{2}{3},-\frac{1}{2}]}$$
$$= [0,2]\cdot[-2,-3/2]$$
$$= [-4,0].$$

Thus, $\widetilde{f}_I(X;A) \neq f_I(X;A)$ but $\widetilde{f}_I(X;A) \subseteq f_I(X;A)$. The reason is that formula $f_I(X;A)$ has three intervals but $\widetilde{f}_I(X;A)$ has only two. Thus, intuition shows that the calculation "errors" accumulated in the former can be larger than that of the latter. This example convinces this observation. Indeed, from the above evaluation using numerical minimization and maximization we know that the correct answer should be [–4,0], and the result of [–6,0] is too conservative in bounding the actual result. Hence, one should always try to reduce the number of intervals involved in an evaluation of a rational interval expression.

As another example, in calculating a power series of intervals

$$f_I(X) = I + X + X^2 + X^3 + \ldots + X^n,$$

the best way is to reformulate it as

$$\widetilde{f}_I(X) = I + X(I + X(I + \ldots + X(I + X) \ldots)$$

before we actually carry out the calculations. The reason is that one needs to apply

$$1 + 2 + 3 + \ldots + n = \frac{n(n+1)}{2}$$

times of the interval X in computing f_I but only n times of X in the formula $\widetilde{f}_I$, where $X(I+X) \subseteq X + X^2$ by Theorem 1.1 (7).

We should remark that the best way to evaluate a rational interval expression $f_I(X_1,\ldots,X_n,A_1,\ldots,A_m)$ is to use the definition directly and apply both minimization and maximization to determine the exact lower and upper bounds for the result:

$$f_I(X_1,\ldots,X_n,A_1,\ldots,A_m) = [\min_{\substack{x_k\in X_k, 1\le k\le n\\ a_l\in A_l, 1\le l\le m}} f_I(x_1,\ldots,x_n,a_1,\ldots,a_m),$$

$$\max_{\substack{x_k \in X_k, 1 \le k \le n \\ a_l \in A_l, 1 \le l \le m}} f_{\rm I}(x_1,\ldots,x_n,a_1,\ldots,a_m)].$$

This is not always possible, however, unless a computer is used to perform the minimization and maximization numerically. Yet, interval arithmetic is much simpler in many cases, even for complicated rational functions consisting of only addition, subtraction, multiplication, and division. Therefore, the above observation is practically important.

G. Interval Matrix Operations

When some entries of a constant matrix are uncertain, namely, when they are intervals rather than exact numbers, the matrix becomes an *interval matrix*.

For example, the following two matrices are interval matrices:

$$A_{\rm I} = \begin{bmatrix} [2,3] & [0,1] \\ [1,2] & [2,3] \end{bmatrix}, \qquad B_{\rm I} = \begin{bmatrix} [0,120] \\ [60,140] \end{bmatrix}.$$

Two $n{\times}m$ interval matrices $A_{\rm I} = [A_{\rm I}(i,j)]$ and $B_{\rm I} = [B_{\rm I}(i,j)]$ are said to be *equal* if $A_{\rm I}(i,j) = B_{\rm I}(i,j)$ for all $1 \le i \le n$ and $1 \le j \le m$. Similarly, $A_{\rm I}$ is said to be *contained* in $B_{\rm I}$, denoted $A_{\rm I} \subseteq B_{\rm I}$, if $A_{\rm I}(i,j) \subseteq B_{\rm I}(i,j)$ for all $1 \le i \le n$ and $1 < j \le m$. In particular, if A is an (ordinary) constant matrix contained in $B_{\rm I}$, we write $A \subseteq B_{\rm I}$.

The fundamental operations of interval matrices are defined as follows:

(1) Addition and Subtraction: Let $A_{\rm I}$ and $B_{\rm I}$ be $n{\times}m$ interval matrices. Then

$$A_{\rm I} \pm B_{\rm I} = [A_{\rm I}(i,j) \pm B_{\rm I}(i,j)].$$

(2) Multiplication: Let $A_{\rm I}$ and $B_{\rm I}$ be $n{\times}r$ and $r{\times}m$ interval matrices, respectively. Then

$$A_{\rm I}\, B_{\rm I} = [\, \sum_{k=1}^{r} A_{\rm I}(i,k)\, B_{\rm I}(k,j)].$$

In particular, if $B_{\rm I} = X$ is an interval, we define

$$A_{\rm I}\, X = X\, A_{\rm I} = [X\, A_{\rm I}(i,j)].$$

Let A and B be two constant matrices and $A_{\rm I}$ and $B_{\rm I}$ be two interval matrices, respectively, of appropriate dimensions. If $A \in A_{\rm I}$ and $B \in B_{\rm I}$, then we have

$$\{\, AB \mid A \in A_{\rm I}, B \in B_{\rm I} \,\} \subseteq \{\, C \mid C \in A_{\rm I}B_{\rm I} \,\}.$$

This relation can be verified by using the inclusion monotonic property of the interval operations. Note that the equality does not hold in general as can be seen from the following example.

Example 1.8. Let

$$A = A_{\rm I} = \begin{bmatrix} 1 & 1 \\ -1 & 1 \end{bmatrix} \qquad \text{and} \qquad B_{\rm I} = \begin{bmatrix} [0,1] \\ [0,1] \end{bmatrix}.$$

Then we have

$$A\, B_{\rm I} = \begin{bmatrix} [0,2] \\ [-1,1] \end{bmatrix}.$$

However, the matrix

$$C = \begin{bmatrix} 2 \\ -1 \end{bmatrix} \in A\,B_{\mathrm{I}},$$

but there is no matrix $B \in B_{\mathrm{I}}$ that satisfies $A\,B = C$. This fact implies that

$$\{ AB \mid A \in A_{\mathrm{I}}, B \in B_{\mathrm{I}} \} \underset{\neq}{\subset} \{ C \mid C \in A_{\mathrm{I}} B_{\mathrm{I}} \}.$$

In the following, we summarize some useful properties of interval matrix operations. First, we point out that the fundamental property of inclusion monotonicity is also valid for interval matrix operations.

Theorem 1.12. Let A, B, C be constant matrices and A_{I}, B_{I}, C_{I} be interval matrices, respectively, of appropriate dimensions. Then

(1) $\{ A \pm B \mid A \in A_{\mathrm{I}}, B \in B_{\mathrm{I}} \} = A_{\mathrm{I}} \pm B_{\mathrm{I}}$;

(2) $\{ A\,B \mid A \in A_{\mathrm{I}} \} = A_{\mathrm{I}} B = B\,A_{\mathrm{I}}$;

(3) $A_{\mathrm{I}} + B_{\mathrm{I}} = B_{\mathrm{I}} + A_{\mathrm{I}}$;

(4) $A_{\mathrm{I}} + (B_{\mathrm{I}} + C_{\mathrm{I}}) = (A_{\mathrm{I}} + B_{\mathrm{I}}) + C_{\mathrm{I}}$;

(5) $A_{\mathrm{I}} + \mathbf{0} = \mathbf{0} + A_{\mathrm{I}} = A_{\mathrm{I}}$, $\quad \mathbf{0}$ = zero matrix;

(6) $A_{\mathrm{I}}\,\mathrm{I} = \mathrm{I}\,A_{\mathrm{I}} = A_{\mathrm{I}}$, $\quad \mathrm{I}$ = identity matrix;

(7) Subdistributive Law:

$$\begin{cases} (A_{\mathrm{I}} + B_{\mathrm{I}}) C_{\mathrm{I}} \subseteq A_{\mathrm{I}} C_{\mathrm{I}} + B_{\mathrm{I}} C_{\mathrm{I}}, \\ C_{\mathrm{I}} (A_{\mathrm{I}} + B_{\mathrm{I}}) \subseteq C_{\mathrm{I}} A_{\mathrm{I}} + C_{\mathrm{I}} B_{\mathrm{I}}; \end{cases}$$

(8) $(A_{\mathrm{I}} + B_{\mathrm{I}})\, C = A_{\mathrm{I}}\, C + B_{\mathrm{I}}\, C$;

(9) $C\, (A_{\mathrm{I}} + B_{\mathrm{I}}) = C\, A_{\mathrm{I}} + C\, B_{\mathrm{I}}$;

(10) Associative and Subassociative Laws:

(a) $A_{\mathrm{I}}\, (B\, C) \subseteq (A_{\mathrm{I}}\, B)\, C$;

(b) $(A\, B_{\mathrm{I}})\, C_{\mathrm{I}} \subseteq A\, (B_{\mathrm{I}}\, C_{\mathrm{I}})$, $\quad$ if $C_{\mathrm{I}} = -C_{\mathrm{I}}$;

(c) $A\, (B_{\mathrm{I}}\, C) = (A\, B_{\mathrm{I}})\, C$;

(d) $A_{\mathrm{I}}\, (B_{\mathrm{I}}\, C_{\mathrm{I}}) = (A_{\mathrm{I}}\, B_{\mathrm{I}})\, C_{\mathrm{I}}$, $\quad$ if $B_{\mathrm{I}} = -B_{\mathrm{I}}$ and $C_{\mathrm{I}} = -C_{\mathrm{I}}$.

Proof. Formulas (1)-(9) can be verified by using the formulas of Theorem 1.1. Here, we only prove the formulas in (10). For simplicity of notation, we assume that all matrices are square with the same dimension, n.

Formula (a):

$$\begin{aligned} A_{\mathrm{I}}\,(B\, C) &= \left[\sum_{j=1}^{n} A_{\mathrm{I}}(i,j) \left[\sum_{l=1}^{n} B_{jl} C_{lk} \right] \right] \\ &\subseteq \left[\sum_{j=1}^{n} \sum_{l=1}^{n} A_{\mathrm{I}}(i,j) B_{jl} C_{lk} \right] \\ &= \left[\sum_{l=1}^{n} \left[\sum_{j=1}^{n} A_{\mathrm{I}}(i,j) B_{jl} \right] C_{lk} \right] \\ &= (A_{\mathrm{I}}\, B)\, C. \end{aligned}$$

Formula (b): Since $C_{\mathrm{I}} = -C_{\mathrm{I}}$, we have

$$
\begin{aligned}
(A B_{\mathrm{I}}) C_{\mathrm{I}} &= \left[\sum_{l=1}^{n}\left[\sum_{k=1}^{n} A_{ik} B_{\mathrm{I}}(k,l)\right] C_{\mathrm{I}}(l,j)\right] \\
&= \left[\sum_{l=1}^{n}\left|\sum_{k=1}^{n} A_{ik} B_{\mathrm{I}}(k,l)\right| C_{\mathrm{I}}(l,j)\right] \\
&\subseteq \left[\sum_{l=1}^{n}\left[\sum_{k=1}^{n} |A_{ik}|| B_{\mathrm{I}}(k,l)|\right] C_{\mathrm{I}}(l,j)\right] \\
&\subseteq \left[\sum_{l=1}^{n}\left[\sum_{k=1}^{n} |A_{ik}|| B_{\mathrm{I}}(k,l)| C_{\mathrm{I}}(l,j)\right]\right] \\
&= \left[\sum_{k=1}^{n} |A_{ik}|\left[\sum_{l=1}^{n} | B_{\mathrm{I}}(k,l)| C_{\mathrm{I}}(l,j)\right]\right] \\
&= \left[\sum_{k=1}^{n} A_{ik}\left[\sum_{l=1}^{n} B_{\mathrm{I}}(k,l) C_{\mathrm{I}}(l,j)\right]\right] \\
&= A\,(B_{\mathrm{I}} C_{\mathrm{I}}).
\end{aligned}
$$

Formula (c):

$$
\begin{aligned}
(A B_{\mathrm{I}}) C &= \left[\sum_{l=1}^{n}\left[\sum_{k=1}^{n} A_{ik} B_{\mathrm{I}}(k,l)\right] C(l,j)\right] \\
&= \left[\sum_{l=1}^{n}\left[\sum_{k=1}^{n} A_{ik} B_{\mathrm{I}}(k,l) C(l,j)\right]\right] \\
&= \left[\sum_{k=1}^{n} A_{ik}\left[\sum_{l=1}^{n} B_{\mathrm{I}}(k,l) C(l,j)\right]\right] \\
&= A\,(B_{\mathrm{I}} C).
\end{aligned}
$$

Formula (d): Since $B_{\mathrm{I}} = -B_{\mathrm{I}}$ and $C_{\mathrm{I}} = -C_{\mathrm{I}}$, we have

$$
\begin{aligned}
A_{\mathrm{I}}(B_{\mathrm{I}} C_{\mathrm{I}}) &= \left[\sum_{k=1}^{n} A_{\mathrm{I}}(i,k)\left[\sum_{l=1}^{n} B_{\mathrm{I}}(k,l) C_{\mathrm{I}}(l,j)\right]\right] \\
&= \left[\sum_{k=1}^{n} A_{\mathrm{I}}(i,k)\left[\sum_{l=1}^{n} | B_{\mathrm{I}}(k,l)| C_{\mathrm{I}}(l,j)\right]\right] \\
&= \left[\sum_{k=1}^{n} |A_{\mathrm{I}}(i,k)|\left[\sum_{l=1}^{n} | B_{\mathrm{I}}(k,l)| C_{\mathrm{I}}(l,j)\right]\right] \\
&= \left[\sum_{l=1}^{n}\left[\sum_{k=1}^{n} |A_{\mathrm{I}}(i,k)|| B_{\mathrm{I}}(k,l)| C_{\mathrm{I}}(l,j)\right]\right] \\
&= \left[\sum_{l=1}^{n}\left[\sum_{k=1}^{n} |A_{\mathrm{I}}(i,k)| B_{\mathrm{I}}(k,l)\right] C_{\mathrm{I}}(l,j)\right]
\end{aligned}
$$

$$= \left[\sum_{l=1}^{n}\left[\sum_{k=1}^{n} A_{\mathrm{I}}(i,k)B_{\mathrm{I}}(k,l)\right]C_{\mathrm{I}}(l,j)\right]$$

$$= (A_{\mathrm{I}}\, B_{\mathrm{I}})\, C_{\mathrm{I}}.$$

We remark that in the formulas (a) and (b) established in part (10) above, we do not have the following identities:

(a') $A_{\mathrm{I}}(B\,C) = (A_{\mathrm{I}}\,B)\,C$

(b') $(A\,B_{\mathrm{I}})\,C_{\mathrm{I}} = A\,(B_{\mathrm{I}}\,C_{\mathrm{I}})$, even if $C_{\mathrm{I}} = -C_{\mathrm{I}}$.

Counterexamples can be easily constructed. For example, we can check with the following matrices:

$$A_{\mathrm{I}} = \begin{bmatrix} [-1,1] & [1,1] \\ [-1,-1] & [0,1] \end{bmatrix}, \qquad B = \begin{bmatrix} 1 & 1 \\ 0 & 1 \end{bmatrix}, \qquad C = \begin{bmatrix} -1 & 0 \\ 1 & -1 \end{bmatrix}.$$

This example shows that (a') does not hold:

$$A_{\mathrm{I}}(B\,C) = \begin{bmatrix} [1,1] & [-2,0] \\ [0,1] & [0,1] \end{bmatrix} \neq \begin{bmatrix} [-1,3] & [-2,0] \\ [0,1] & [0,1] \end{bmatrix} = (A_{\mathrm{I}}\,B)\,C.$$

Theorem 1.13. Let A_{I}^1, A_{I}^2, B_{I}^1, and B_{I}^2 be interval matrices of the same dimension, with

$$A_{\mathrm{I}}^1 \subseteq B_{\mathrm{I}}^1 \qquad \text{and} \qquad A_{\mathrm{I}}^2 \subseteq B_{\mathrm{I}}^2,$$

and let X and Y be intervals with $X \subseteq Y$. Then the relations

$$A_{\mathrm{I}}^1 * A_{\mathrm{I}}^2 \subseteq B_{\mathrm{I}}^1 * B_{\mathrm{I}}^2$$

and

$$X A_{\mathrm{I}}^1 \subseteq Y B_{\mathrm{I}}^1$$

hold for $* \in \{+, -, \cdot\}$.

This result can be proved by using Theorem 1.2.

As a special case, we have

$$A \in A_{\mathrm{I}},\quad B \in B_{\mathrm{I}} \quad \Rightarrow \quad A * B \in A_{\mathrm{I}} * B_{\mathrm{I}}, \qquad * \in \{+, -, \cdot\}$$

and

$$x \in X, \quad A \in A_{\mathrm{I}} \quad \Rightarrow \quad x\,A \in X A_{\mathrm{I}}.$$

We next recall the concepts of *width* and *absolute value* for an interval from Definition 1.1, which are now used to define the width and absolute value of an interval matrix.

Definition 1.5. Let $A_{\mathrm{I}} = [A_{\mathrm{I}}(i,j)]$ be an interval matrix. The real nonnegative matrix

$$\mathrm{w}(A_{\mathrm{I}}) = [\, \mathrm{w}(A_{\mathrm{I}}(i,j))\,]$$

is called the *width* of A_{I}, and the real nonnegative matrix

$$|A_{\mathrm{I}}| = [\, |A_{\mathrm{I}}(i,j)|\,]$$

is called the *absolute value* of A_{I}.

Some basic properties of width and absolute value of an interval matrix are now summarized in the following theorem. A proof can be given by comparing it with Exercises *P1.1* and *P1.2*. In this theorem, we use the standard *partial ordering*, defined for constant matrices by $A \le B$, if and only if $A(i,j) \le B(i,j)$ for all entries of A and B.

Theorem 1.14. Let A_I and B_I be interval matrices of the same dimension. Then

(1) $A_I \subseteq B_I \Rightarrow w(A_I) \le w(B_I)$;
(2) $w(A_I + B_I) = w(A_I) + w(B_I)$;
(3) $w(A_I) = \sup_{A,\tilde{A} \in A_I} |A - \tilde{A}|$, $\quad (|A - \tilde{A}| = [|A_{ij} - \tilde{A}_{ij}|])$;
(4) $|A_I| = \sup_{A \in A_I} |A|$, $\quad (|A| = [|A_{ij}|])$;
(5) $A_I \subseteq B_I \Rightarrow |A_I| \le |B_I|$;
(6) $|A_I| \ge \mathbf{0}$, where $|A_I| = \mathbf{0}$ if and only if $A_I = \mathbf{0}$;
(7) $|A_I + B_I| \le |A_I| + |B_I|$;
(8) $|a A_I| = |A_I a| = |a|\,|A_I|$, ($a$ = real number);
(9) $|A_I B_I| \le |A_I|\,|B_I|$;
(10) $w(A_I B_I) \le w(A_I)\,|B_I| + |A_I|\,w(B_I)$;
(11) $w(A_I B_I) \ge |A_I|\,w(B_I)$ and $w(A_I B_I) \ge |B_I|\,w(A_I)$;
(12) $w(a A_I) = |a|\,w(A_I)$, (a = real number);
(13) $w(A A_I) = |A|\,w(A_I)$ and $w(A_I A) = w(A_I)\,|A|$;
(14) $0 \in A_I \Rightarrow |A_I| \le w(A_I) \le 2\,|A_I|$;
(15) $A_I = -A_I \Rightarrow A_I B_I = A_I |B_I|$;
(16) $0 \in A_I$ and $0 \notin B_I \Rightarrow w(A_I B_I) = w(A_I)\,|B_I|$.

We next recall the measure theory of intervals, particularly the concept of *distance* between two intervals given in Definition 1.3. Using this distance, we can define the distance between two interval matrices.

Definition 1.6. Let A_I and B_I be two interval matrices. The real nonnegative matrix

$$D(A_I,B_I) = [\, d(A_I(i,j),B_I(i,j)) \,]$$

is called the *distance* between A_I and B_I.

The following results can be easily verified.

Theorem 1.15. Let A_I, B_I, C_I, and D_I be interval matrices of appropriate dimensions. Then

(1) $D(A_I,B_I) = 0$ if and only if $A_I = B_I$;
(2) $D(A_I,B_I) \le D(A_I,C_I) + D(C_I,B_I)$;
(3) $D(A_I+C_I,B_I+C_I) = D(A_I,B_I)$;
(4) $D(A_I+B_I,C_I+D_I) \le D(A_I,C_I) + D(B_I,D_I)$;
(5) $D(A_IB_I,A_IC_I) \le |A_I|\,D(B_I,C_I)$.

With the help of a distance, we can then use a monotonic matrix norm (e.g., the ordinary spectral norm), $\|\times\|$, to define a *metric* for a family of interval matrices. For instance, we can use $\|D(A_I,B_I)\|$ as a metric. Under this metric, we can introduce the concept of normed linear space for interval matrices, which can be a complete metric space. Consequently, the concept of *convergence* can also be introduced, which is equivalent to the convergence of individual entries of the sequence of interval matrices. More precisely, we have the following.

Definition 1.7. A sequence $\{A_{\mathrm{I}}^{(n)}\}_{n=1}^{\infty}$ of $p{\times}m$ interval matrices is said to *converge* to an interval matrix A_{I}, denoted

$$\lim_{n\to\infty} A_{\mathrm{I}}^{(n)} = A_{\mathrm{I}},$$

if for all entries of $A_{\mathrm{I}}^{(n)}$, we have

$$\lim_{n\to\infty} A_{\mathrm{I}}^{(n)}(i,j) = A_{\mathrm{I}}(i,j), \quad 1 \le i \le p,\ 1 \le j \le m.$$

As a consequence of Theorems 1.6 and 1.7, we have the following.

Corollary 1.4. Let $\{A_{\mathrm{I}}^{(n)}\}_{n=1}^{\infty}$ be a sequence of $n{\times}m$ interval matrices such that

$$A_{\mathrm{I}}^{(1)} \supseteq A_{\mathrm{I}}^{(2)} \supseteq A_{\mathrm{I}}^{(3)} \supseteq \cdots$$

Then $\lim_{n\to\infty} A_{\mathrm{I}}^{(n)} = A_{\mathrm{I}}$, where

$$A_{\mathrm{I}}(i,j) = \bigcap_{n=1}^{\infty} A_{\mathrm{I}}^{(n)}(i,j), \quad 1 \le i \le p,\ 1 \le j \le m.$$

Corollary 1.5. The interval matrix operations $\{+,-,\cdot\}$ introduced in this section are continuous functions of interval matrices.

Corollary 1.6. For interval matrices A_{I}, B_{I}, C_{I}, and D_{I}, we have the following inclusion monotonic property:

$$A_{\mathrm{I}} \subseteq C_{\mathrm{I}} \text{ and } B_{\mathrm{I}} \subseteq D_{\mathrm{I}} \Rightarrow A_{\mathrm{I}} \cap B_{\mathrm{I}} \subseteq C_{\mathrm{I}} \cap D_{\mathrm{I}}.$$

Moreover, this intersection operation is a continuous function of interval matrices.

H. Interval Matrix Equations and Interval Matrix Inversion

In this section, we consider the problem of solving an interval matrix equation, namely, finding a solution matrix X_{I} such that for given interval matrices A_{I} and B_{I}, we have

$$A_{\mathrm{I}} X_{\mathrm{I}} = B_{\mathrm{I}}.$$

Usually, B_{I} is an interval vector, and so is X_{I} (if it exists). Let

$$S = \{\, X \mid AX = B,\ A \in A_{\mathrm{I}},\ B \in B_{\mathrm{I}} \,\}$$

be the solution set of the problem. If the problem has no solutions, then $S = \varnothing$. To have some idea about this general problem of solving an interval matrix equation, we first consider the following specific example.

Example 1.9. Let

$$A_{\mathrm{I}} = \begin{bmatrix} [2,3] & [0,1] \\ [1,2] & [2,3] \end{bmatrix} \quad \text{and} \quad B_{\mathrm{I}} = \begin{bmatrix} [0,120] \\ [60,240] \end{bmatrix}.$$

Find $X_{\mathrm{I}} = \begin{bmatrix} X_{\mathrm{I}}(1) \\ X_{\mathrm{I}}(2) \end{bmatrix}$ with $X_{\mathrm{I}}(1) \ge 0$ and $X_{\mathrm{I}}(2) \ge 0$, such that $A_{\mathrm{I}}X_{\mathrm{I}} = B_{\mathrm{I}}$.

We first write

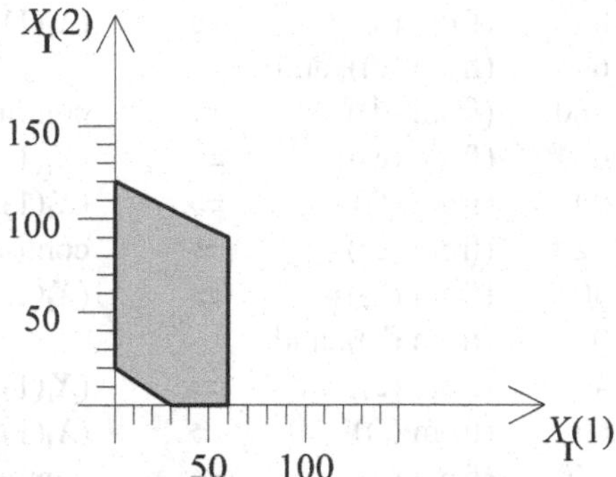

Figure 1.7 The polyhedron solution set of Example 1.9 with restrictions.

$$\begin{cases} [2,3]X_I(1)+[0,1]X_I(2)=[0,120] \\ [1,2]X_I(1)+[2,3]X_I(2)=[60,240], \end{cases}$$

which implies that

$$0 \le 2\,X_I(1) \le 120,$$
$$0 \le 3\,X_I(1) + X_I(2) \le 120,$$
$$60 \le X_I(1) + 2\,X_I(2) \le 240,$$
$$60 \le 2\,X_I(1) + 3\,X_I(2) \le 240.$$

If $X_I = X$ is a constant vector in the solution set S, then it must satisfy the property that the interval on the left-hand side of the equation $A_I X_I = B_I$ intersects the interval on the right-hand side of the equation. It follows from the first inequality that

$$X_I(1) \ge 0 \qquad \text{and} \qquad X_I(1) \le 60.$$

Since $X_I(2) \ge 0$ by assumption, we see that the second inequality

$$0 \le 3\,X_I(1) + X_I(2)$$

is satisfied. We then look at all the extremal cases (boundary lines):

(a) $X_I(1) = 0$
(b) $X_I(2) = 0$
(c) $X_I(1) = 60$
(d) $3X_I(1) + X_I(2) = 120$
(e) $X_I(1) + 2X_I(2) = 60$
(f) $X_I(1) + 2X_I(2) = 240$
(g) $2X_I(1) + 3X_I(2) = 60$
(h) $2X_I(1) + 3X_I(2) = 240$,

which together yield the following results (note that $X_I(1) = X_I(2) = 0$ is not a solution):

(i) $X_I(1) = 0$ (from (a)), and
$X_I(2) = 120$ (from (d)) $\Rightarrow$ $(X_I(1),X_I(2)) = (0,120)$;
$X_I(2) = 30$ (from (e)) $\Rightarrow$ $(X_I(1),X_I(2)) = (0,30)$;
$X_I(2) = 120$ (from (f)) $\Rightarrow$ $(X_I(1),X_I(2)) = (0,120)$;
$X_I(2) = 20$ (from (g)) $\Rightarrow$ $(X_I(1),X_I(2)) = (0,20)$;

$X_I(2) = 80$ (from (h)) $\Rightarrow$ $(X_I(1),X_I(2)) = (0,80)$;

(ii) $X_I(1) = 60$ (from (c)), and

$X_I(2) = -60$ (from (d)) $\Rightarrow$ contradicts $X_I(2) \geq 0$;

$X_I(2) = 0$ (from (e)) $\Rightarrow$ $(X_I(1),X_I(2)) = (60,0)$;

$X_I(2) = 90$ (from (f)) $\Rightarrow$ $(X_I(1),X_I(2)) = (60,90)$;

$X_I(2) = -20$ (from (g)) $\Rightarrow$ contradicts $X_I(2) \geq 0$;

$X_I(2) = 40$ (from (h)) $\Rightarrow$ $(X_I(1),X_I(2)) = (60,40)$;

(iii) $X_I(2) = 0$ (from (b)), and

$X_I(1) = 40$ (from (e)) $\Rightarrow$ $(X_I(1),X_I(2)) = (40,0)$;

$X_I(1) = 60$ (from (f)) $\Rightarrow$ $(X_I(1),X_I(2)) = (60,0)$;

$X_I(1) = 240$ (from (g)) $\Rightarrow$ contradicts $X_I(1) \leq 60$;

$X_I(1) = 30$ (from (h)) $\Rightarrow$ $(X_I(1),X_I(2)) = (30,0)$;

$X_I(1) = 120$ (from (h)) $\Rightarrow$ contradicts $X_I(1) \leq 60$.

The solution set S is located inside all these boundary lines, as shown in Figure 1.7.

If we do not restrict the solution set to be with $X_I(1) \geq 0$ and $X_I(2) \geq 0$, then the solution set for the interval matrix equation $A_I X_I = B_I$, namely,

$$\begin{bmatrix} [2,3] & [0,1] \\ [1,2] & [2,3] \end{bmatrix} \begin{bmatrix} X_I(1) \\ X_I(2) \end{bmatrix} = \begin{bmatrix} [0,120] \\ [60,140] \end{bmatrix}$$

is visualized by Figure 1.8, which is, however, very difficult to find without using a computer.

In general, a solution set S of an interval matrix equation $A_I X_I = B_I$ is not a convex bounded set, but its intersection with each othrant of R^n is a convex polyhedron, where n is the dimension of the square matrix A_I (or the dimension of the interval vectors X_I and B_I).

Now, let us consider again the interval matrix equation

$$A_I X_I = B_I,$$

where A_I and B_I are given $n \times m$ and $n \times 1$ interval matrix and vector, respectively, and X_I is to be determined.

The interval matrix A_I is said to be *regular* if every $A \in A_I$ is nonsingular.

In the above example, the matrix

$$A_I = \begin{bmatrix} [2,3] & [0,1] \\ [1,2] & [2,3] \end{bmatrix}$$

is regular. Thus, we may expect to find X_I by inverting A_I as

$$X_I = [A_I]^{-1} B_I,$$

as usually done in linear algebra. In so doing, we have

$$\det[A_I] = [2,3][2,3] - [0,1][1,2] = [4,9] - [0,2] = [2,9],$$

so that

$$[A_I]^{-1} = \frac{\text{adj}[A_I]}{\det[A_I]} = \frac{\begin{bmatrix} [2,3] & [-1,0] \\ [-2,-1] & [2,3] \end{bmatrix}}{[2,9]}$$

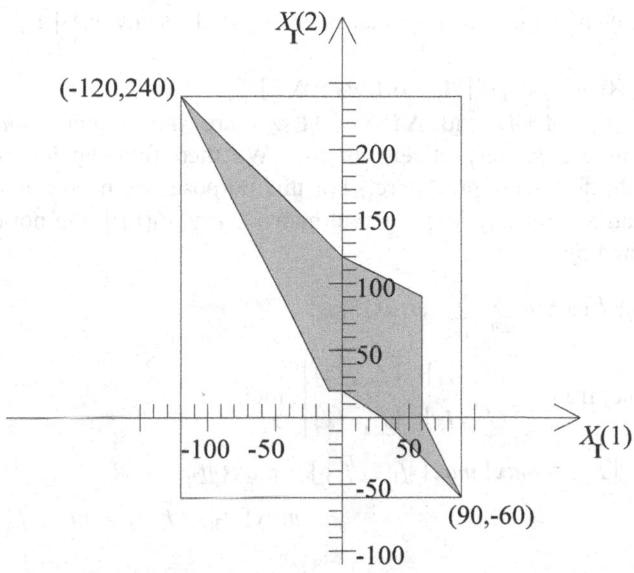

Figure 1.8 The polyhedron solution set of Example 1.9 without restrictions.

$$= \begin{bmatrix} [2/9,3/2] & [-1/2,0] \\ [-1,-1/9] & [2/9,3/2] \end{bmatrix}.$$

Consequently, we have

$$X_I = [A_I]^{-1} B_I = \begin{bmatrix} [-120,180] \\ [-320/3,360] \end{bmatrix}.$$

Note, however, that a better solution can be obtained by using another approach (to be discussed below), that is,

$$X_I^* = \begin{bmatrix} [-120,90] \\ [-60,240] \end{bmatrix} \in X_I.$$

This means that the direct matrix inversion method $[A_I] = \mathrm{adj}[A_I] \,/\, \det[A_I]$ gives a rather conservative interval matrix solution.

It is very important to point out that either X_I or X_I^* shown above only gives an interval matrix solution that contains all the true solutions (namely, the solution set S shown in Figures 1.7 or 1.8). In other words, both X_I and X_I^* are intervals that cover all true solutions in which X_I^* is less conservative than X_I.

We recall from Example 1.7 that the more interval operations we use, the more conservative the result will be. Hence, to find a less conservative solution for the inverse matrix $[A_I]^{-1}$ for a regular interval matrix A_I, we should look for a method that uses less interval operations. The so-called *Hansen inverse method* provides one of such good approaches.

For a given regular interval matrix A_I, to find its inverse $[A_I]^{-1}$, we first write

$$A_I = [\underline{A}, \overline{A}] = [A_0 - \Delta A, A_0 + \Delta A],$$

where $A_0 = \frac{1}{2}[\overline{A} + \underline{A}]$ and $\Delta A = \frac{1}{2}[\overline{A} - \underline{A}]$ are the *nominal matrix* and *perturbation matrix*, respectively, of A_I. We then find the inverse matrix $[A_I]^{-1}$ via the following procedure. For this purpose, we need a *norm* for an interval matrix. For any $n \times m$ interval matrix $F_I = [F_I(i,j)]$, the norm that we use is defined by

$$\| F_I \| = \max_{1 \le i \le n} \sum_{j=1}^{m} | F_I(i,j) |.$$

For example, if $F_I = \begin{bmatrix} [\underline{f}_1, \bar{f}_1] & [\underline{f}_2, \bar{f}_2] \\ [\underline{f}_3, \bar{f}_3] & [\underline{f}_4, \bar{f}_4] \end{bmatrix}$, then

$$\| F_I \| = \max\{ \max\{ |\underline{f}_1|, |\bar{f}_1| \} + \max\{ |\underline{f}_2|, |\bar{f}_2| \}, \\ \max\{ |\underline{f}_3|, |\bar{f}_3| \} + \max\{ |\underline{f}_4|, |\bar{f}_4| \} \}.$$

The Hansen Matrix Inversion Algorithm

Step 1. Choose $A_r = A_0 = \frac{1}{2}[\overline{A} + \underline{A}] \in A_I$.

Step 2. Compute $G = A_r^{-1}$.
Note that G is contained in $[A_I]^{-1}$ by the inclusion monotonic property.

Step 3. Introduce the error matrix $E_r = I - A_r G$. Obviously, $E_r = 0$. But we note that

$$E_r = I - A_r G \in I - A_I G := E_I,$$

where $E_I \neq \mathbf{0}$. If $\| E_I \| < 1$, then we have a convergent series formula:

$$[I - E_I]^{-1} = I + E_I + E_I^2 + E_I^3 + \cdots.$$

Step 4. Compute $S_I^{(m)}$ and $P_I^{(m)}$ defined as follows.
First, we observe that since $E_r = 0$, we formally have

$$\begin{aligned} A_r^{-1} &= G[I - E_r]^{-1} \\ &= G[I + E_r + E_r^2 + E_r^3 + \cdots + E_r^m + E_r^{m+1} + E_r^{m+2} + \cdots] \\ &= G[(I + E_r + \cdots + E_r^m) + E_r^{m+1}(I + E_r + \cdots)] \\ &= G[I + E_r(I + E_r(I + \cdots + E_r(I + E_r)\cdots)) + E_r^{m+1}(I - E_r)^{-1}] \\ &= G[\, S_r^{(m)} + E_r^{m+1}(I - E_r)^{-1} \,] \\ &= G[\, S_r^{(m)} + R_r^{(m+1)} \,], \end{aligned}$$

where

$$\| R_r^{(m+1)} \| = \| E_r^{m+1}(I - E_r)^{-1} \| \le \frac{\|E_r\|^{m+1}}{1 - \|E_r\|}.$$

Since $E_r \in E_{\mathrm{I}}$, by the inclusion monotonic property we have $R_r^{(m+1)} \in R_{\mathrm{I}}^{(m+1)}$ and $\| R_r^{(m+1)} \| \leq \| R_{\mathrm{I}}^{(m+1)} \|$, where

$$\| R_{\mathrm{I}}^{(m+1)} \| = \| E_{\mathrm{I}}^{m+1} (\mathrm{I}-E_{\mathrm{I}})^{-1} \| \leq \frac{\|E_{\mathrm{I}}\|^{m+1}}{1-\|E_{\mathrm{I}}\|}.$$

This inequality is what we need later.

Then, we construct an $n \times n$ interval matrix $P_{\mathrm{I}}^{(m)}$ with identical entries of the form $[-P,P]$, namely,

$$P_{\mathrm{I}}^{(m)} = \begin{bmatrix} [-P,P] & \cdots & [-P,P] \\ \vdots & \ddots & \vdots \\ [-P,P] & \cdots & [-P,P] \end{bmatrix},$$

such that $R_{\mathrm{I}}^{(m+1)} \subseteq P_{\mathrm{I}}^{(m)}$. For instance, if

$$R_{\mathrm{I}}^{(m+1)} = \begin{bmatrix} [-0.1,0.1] & [0.1,0.2] \\ [-0.2,0.4] & [0.1,0.2] \end{bmatrix},$$

then

$$\| R_{\mathrm{I}}^{(m+1)} \| = max\{ 0.1+0.2, 0.4+0.2 \} = 0.6;$$

we can pick

$$P_{\mathrm{I}}^{(m)} = \begin{bmatrix} [-0.6,0.6] & [-0.6,0.6] \\ [-0.6,0.6] & [-0.6,0.6] \end{bmatrix}.$$

In general, we can pick $\mathrm{P} = \| E_{\mathrm{I}} \|^{m+1}/(1-\| E_{\mathrm{I}} \|)$ since $\| R_{\mathrm{I}}^{(m+1)} \| \leq \| E_{\mathrm{I}} \|^{m+1}/(1-\| E_{\mathrm{I}} \|)$. Thus we can rewrite

$$\begin{aligned} A_r^{-1} &= G[\, S_r^{(m)} + R_r^{(m+1)} \,] \\ &\in G[\, S_{\mathrm{I}}^{(m)} + R_{\mathrm{I}}^{(m+1)} \,] \\ &\subseteq G[\, S_{\mathrm{I}}^{(m)} + P_{\mathrm{I}}^{(m)} \,] \\ &= M_{\mathrm{I}}^{(m)}, \end{aligned}$$

where

$$S_{\mathrm{I}}^{(m)} = \mathrm{I} + E_{\mathrm{I}}(\mathrm{I} + E_{\mathrm{I}} (\mathrm{I}+\cdots+E_{\mathrm{I}} (\mathrm{I} + E_{\mathrm{I}})\cdots)).$$

Step 5. Answer: The desired $[A_{\mathrm{I}}]^{-1}$ is $M_{\mathrm{I}}^{(m)}$.

Note that the inverse matrix so obtained is usually much less conservative. In summary, we have an effective formula for the inverse of a regular interval matrix A_{I}:

$$[A_{\mathrm{I}}]^{-1} = G[\, S_{\mathrm{I}}^{(m)} + P_{\mathrm{I}}^{(m)} \,],$$

where

$$S_{\mathrm{I}}^{(m)} = \mathrm{I} + E_{\mathrm{I}} (\mathrm{I} + E_{\mathrm{I}} (\mathrm{I}+\cdots+E_{\mathrm{I}} (\mathrm{I} + E_{\mathrm{I}})\cdots)),$$

$$P_{\mathrm{I}}^{(m)} = \begin{bmatrix} [-P,P] & \cdots & [-P,P] \\ \vdots & \ddots & \vdots \\ [-P,P] & \cdots & [-P,P] \end{bmatrix},$$

$$P = \frac{\|E_{\mathrm{I}}\|^{m+1}}{1-\|E_{\mathrm{I}}\|},$$

with

$$G = A_r^{-1} \ (A_r \in A_{\mathrm{I}}),$$

$$E_{\mathrm{I}} = \mathrm{I} - A_{\mathrm{I}}G,$$

$$\| E_{\mathrm{I}} \| = \max_{1 \le i \le n} \sum_{j=1}^{m} | E_{\mathrm{I}}(i,j) |.$$

Example 1.10. Given

$$A_{\mathrm{I}} = \begin{bmatrix} [0.999,1.01] & [-0.001,0.001] \\ [-0.001,0.001] & [0.999,1.01] \end{bmatrix}.$$

Find $[A_{\mathrm{I}}]^{-1}$.

Step 1. We pick

$$A_r = \frac{1}{2}[\underline{A}_{\mathrm{I}}, \overline{A}_{\mathrm{I}}] = \begin{bmatrix} 1.0045 & 0 \\ 0 & 1.0045 \end{bmatrix},$$

where

$$\underline{A}_{\mathrm{I}} = \begin{bmatrix} 0.999 & -0.001 \\ -0.001 & 0.999 \end{bmatrix}$$

and

$$\overline{A}_{\mathrm{I}} = \begin{bmatrix} 1.010 & 0.001 \\ 0.001 & 1.010 \end{bmatrix}.$$

Step 2. Compute

$$G = A_r^{-1} = \begin{bmatrix} 0.9955201 & 0 \\ 0 & 0.9955201 \end{bmatrix}.$$

Step 3. Compute

$$E_r = \mathrm{I} - A_r G = 0,$$

$$E_{\mathrm{I}} = \mathrm{I} - A_{\mathrm{I}}G =$$

$$\begin{bmatrix} [-5.475283\times10^{-3}, 5.475402\times10^{-3}] & [-9.955202\times10^{-4}, 9.955202\times10^{-4}] \\ [-9.955202\times10^{-4}, 9.955202\times10^{-4}] & [-5.475283\times10^{-3}, 5.475402\times10^{-3}] \end{bmatrix},$$

and

$$\| E_r \| = 0 < 1,$$

$$\begin{aligned} \| E_{\mathrm{I}} \| = \max\{ & \max\{ |-5.475283\times10^{-3}|, |5.475283\times10^{-3}| \} \\ & + \max\{ |-9.955202\times10^{-4}|, |9.955202\times10^{-4}| \}, \\ & \max\{ |-5.475283\times10^{-3}|, |5.475283\times10^{-3}| \} + \\ & + \max\{ |-9.955202\times10^{-4}|, |9.955202\times10^{-4}| \} \} \end{aligned}$$

$$= 6.470881 \times 10^{-3} < 1.$$

Step 4. Compute

$$P = \frac{\|E_{\mathbf{I}}\|^{m+1}}{1-\|E_{\mathbf{I}}\|}.$$

If we pick $m = 1$, say, then we have

$$P = \frac{(1.294184\times10^{-2})^2}{1-1.294184\times10^{-2}} = 4.2145 \times 10^{-5},$$

and so we have

$$P_{\mathbf{I}}^{(1)} =$$

$$\begin{bmatrix} [-4.2145\times10^{-5}, 4.2145\times10^{-5}] & [-4.2145\times10^{-5}, 4.2145\times10^{-5}] \\ [-4.2145\times10^{-5}, 4.2145\times10^{-5}] & [-4.2145\times10^{-5}, 4.2145\times10^{-5}] \end{bmatrix}.$$

Also, we have

$$S_{\mathbf{I}}^{(1)} = \mathrm{I} + E_{\mathbf{I}} =$$

$$\begin{bmatrix} [0.994524, 1.005475] & [-9.955202\times10^{-4}, 9.955202\times10^{-4}] \\ [-9.955202\times10^{-4}, 9.955202\times10^{-4}] & [0.994524, 1.005475] \end{bmatrix}.$$

Step 5. Answer:

$$[A_{\mathbf{I}}]^{-1} = M_{\mathbf{I}}^{(1)} = G[\, S_{\mathbf{I}}^{(1)} + P_{\mathbf{I}}^{(1)}\,]$$

$$= \begin{bmatrix} [0.990027, 1.001013] & [-0.001033, 0.001033] \\ [-0.001033, 0.001033] & [0.990027, 1.001013] \end{bmatrix}.$$

In this example, if we pick $m = 2$, then, we will obtain

$$[A_{\mathbf{I}}]^{-1} = M_{\mathbf{I}}^{(2)} =$$

$$\begin{bmatrix} [0.990038, 1.00100] & [-0.001021, 0.001021] \\ [-0.001021, 0.001021] & [0.990038, 1.00100] \end{bmatrix}.$$

Obviously, $M_{\mathbf{I}}^{(2)} \subset M_{\mathbf{I}}^{(1)}$, giving a less conservative result to the inverse matrix. In the limit, we will have

$$M_{\mathbf{I}}^{(1)} \supset M_{\mathbf{I}}^{(2)} \supset \cdots \supset [A_{\mathbf{I}}]^{-1}.$$

IV. OPERATIONS ON FUZZY SETS

A. Fuzzy Subsets

Let S_f be a fuzzy subset defined in a universe set $\mathbf{S}$, together with a membership function $\mu_{S_f}(s)$, as shown in Figure 1.9.

Definition 1.8. The two subsets, S_α and $S_{\bar{\alpha}}$, of S_f defined by

$$S_\alpha = \{\, s \in S_f \mid \mu_{S_f}(s) > \alpha \,\} \qquad \alpha \in [0,1)$$

$$S_{\bar{\alpha}} = \{\, s \in S_f \mid \mu_{S_f}(s) \geq \alpha \,\} \qquad \alpha \in (0,1]$$

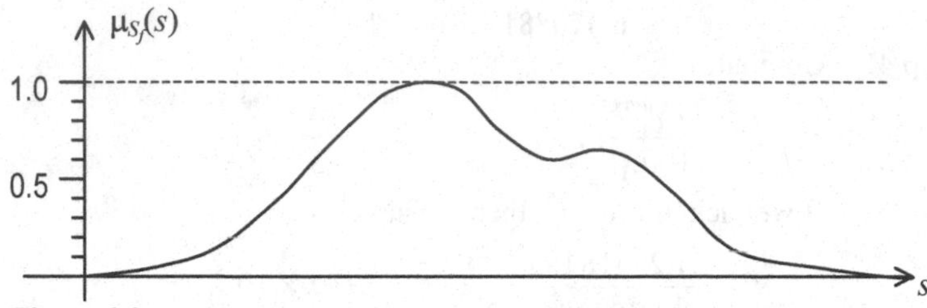

Figure 1.9 A membership function associated with a fuzzy subset.

are called the *strong* α-*cut* and *weak* α-*cut*, respectively.

The concept of α-cuts is visualized in Figure 1.10.

We remark that the weak α-cut is also called the α *level-set*, which is easier to deal with in general. We also remark that if the membership function is continuous, then the distinction of strong and weak α-cuts is not necessary in applications. We only use α level-sets below.

Definition 1.9. A fuzzy subset S_f of **S** = R is *convex* if and only if every ordinary subset (α level-set)

$$S_{\bar{\alpha}} = \{ s \in S_f \mid \mu_{S_f}(s) \geq \alpha \} \qquad \alpha \in (0,1]$$

is convex, namely, for any $s_1, s_2 \in S_f$ and any $\lambda \in [0,1]$,

$$\mu_{S_f}(\lambda s_1+(1-\lambda)s_2) \geq \min\{ \mu_{S_f}(s_1), \mu_{S_f}(s_2) \}.$$

We remark that in Figure 1.10, $S_{0.8}$ is convex but $S_{0.6}$ is not. If **S** = R and the membership function is continuous, then the α level-set of a convex fuzzy set is a closed interval.

To proceed further, we need some new notation. For two real numbers s_1 and s_2, we define

$$\begin{cases} s_1 \wedge s_2 = \min\{s_1, s_2\}, \\ s_1 \vee s_2 = \max\{s_1, s_2\}. \end{cases}$$

Recall also that for a subset S of R, the ordinary *characteristic function* is defined by

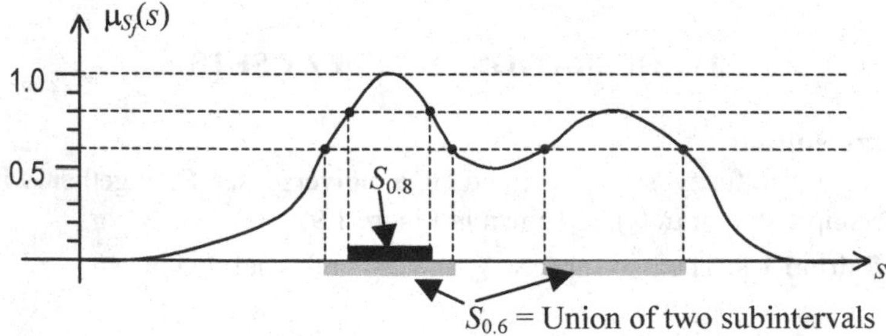

Figure 1.10 Examples of α-cuts.

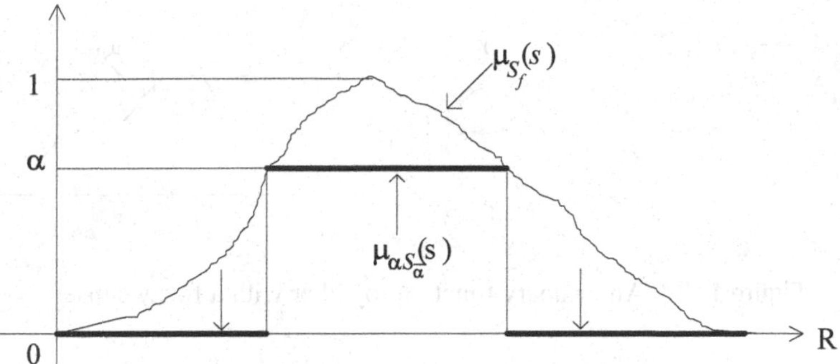

Figure 1.11 Graph of the function $\mu_{\alpha S_{\bar{\alpha}}}(s)$.

$$\mathbf{X}_S(s) = \begin{cases} 1 & \text{if } s \in S, \\ 0 & \text{if } s \notin S. \end{cases}$$

Theorem 1.16. Let S_f be a fuzzy subset with the membership function $\mu_{S_f}(s)$, and let $\mathbf{X}_S$ be the characteristic function of the set S. Then for any $s \in S_f$, we have

$$\mu_{S_f}(s) = \sup_{\alpha \in (0,1]} \{ \alpha \wedge \mathbf{X}_{S_{\bar{\alpha}}}(s) \}$$

and

$$\mu_{S_f}(s) = \sup_{\alpha \in [0,1)} \{ \alpha \wedge \mathbf{X}_S(s) \}$$

Proof. Since

$$\mathbf{X}_{S_{\bar{\alpha}}}(s) = \begin{cases} 1 & \text{if } s \in S_{\bar{\alpha}} \ (i.e., \mu_{S_f}(s) \geq \alpha), \\ 0 & \text{if } s \notin S_{\bar{\alpha}} \ (i.e., \mu_{S_f}(s) < \alpha), \end{cases}$$

we have, for each $s \in S_f$,

$$\begin{aligned} &\sup_{\alpha \in (0,1]} \{ \alpha \wedge \mathbf{X}_{S_{\bar{\alpha}}}(s) \} \\ &= \sup_{\alpha \in (0,\mu_{S_f}(s)]} \{ \alpha \wedge \mathbf{X}_{S_{\bar{\alpha}}}(s) \} \vee \sup_{\alpha \in (\mu_{S_f}(s),1]} \{ \alpha \wedge \mathbf{X}_{S_{\bar{\alpha}}}(s) \} \\ &= \sup_{\alpha \in (0,\mu_{S_f}(s)]} \{ \alpha \wedge 1 \} \vee \sup_{\alpha \in (\mu_{S_f}(s),1]} \{ \alpha \wedge 0 \} \\ &= \sup_{\alpha \in (0,\mu_{S_f}(s)]} \{ \alpha \} \vee 0 \\ &= \mu_{S_f}(s). \end{aligned}$$

The other equality can be similarly proved.

Now, we introduce a new fuzzy subset, $\alpha S_{\bar{\alpha}}$, by defining

$$\alpha S_{\bar{\alpha}} = \{ s \in S_f \mid \mu_{\alpha S_{\bar{\alpha}}}(s) = \alpha \wedge \mathbf{X}_{S_{\bar{\alpha}}}(s) \}.$$

Then, it can be verified that

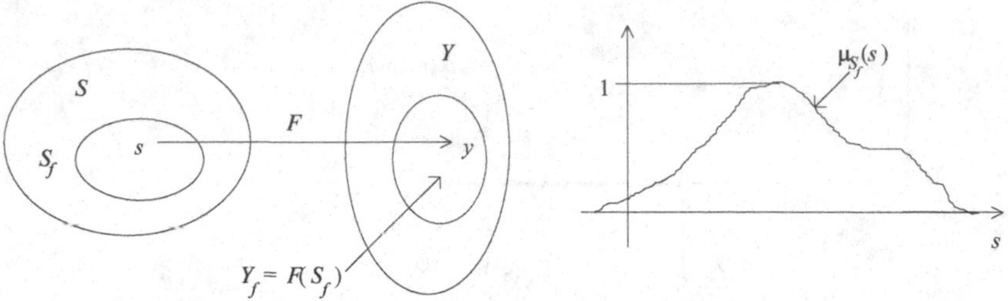

Figure 1.12 An ordinary function together with a fuzzy subset.

$$S_f = \bigcup_{\alpha \in (0,1]} \alpha S_{\bar{\alpha}},$$

which implies that the fuzzy subset has been completely decomposed into the union of $\alpha S_{\bar{\alpha}}$ for all $\alpha \in (0,1]$. This is the *Resolution Principle.*

A few remarks are in order. First, for the fuzzy subset $\alpha S_{\bar{\alpha}}$ defined above, we have

$$\mu_{\alpha S_{\bar{\alpha}}}(s) = \alpha \wedge \mathbf{X}_{S_{\bar{\alpha}}}(s) = \begin{cases} \alpha, & \text{if } s \in S_{\bar{\alpha}}, \\ 0, & \text{if } s \notin S_{\bar{\alpha}}. \end{cases}$$

This is visualized in Figure 1.11.

Second, it is clear from Figure 1.11 that

$$\alpha_1 < \alpha_2 \quad \Rightarrow \alpha_1 S_{\bar{\alpha}_1} \supset \alpha_2 S_{\bar{\alpha}_2}.$$

Third, there is a one-to-one correspondence:

$$\mu_{S_f}(s) \xleftrightarrow{1-1} \alpha S_{\bar{\alpha}} \qquad \alpha \in (0,1],$$

which means that a fuzzy subset can be defined and described by α-cuts only (without using a membership function). This is the so-called *Representation Theorem.*

Next, we introduce the important *Extension Principle.*

Consider an ordinary function $F{:}S \to Y$, mapping from an ordinary set S to another ordinary set Y. Let S_f be a fuzzy subset of S with an associated membership function $\mu_{S_f}(s)$, as shown in Figure 1.12.

Since S_f is a fuzzy subset, the image of S_f under F, namely, the subset $F(S_f)$ defined by

$$Y_f = F(S_f) = \{\, y \in Y \mid y = F(s),\ s \in S_f \,\},$$

is also a fuzzy subset. The question is: What is the membership function $\mu_{Y_f}(y)$ associated with the fuzzy subset Y_f? The answer is: For any fixed $y \in Y$, there are some $s \in S_f$ satisfying $y = f(s)$, and so we can define

$$\mu_{Y_f}(y) = \sup_{s:F(s)=y} \mu_{Y_f}(s).$$

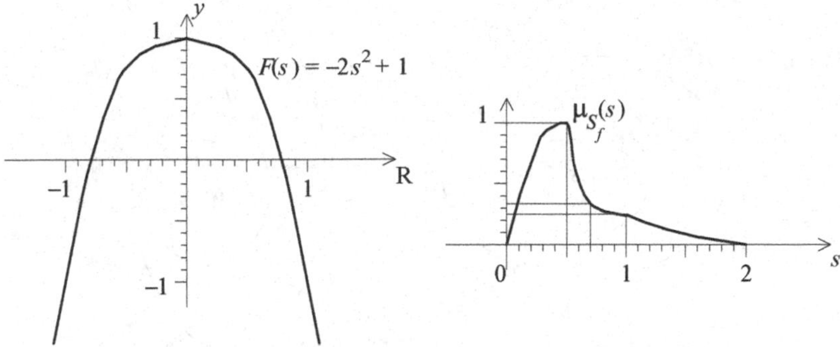

Figure 1.13 An example of the Extension Principle.

In doing so, we have extended the domain and range of an ordinary function from ordinary sets to fuzzy subsets. This is the important *Extension Principle*, where f is called an *extended function*.

Example 1.11. Consider the ordinary function

$$y = F(s) = -2s^2 + 1,$$

with domain $S = \mathrm{R}$ and range $Y = (-\infty,1]$. Suppose that $S_f = [0,2]$ is a fuzzy subset with the membership function shown in Figure 1.13.

The fuzzy subset $Y_f = F(S_f)$ is given via the interval arithmetic by

$$\begin{aligned} Y_f &= F(S_f) = F([0,2]) = -2[0,2]\cdot[0,2] + 1 \\ &= -2[0,4] + 1 = [-8,0] + 1 \\ &= [-7,1]. \end{aligned}$$

The membership function $\mu_{Y_f}(s)$ associated with Y_f is determined as follows. Let y run through from -7 to 1. For each y, find the corresponding $s \in S_f$ satisfying $y = F(s)$. Then for all such s, calculate $\mu_{Y_f}(s)$: if there is only one s satisfying $y = F(s)$, then

$$\mu_{Y_f}(y) = \mu_{Y_f}(s)$$

at this s; if there are more than one s satisfying $y = F(s)$, then

$$\mu_{Y_f}(y) = \sup_{s:F(s)=y} \mu_{S_f}(s).$$

In this example, it is clear that for any $y \in [-7,1]$, there is always one $s \in [0,2]$ satisfying $y = F(s) = -2s^2 + 1$. Hence, it can be easily verified that the

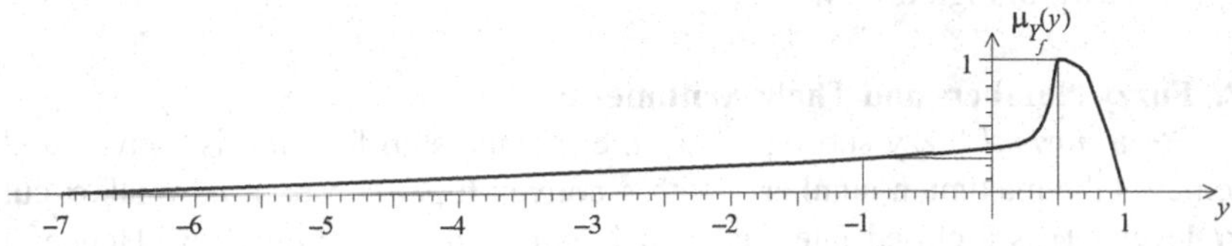

Figure 1.14 The resulting membership function of Example 1.11.

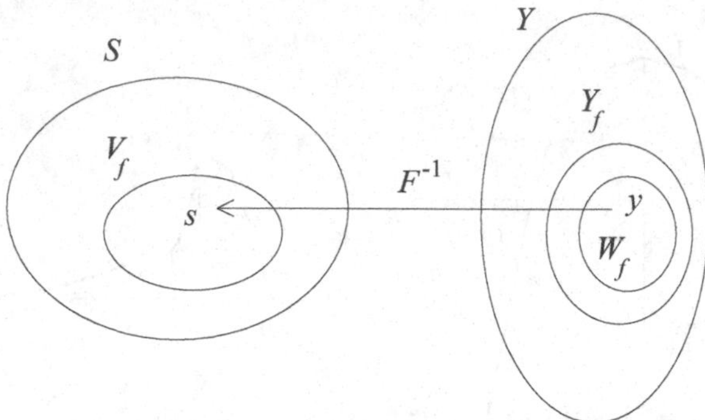

Figure 1.15 The inverse image of an ordinary function defined on a fuzzy subset.

membership function $\mu_{Y_f}(y)$ is as shown in Figure 1.14. For example, when $s = 1/2$, $\mu_{S_f}(1/2) = 1$ from Figure 1.13. For this value of s, we have $y = -2(1/2)^2 + 1 = 1/2$. Hence, we let $\mu_{Y_f}(1/2) = \mu_{S_f}(1/2) = 1$, which yields the point $(y, \mu_{Y_f}(y)) = (1/2,1)$ in Figure 1.14. As another example, when $s = 2$, we have $\mu_{S_f}(0) = 0$ from Figure 1.13. Then, $y = -2(2)^2 + 1 = -7$. At this value, we let $\mu_{Y_f}(-7) = \mu_{S_f}(2) = 0$, which yields the point $(y, \mu_{Y_f}(y)) = (-7,0)$ in Figure 1.14. In the next section, we will show that there is an easier routine procedure to do such calculations.

If the ordinary function $y = F(s)$ is one-to-one and onto, then its inverse function exists: $s = F^{-1}(y)$. Using the fuzzy subset Y_f and its associated membership function $\mu_{Y_f}(y)$ introduced above, we can define the *inverse image* of a fuzzy subset $W_f \subseteq Y_f$ (see Figure 1.15) as follows. Let

$$V_f = F^{-1}(W_f) = \{\, s \in S \mid s = F^{-1}(y),\, y \in W_f\}.$$

The question is, again, what is the membership function $\mu_{V_f}(s)$ associated with the fuzzy subset V_f? To answer this question, we define

$$\mu_{V_f}(s) = \mu_{W_f}(s)\,|_{y=f(s)} = \mu_{Y_f}(y)\,|_{y=f(s)},$$

where we don't have to consider the "sup" on the right-hand side since F is one-to-one, namely, there is always one and only one corresponding value.

If we consider Example 1.11 again, where $y = F(s) = -2s^2 + 1$, which is one-to-one from $[0,2]$ to $[-7,1]$, we see that the inverse image has the membership function $\mu_{S_f}(s)$ shown in Figure 1.13, which is obtained from $\mu_{V_f}(y)$ shown in Figure 1.14.

B. Fuzzy Numbers and Their Arithmetic

For a *normal* fuzzy subset, where the membership function is convex and achieves the maximum number 1 with a convex fuzzy subset, if its weak α-cut (α level-set) is a closed interval, then it is called a *fuzzy number*. Hence, a fuzzy number is a convex fuzzy subset, which has a normalized membership function and represents an interval of confidence.

We remark that the fuzzy subset, with subset $[0,\infty)$ and the associate fuzzy membership function $\mu(x) = 1 - e^{-x}$, is considered to be normal. Although this membership function does not attain the maximum value 1, it approaches 1 as a limit. We consider all such fuzzy subsets as normal in this book.

For the arithmetic of fuzzy numbers, there is a general rule.

The General Rule. Let S_x and S_y be two fuzzy subsets of the universe set S, $Z \subseteq R$, and consider a two-variable extended function

$$F: S_x \times S_x \to Z.$$

Let S_z be the image of F, which is a fuzzy subset of Z as discussed above, and $\mu_{S_x}(x)$, $\mu_{S_y}(y)$, and $\mu_{S_z}(z)$ be the associate membership functions. Given $\mu_{S_x}(x)$ and $\mu_{S_y}(y)$, we define

$$\mu_{S_z}(z) = \sup_{z=F(x,y)} \{ \mu_{S_x}(x) \wedge \mu_{S_y}(y) \}.$$

The reason for choosing the smaller value in this definition is that when one has two different degrees of confidence about two events, then the confidence about both events together is lower. Using the α-cut notation, this is equivalent to the following:

$$\begin{aligned}(S_z)_{\overline{\alpha}} &= F(\ (S_x)_{\overline{\alpha}},\ (S_y)_{\overline{\alpha}}\ ) \\ &= \{ z \in S_z \mid z = F(x,y), x \in (S_x)_{\overline{\alpha}}, y \in (S_y)_{\overline{\alpha}} \}.\end{aligned}$$

Now, given two fuzzy numbers $\tilde{x}$ and $\tilde{y}$, which are two normal and convex fuzzy subsets $S_{\tilde{x}}$ and $S_{\tilde{y}}$ with membership functions $\mu_{S_{\tilde{x}}}$ and $\mu_{S_{\tilde{y}}}$, respectively, consider a two-variable function

$$F: X \times Y \to Z$$

defined by $z = F(x,y)$, $x \in S_{\tilde{x}}$, $y \in S_{\tilde{y}}$, via the general operation rule. We only discuss the fuzzy number arithmetic for addition, subtraction, multiplication, division, minimum, and maximum operations. These results will be needed in the next chapter.

(1) Addition. Let $z = F(x,y) = x + y$. Then $\tilde{z} = \tilde{x} + \tilde{y}$ with

$$S_{\tilde{z}} = \{ z \in Z \mid z = x + y, x \in S_{\tilde{x}}, y \in S_{\tilde{y}} \}$$

and

$$\mu_{S_{\tilde{z}}}(z) = \sup_{z=x+y} \{ \mu_{S_{\tilde{x}}}(x) \wedge \mu_{S_{\tilde{y}}}(y) \}.$$

In the α-cut notation:

$$(S_{\tilde{z}})_{\overline{\alpha}} = F(\ (S_{\tilde{x}})_{\overline{\alpha}},\ (S_{\tilde{y}})_{\overline{\alpha}}\ ) = (S_{\tilde{x}})_{\overline{\alpha}} + (S_{\tilde{y}})_{\overline{\alpha}}.$$

Example 1.12. Let $\tilde{x}$ and $\tilde{y}$ be such that

$$S_{\tilde{x}} = [-5,1], \qquad S_{\tilde{y}} = [-5,12],$$

with associate membership functions shown in Figure 1.16:

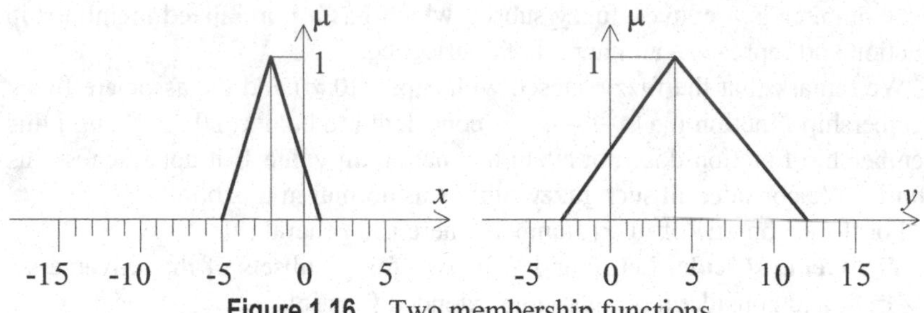

Figure 1.16 Two membership functions.

$$\mu_{S_{\tilde{x}}}(x) = \begin{cases} \frac{x}{3}+\frac{5}{3}, & -5\leq x\leq -2, \\ -\frac{x}{3}+\frac{1}{3}, & -2\leq x\leq 1, \end{cases}$$

and

$$\mu_{S_{\tilde{y}}}(y) = \begin{cases} 0 & -5\leq y\leq -3, \\ \frac{y}{7}+\frac{3}{7} & -3\leq y\leq 4, \\ -\frac{y}{8}+\frac{12}{8} & 4\leq y\leq 12. \end{cases}$$

Then, by the general operation rule we have

$$S_{\tilde{z}} = S_{\tilde{x}} + S_{\tilde{y}} = [-5,1] + [-5,12] = [-10,13]$$

and, by comparing $\mu_{S_{\tilde{x}}}(x)$ and $\mu_{S_{\tilde{y}}}(y)$ pointwise, we obtain

$$\mu_{S_{\tilde{z}}}(z) = \sup_{z=x+y} \{ \mu_{S_{\tilde{x}}}(x) \wedge \mu_{S_{\tilde{y}}}(y) \}$$

$$= \begin{cases} 0, & -10 \leq z \leq -8, \\ \frac{z}{10}+\frac{8}{10}, & -8 \leq z \leq 2, \\ -\frac{z}{11}+\frac{13}{11}, & 2 \leq z \leq 13. \end{cases}$$

Here, it is clear that the general "sup" rule does not yield the explicit formulas easily. In contrast, this resulting explicit formula of $\mu_{S_{\tilde{z}}}(z)$ can be easily obtained by using the equivalent α-cut operation as follows.

In the α-cut notation, for any α value, the α-cut of $S_{\tilde{x}}$ is obtained by letting $\alpha = x/3 + 5/3$ and $\alpha = -x/3 + 1/3$, respectively, which give $x_1 = 3\alpha - 5$ and $x_2 = -3\alpha + 1$, as shown in (the enlarged) Figure 1.17. Hence, the projection interval is

$$(S_{\tilde{x}})_{\overline{\alpha}} = [x_1,x_2] = [3\alpha - 5, -3\alpha + 1].$$

Similarly,

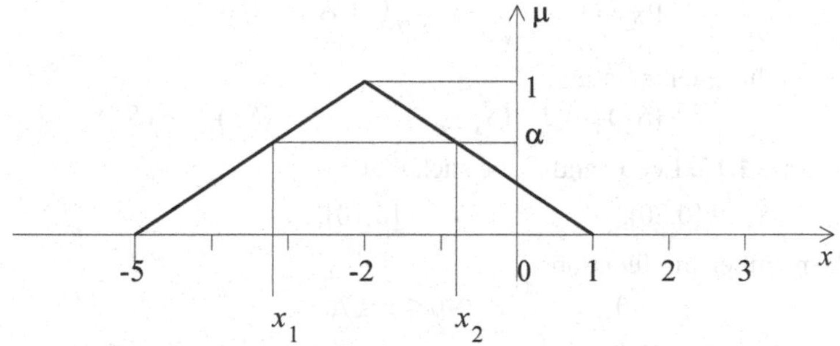

Figure 1.17 The α-cut of the membership function $\mu_{S_{\tilde{x}}}(x)$.

$$(S_{\tilde{y}})_{\overline{\alpha}} = [7\alpha - 3, -8\alpha + 12],$$

so that

$$(S_{\tilde{z}})_{\overline{\alpha}} = (S_{\tilde{x}})_{\overline{\alpha}} + (S_{\tilde{y}})_{\overline{\alpha}} = [10\alpha - 8, -11\alpha + 13].$$

Setting $z_1 = 10\alpha - 8$ and $z_2 = -11\alpha + 13$ gives $\alpha = z_1/10 + 8/10$ and $\alpha = -z_2/11 + 12/11$, which yield the membership function shown in Figure 1.18.

Taking $S_{\tilde{z}} = [-10, 13]$ into account, we finally arrive at

$$\mu_{S_{\tilde{z}}}(z) = \begin{cases} 0, & -10 \le z \le -8, \\ \dfrac{z}{10} + \dfrac{8}{10}, & -8 \le z \le 2, \\ -\dfrac{z}{11} + \dfrac{13}{11}, & 2 \le z \le 13. \end{cases}$$

(2) Subtraction. Let $z = F(x,y) = x - y$. Then $\tilde{z} = \tilde{x} - \tilde{y}$ with

$$S_{\tilde{z}} = \{ z \in Z \mid z = x - y, x \in S_{\tilde{x}}, y \in S_{\tilde{y}} \}$$

and

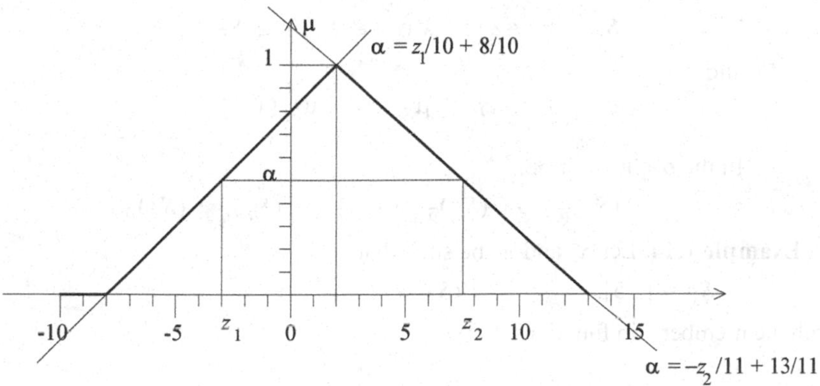

Figure 1.18 The resulting membership function of Example 1.12.

$$\mu_{S_{\tilde{z}}}(z) = \sup_{z=x-y} \{ \mu_{S_{\tilde{x}}}(x) \wedge \mu_{S_{\tilde{y}}}(y) \}.$$

In the α-cut notation:

$$(S_{\tilde{z}})_{\overline{\alpha}} = F((S_{\tilde{x}})_{\overline{\alpha}}, (S_{\tilde{y}})_{\overline{\alpha}}) = (S_{\tilde{x}})_{\overline{\alpha}} - (S_{\tilde{y}})_{\overline{\alpha}} .$$

Example 1.13. Let $\tilde{x}$ and $\tilde{y}$ be such that

$$S_{\tilde{x}} = [0,20], \qquad S_{\tilde{y}} = [0,10],$$

with the membership functions

$$\mu_{S_{\tilde{x}}}(x) = \begin{cases} 0, & 0 \le x \le 7, \\ \dfrac{x}{7} - 1, & 7 \le x \le 14, \\ -\dfrac{x}{5} + \dfrac{19}{5}, & 14 \le x \le 19, \\ 0, & 19 \le x \le 20, \end{cases}$$

and

$$\mu_{S_{\tilde{y}}}(y) = \begin{cases} 0, & 0 \le y \le 3, \\ \dfrac{y}{2} - \dfrac{3}{2}, & 3 \le y \le 5, \\ -\dfrac{y}{5} + 2, & 5 \le y \le 10. \end{cases}$$

Then we have, via the interval arithmetic,

$$S_{\tilde{z}} = S_{\tilde{x}} - S_{\tilde{y}} = [-10,20],$$

with

$$\mu_{S_{\tilde{z}}}(z) = \begin{cases} 0, & -10 \le z \le -3, \\ \dfrac{z}{12} + \dfrac{3}{12} & -3 \le z \le 9, \\ -\dfrac{z}{7} + \dfrac{16}{7} & 9 \le z \le 16, \\ 0, & 16 \le z \le 20. \end{cases}$$

(3) Multiplication. Let $z = F(x,y) = x \times y$. Then $\tilde{z} = \tilde{x} \times \tilde{y}$ with

$$S_{\tilde{z}} = \{ z \in Z \mid z = x\,y, x \in S_{\tilde{x}}, y \in S_{\tilde{y}} \}$$

and

$$\mu_{S_{\tilde{z}}}(z) = \sup_{z=x\times y} \{ \mu_{S_{\tilde{x}}}(x) \wedge \mu_{S_{\tilde{y}}}(y) \}.$$

In the α-cut notation:

$$(S_{\tilde{z}})_{\overline{\alpha}} = F((S_{\tilde{x}})_{\overline{\alpha}}, (S_{\tilde{y}})_{\overline{\alpha}}) = (S_{\tilde{x}})_{\overline{\alpha}} \cdot (S_{\tilde{y}})_{\overline{\alpha}} .$$

Example 1.14. Let $\tilde{x}$ and $\tilde{y}$ be such that

$$S_{\tilde{x}} = [2,5], \qquad S_{\tilde{y}} = [3,6],$$

with the membership functions

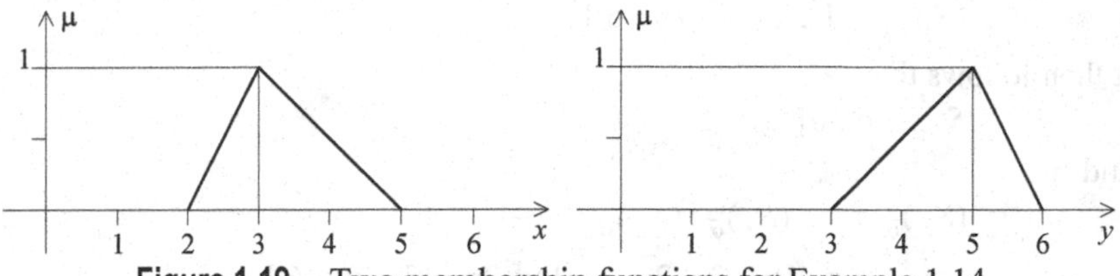

Figure 1.19 Two membership functions for Example 1.14.

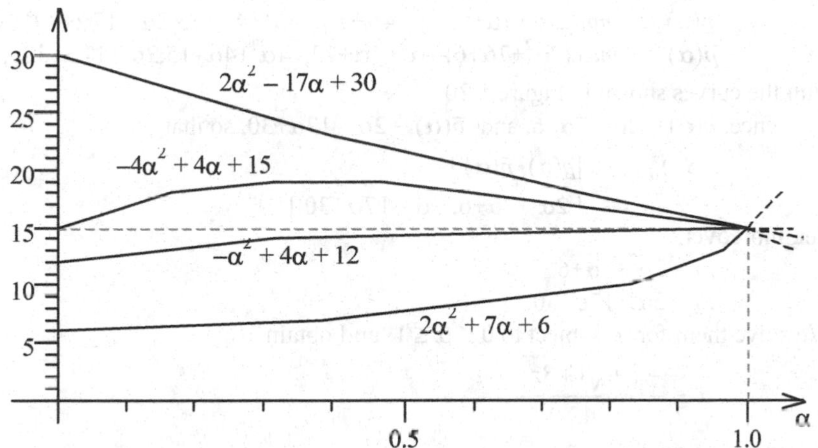

Figure 1.20 The intermediate membership functions of Example 1.14.

$$\mu_{S_{\tilde{x}}}(x) = \begin{cases} x-2, & 2 \le x \le 3, \\ -\dfrac{x}{2}+\dfrac{5}{2}, & 3 \le x \le 5, \end{cases}$$

and

$$\mu_{S_{\tilde{y}}}(y) = \begin{cases} \dfrac{y}{2}-\dfrac{3}{2}, & 3 \le y \le 5, \\ -y+6, & 5 \le y \le 6, \end{cases}$$

as shown in Figure 1.19.

In the α-cut notation, for any α value, letting

$$\alpha = x - 2 \qquad \text{and} \qquad \alpha = -\frac{x}{2} + \frac{5}{2}$$

gives

$$x_1 = \alpha + 2 \qquad \text{and} \qquad x_2 = -2\alpha + 5,$$

so that

$$(S_{\tilde{x}})_{\overline{\alpha}} = [\, \alpha + 2, -2\alpha + 5 \,].$$

Similarly,

$$(S_{\tilde{y}})_{\overline{\alpha}} = [\, 2\alpha + 3, -\alpha + 6 \,].$$

It then follows that

$$S_{\tilde{z}} = [6,30]$$

and

$$\begin{aligned}(S_{\tilde{z}})_{\overline{\alpha}} &= (S_{\tilde{x}})_{\overline{\alpha}} \times (S_{\tilde{y}})_{\overline{\alpha}} \\ &= [\alpha + 2, -2\alpha + 5\,] \times [\, 2\alpha + 3, -\alpha + 6\,] \\ &= [\underline{p}(\alpha), \overline{p}(\alpha)\,],\end{aligned}$$

where

$$\begin{aligned}\underline{p}(\alpha) &= min\{2\alpha^2+7\alpha+6, -\alpha^2+4\alpha+12, -4\alpha^2+4\alpha+15, 2\alpha^2-17\alpha+30\}, \\ \overline{p}(\alpha) &= max\{2\alpha^2+7\alpha+6, -\alpha^2+4\alpha+12, -4\alpha^2+4\alpha+15, 2\alpha^2-17\alpha+30\},\end{aligned}$$

with the curves shown in Figure 1.20.

Hence, $\underline{p}(\alpha) = 2\alpha^2+7\alpha+6$, and $\overline{p}(\alpha) = 2\alpha^2-17\alpha+30$, so that

$$\begin{aligned}(S_{\tilde{z}})_{\overline{\alpha}} &= [\underline{p}(\alpha), \overline{p}(\alpha)\,] \\ &= [\, 2\alpha^2+7\alpha+6,\ 2\alpha^2-17\alpha+30\,].\end{aligned}$$

Let, moreover,

$$z_1 = 2\alpha^2+7\alpha+6,$$
$$z_2 = 2\alpha^2-17\alpha+30.$$

We solve them for α, subject to $0 \le \alpha \le 1$, and obtain

$$\alpha = \frac{-7+\sqrt{1+8z_1}}{4},$$

or

$$\alpha = \frac{17-\sqrt{49+8z_2}}{4}.$$

Consequently, we have

$$\mu_{S_{\tilde{z}}}(z) = \begin{cases} \dfrac{-7+\sqrt{1+8z}}{4}, & 6 \le z \le 15, \\ \dfrac{17-\sqrt{49+8z}}{4}, & 15 \le z \le 30, \end{cases}$$

as shown in Figure 1.21.

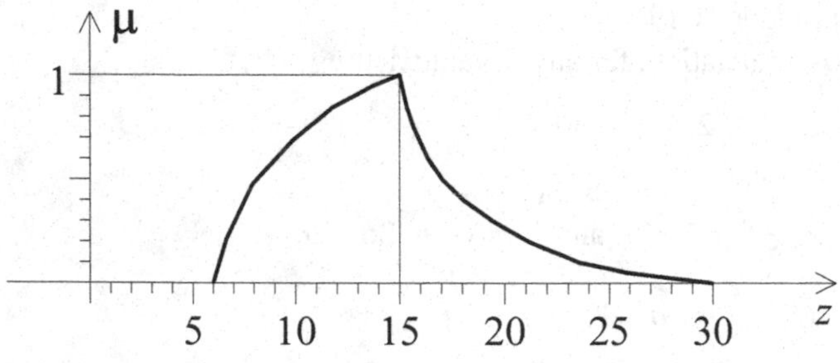

Figure 1.21 The resulting membership function of Example 1.14.

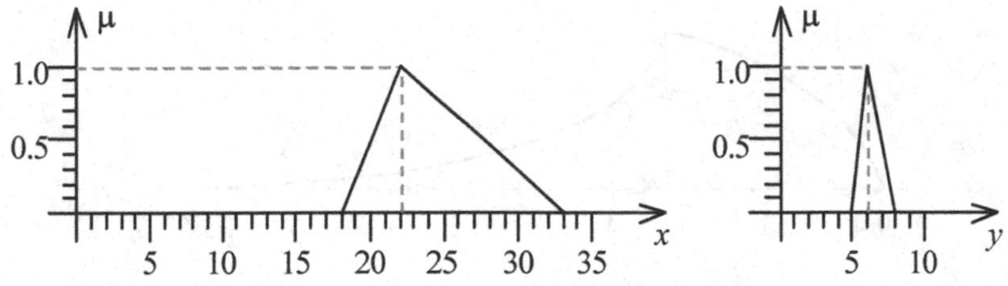

Figure 1.22 The two membership functions for Example 1.15.

(4) Division. Let $z = F(x,y) = \dfrac{x}{y}$. Then $\tilde{z} = \dfrac{\tilde{x}}{\tilde{y}}$ with

$$S_{\tilde{z}} = \{ z \in Z \mid z = x/y,\ x \in S_{\tilde{x}},\ y \in S_{\tilde{y}} \}$$

and

$$\mu_{S_{\tilde{z}}}(z) = \sup_{z=x/y} \{ \mu_{S_{\tilde{x}}}(x) \wedge \mu_{S_{\tilde{y}}}(y) \}.$$

In the α-cut notation:

$$(S_{\tilde{z}})_{\overline{\alpha}} = F(\ (S_{\tilde{x}})_{\overline{\alpha}}\ ,\ (S_{\tilde{y}})_{\overline{\alpha}}\ ) = \frac{(S_{\tilde{x}})_{\overline{\alpha}}}{(S_{\tilde{y}})_{\overline{\alpha}}}.$$

Example 1.15. Let $\tilde{x}$ and $\tilde{y}$ be such that

$$S_{\tilde{x}} = [18,33], \qquad S_{\tilde{y}} = [5,8],$$

with the membership functions

$$\mu_{S_{\tilde{x}}}(x) = \begin{cases} \dfrac{x}{4} - \dfrac{18}{4}, & 18 \le x \le 22, \\ -\dfrac{x}{11} + 3, & 22 \le x \le 33, \end{cases}$$

and

$$\mu_{S_{\tilde{y}}}(y) = \begin{cases} y - 5, & 5 \le y \le 6, \\ -\dfrac{y}{2} + 4, & 6 \le y \le 8, \end{cases}$$

as shown in Figure 1.22.

In the α-cut notation, let

$$\alpha = \frac{x}{4} - \frac{18}{4} \qquad \text{and} \qquad \alpha = -\frac{x}{11} + 3.$$

We obtain

$$x_1 = 4\alpha + 18 \qquad \text{and} \qquad x_2 = -11\alpha + 33,$$

so that

$$(S_{\tilde{x}})_{\overline{\alpha}} = [\ 4\alpha + 18, -11\alpha + 33\ ].$$

Similarly,

$$(S_{\tilde{y}})_{\overline{\alpha}} = [\ \alpha + 5, -2\alpha + 8\ ].$$

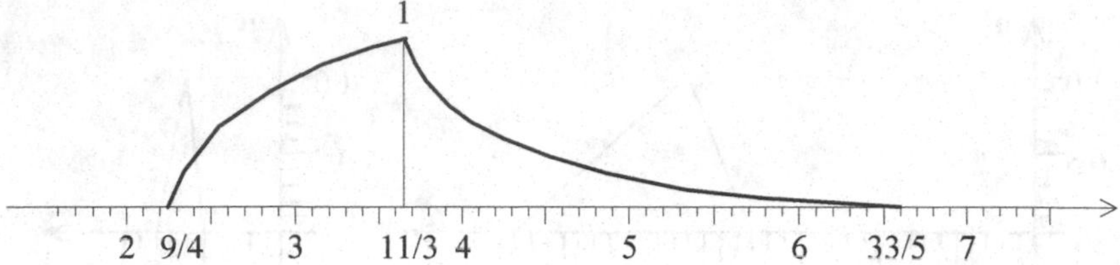

Figure 1.23 The resulting membership function of Example 1.15.

Hence,

$$(S_{\tilde{z}})_{\bar{\alpha}} = \frac{(S_{\tilde{x}})_{\bar{\alpha}}}{(S_{\tilde{y}})_{\bar{\alpha}}} = \frac{[4\alpha+18,-11\alpha+33]}{[\alpha+5,-2\alpha+8]}$$

$$= \left[\frac{4\alpha+18}{-2\alpha+8}, \frac{-11\alpha+33}{\alpha+5}\right].$$

Next, letting

$$z_1 = \frac{4\alpha+18}{-2\alpha+8} \quad \text{and} \quad z_2 = \frac{-11\alpha+33}{\alpha+5}$$

gives

$$\alpha = \frac{8z_1-18}{2z_1+4} \quad \text{and} \quad \alpha = \frac{-5z_2+33}{z_2+11},$$

so that

$$\mu_{S_{\tilde{z}}}(z) = \begin{cases} \dfrac{8z-18}{2z+4}, & \dfrac{9}{4} \le z \le \dfrac{11}{3}, \\ \dfrac{-5z+33}{z+11}, & \dfrac{11}{3} \le z \le \dfrac{33}{5}, \end{cases}$$

as shown in Figure 1.23.

(5) Minimum and Maximum. Let $\tilde{x}$ and $\tilde{y}$ be two fuzzy numbers with $S_{\tilde{x}} = [a,b]$ and $S_{\tilde{y}} = [c,d]$, and with membership functions $\mu_{S_{\tilde{x}}}(x)$ and $\mu_{S_{\tilde{y}}}(y)$, respectively. For any α value, let us denote

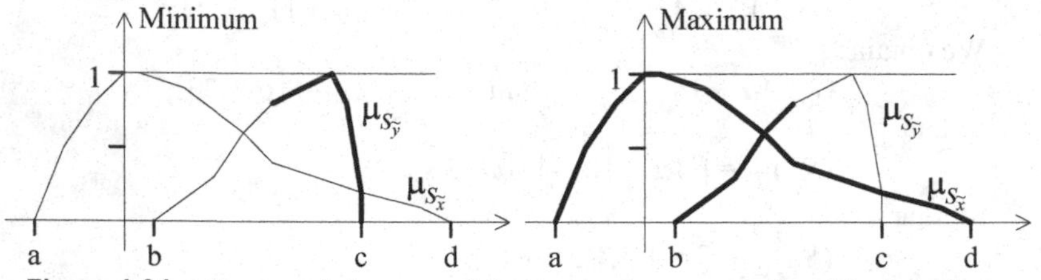

Figure 1.24 Fuzzy minimum and fuzzy maximum membership functions.

$$(S_{\tilde{x}})_{\bar{\alpha}} = [\underline{x}(\alpha), \bar{x}(\alpha)] \qquad \text{and} \qquad (S_{\tilde{y}})_{\bar{\alpha}} = [\underline{y}(\alpha), \bar{y}(\alpha)].$$

Then, the *fuzzy minimum* of $\tilde{x}$ and $\tilde{y}$ is defined to be $\tilde{z} = \tilde{x} \wedge \tilde{y}$, with

$$S_{\tilde{z}} = [\, min\{a,c\}, min\{b,d\} \,]$$

and

$$\begin{aligned} \mu_{S_{\tilde{z}}}(z) &= \sup_{z=min\{x,y\}} \{ \mu_{S_{\tilde{x}}}(x) \wedge \mu_{S_{\tilde{y}}}(y) \} \\ &= \bigvee_{z=x\wedge y} \{ \mu_{S_{\tilde{x}}}(x) \wedge \mu_{S_{\tilde{y}}}(y) \} \end{aligned}$$

In the α-cut notation:

$$\begin{aligned} (S_{\tilde{z}})_{\bar{\alpha}} &= (S_{\tilde{x}})_{\bar{\alpha}} \wedge (S_{\tilde{y}})_{\bar{\alpha}} \\ &= [\, \underline{x}(\alpha) \wedge \underline{y}(\alpha), \bar{x}(\alpha) \wedge \bar{y}(\alpha) \,]. \end{aligned}$$

Similarly, the *fuzzy maximum* of $\tilde{x}$ and $\tilde{y}$ is defined as $\tilde{z} = \tilde{x} \vee \tilde{y}$, with

$$S_{\tilde{z}} = [\, max\{a,c\}, max\{b,d\} \,]$$

and

$$\mu_{S_{\tilde{z}}}(z) = \bigvee_{z=x\vee y} \{ \mu_{S_{\tilde{x}}}(x) \wedge \mu_{S_{\tilde{y}}}(y) \}.$$

In the α-cut notation:

$$\begin{aligned} (S_{\tilde{z}})_{\bar{\alpha}} &= (S_{\tilde{x}})_{\bar{\alpha}} \vee (S_{\tilde{y}})_{\bar{\alpha}} \\ &= [\, \underline{x}(\alpha) \vee \underline{y}(\alpha), \bar{x}(\alpha) \vee \bar{y}(\alpha) \,]. \end{aligned}$$

These two extrema are visualized in Figure 1.24. Graphically, if we draw horizontal lines to cut the two membership functions, then the resulting membership functions have all left-hand or right-hand crossing points, respectively.

Example 1.16. Let $\tilde{x}$ and $\tilde{y}$ be such that

$$S_{\tilde{x}} = [-2,6], \qquad S_{\tilde{y}} = [-4,5],$$

with the membership functions

$$\mu_{S_{\tilde{x}}}(x) = \begin{cases} \dfrac{x}{2} + 1, & -2 \le x \le 0, \\ -\dfrac{x}{6} + 1, & 0 \le x \le 6, \end{cases}$$

$$\mu_{S_{\tilde{y}}}(y) = \begin{cases} \dfrac{y}{7} + \dfrac{4}{7}, & -4 \le y \le 3, \\ -\dfrac{y}{2} + \dfrac{5}{2}, & 3 \le y \le 5, \end{cases}$$

as shown in Figure 1.25.

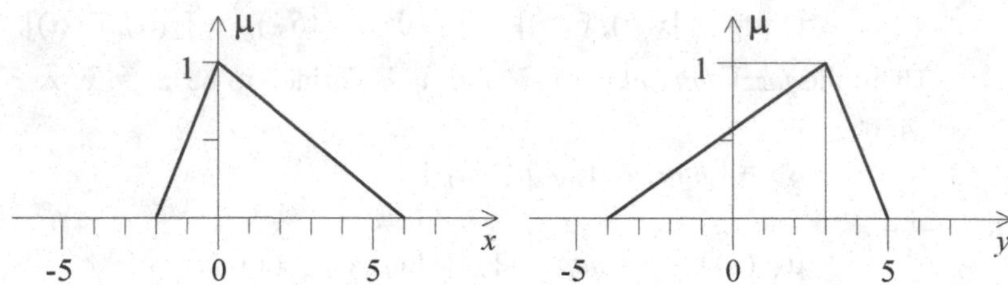

Figure 1.25 Two membership functions in Example 1.16.

In the α-cut notation, following the procedure described above, we obtain

$$(S_{\tilde{x}})_{\overline{\alpha}} = [\, 2\alpha - 2, -6\alpha + 6 \,],$$

and

$$(S_{\tilde{y}})_{\overline{\alpha}} = [\, 7\alpha - 4, -2\alpha + 5 \,],$$

so that

$$\begin{aligned}(S_{\tilde{z}})_{\overline{\alpha}} &= (S_{\tilde{x}})_{\overline{\alpha}} \wedge (S_{\tilde{y}})_{\overline{\alpha}} \\ &= [\, (2\alpha - 2) \wedge (7\alpha - 4), (-6\alpha + 6) \wedge (-2\alpha + 5) \,] \\ &= \begin{cases} [7\alpha - 4, -2\alpha + 5], & 0 \le \alpha \le 1/4, \\ [7\alpha - 4, -6\alpha + 6], & 1/4 \le \alpha \le 2/5, \\ [2\alpha - 2, -6\alpha + 6], & 2/5 \le \alpha \le 1, \end{cases}\end{aligned}$$

in which by matching $2\alpha - 2 = 7\alpha - 4$ and $-6\alpha + 6 = -2\alpha + 5$ we can find the switching points $\alpha = 1/4$ and $\alpha = 2/5$, respectively. It then yields $S_{\tilde{z}} = [-4,5]$, with

$$\mu_{S_{\tilde{z}}}(z) = \begin{cases} -\dfrac{z}{7} + \dfrac{4}{7}, & -4 \le z \le -\dfrac{6}{5}, \\ \dfrac{z}{2} + 1, & -\dfrac{6}{5} \le z \le 0, \\ -\dfrac{z}{6} + 1, & 0 \le z \le \dfrac{9}{2}, \\ -\dfrac{z}{2} + \dfrac{5}{2}, & \dfrac{9}{2} \le z \le 5, \end{cases}$$

as shown in Figure 1.26. Graphically, if we draw horizontal lines to cut the two membership functions, then the resulting membership function has all left-hand crossing points.

Similarly, it can be verified that

$$\begin{aligned}(S_{\tilde{z}})_{\overline{\alpha}} &= (S_{\tilde{x}})_{\overline{\alpha}} \vee (S_{\tilde{y}})_{\overline{\alpha}} \\ &= [\, (2\alpha - 2) \vee (7\alpha - 4), (-6\alpha + 6) \vee (-2\alpha + 5) \,] \\ &= \begin{cases} [2\alpha - 2, -6\alpha + 6], & 0 \le \alpha \le 1/4, \\ [2\alpha - 2, -2\alpha + 5], & 1/4 \le \alpha \le 2/5, \\ [7\alpha - 4, -2\alpha + 5], & 2/5 \le \alpha \le 1, \end{cases}\end{aligned}$$

and

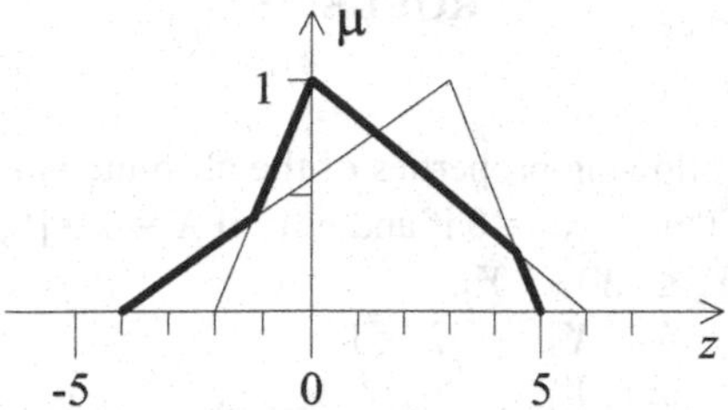

Figure 1.26 The resulting fuzzy minimum membership function.

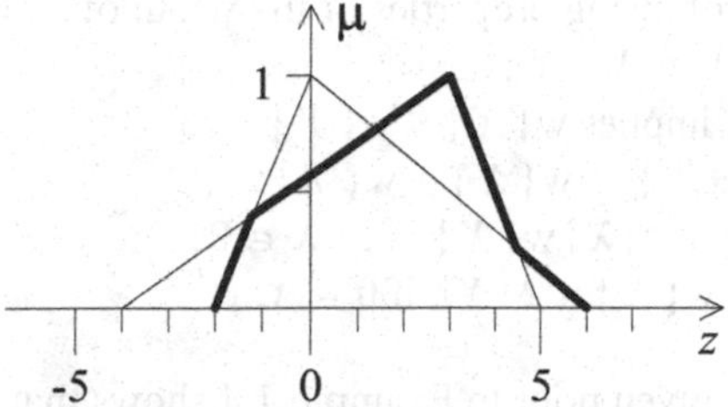

Figure 1.27 The resulting fuzzy maximum membership function.

$$\mu_{S_{\tilde{z}}}(z) = \begin{cases} \dfrac{z}{2}+1, & -2 \le z \le -\dfrac{6}{5}, \\ \dfrac{z}{7}+\dfrac{4}{7}, & -\dfrac{6}{5} \le z \le 3, \\ -\dfrac{z}{2}+\dfrac{5}{2}, & 3 \le z \le \dfrac{9}{2}, \\ -\dfrac{z}{6}+1, & \dfrac{9}{2} \le z \le 6, \end{cases}$$

as shown in Figure 1.27. Graphically, if we draw horizontal lines to cut the two membership functions, then the resulting membership function has all right-hand crossing points.

PROBLEMS

P1.1 Verify the following properties of the absolute value of an interval:

(1) $|X| \geq 0$ and $|X| = 0$ if and only if $X = 0 = [0,0]$;

(2) $|X + Y| \leq |X| + |Y|$;

(3) $|\lambda X| = |\lambda||X|$, $\quad \lambda \in R$;

(4) $|XY| = |X||Y|$;

(5) $X \subseteq Y$ implies $|X| \leq |Y|$.

P1.2 Verify the following properties of the width of an interval:

(1) $w\{X\} = |X - X|$;

(2) $X \subseteq Y$ implies $w\{X\} \leq w\{Y\}$;

(3) $w\{X \pm Y\} = w\{X\} + w\{Y\}$;

(4) $w\{\lambda X\} = |\lambda| w\{X\}$, $\quad \lambda \in R$;

(5) $|X| \leq w\{X\} \leq 2|X|$, if $0 \in X$.

P1.3 The remark given prior to Example 1.4 shows that

$$Z(x+y) \neq Zx + Zy$$

for some points x and y. Theorem 1.1, part 7(c), shows that generally we have

$$Z(x+y) \subseteq Zx + Zy.$$

However, someone tries to show the equality by the following argument:

$$\begin{aligned} Z(x+y) &= \{ z\cdot(x+y) \mid z \in Z \} \\ &= \{ z\cdot x + z\cdot y \mid z \in Z \} \\ &= Zx + Zy. \end{aligned}$$

What is wrong with this proof?

P1.4 Give an example to show that for a constant matrix A and two interval matrices B_I and C_I of appropriate dimensions,

$$(AB_I)C_I \neq A(B_I C_I)$$

even if $C_I = -C_I$. This shows that the equality in Theorem 1.12, part (10), formula (b), does not hold in general.

P1.5 Consider the circuit

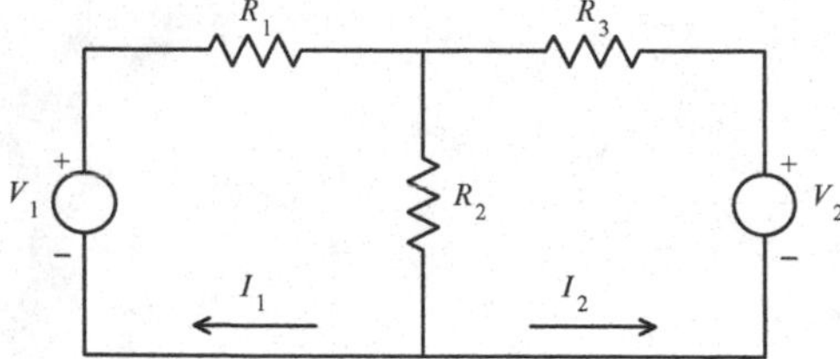

In the normal case we have

$$\begin{bmatrix} R_1+R_2 & R_2 \\ R_2 & R_2+R_3 \end{bmatrix}\begin{bmatrix} I_1 \\ I_2 \end{bmatrix}=\begin{bmatrix} V_1 \\ V_2 \end{bmatrix}.$$

Now, assume that $V_1 = 1$, $V_2 = 0$, but all the resistors have uncertainties, so that

$$R_1 = R_2 = R_3 = R_I = [\,0.98, 1.02\,].$$

Find the interval solution of the current vector by using both the direct matrix inversion method

$$\begin{bmatrix} I_1 \\ I_2 \end{bmatrix} = [R_I]^{-1}\begin{bmatrix} V_1 \\ V_2 \end{bmatrix} = \frac{\text{adj}[R_I]}{\det[R_I]}\begin{bmatrix} V_1 \\ V_2 \end{bmatrix}$$

and the Hansen inverse method. Compare your results.

P1.6 Verify Theorems 1.14 and 1.15.

P1.7* Verify the Resolution Principle

$$S_f = \bigcup_{\alpha\in(0,1]} \alpha S_{\overline{\alpha}},$$

which is formulated and discussed in Section IV.A.

P1.8 Consider the ordinary real function

$$y = F(s) = -2s^2 + 1$$

defined on a fuzzy set $S_f = [-2,2]$ with the membership function

$$\mu_{S_f}(s) = 1 - \tfrac{1}{4}s^2,\; s \in [-2,2].$$

Find the fuzzy set Y_f and its associate membership function $\mu_{Y_f}(y)$.

P1.9 Let $S_x = [1,5]$ and $S_y = [2,6]$ with fuzzy membership functions

$$\mu_{S_x} = \begin{cases} (x-1), & 1 \le x \le 2, \\ (5-x)/3, & 2 \le x \le 5, \end{cases}$$

and

$$\mu_{S_y} = \begin{cases} (y-2)/2, & 2 \le y \le 4, \\ (6-y)/2, & 4 \le y \le 6, \end{cases}$$

respectively. Compute the following six operations to obtain S_z and μ_{S_z}:

$$z = F(x,y) = x * y \text{ for } \quad * \in \{\,+, -, \cdot, /, \max, \min\,\}.$$

P1.10 Consider Figure 1.20. Can these four curves get crossed in a general case? If so, show one example; if not, explain why (a few sentences only).

REFERENCES

[1] G. Alefeld and J. Herzberger. *Introduction to Interval Computations*. New York, NY: Academic Press (1983).

[2] A. Deif. *Sensitivity Analysis in Linear Systems*. New York, NY: Springer-Verlag (1992).

[3] E. Hansen. "Interval Arithmetic in Matrix Computations, Part I." *SIAM Journal of Numerical Analysis*, Vol. 2 (1965), pp 308-320.

[4] E. Hansen. "On Linear Algebraic Equations with Interval Coefficients." *Topics in Interval Analysis* (E. Hansen, ed.). Oxford, England: Claredon Press (1969).

[5] E. Hansen and R. Smith. "Interval Arithmetic in Matrix Computations, Part II." *SIAM Journal of Numerical Analysis*, Vol. 4 (1967), pp 1-9.

[6] A. Kaufmann and M. M. Gupta. *Introduction to Fuzzy Arithmetic: Theory and Applications*. New York, NY: Van Nostrand Reinhold (1991).

[7] R. B. Kearfott. *Rigorous Global Search: Continuous Problems*. Boston, MA: Kluwer Academic Publishers (1996).

[8] G. J. Klir, U. H. St. Clair, and B. Yuan. *Fuzzy Set Theory: Foundations and Applications*. Upper Saddle River, NJ: Prentice-Hall PTR (1997).

[9] R. E. Moore. *Methods and Applications of Interval Analysis*. Philadelphia, PA: SIAM Press (1979).

CHAPTER 2

Fuzzy Logic Theory

Fuzzy logic is a logic. *Logic* refers to the study of methods and principles of human reasoning.

Classical logic, as common practice, deals with propositions (e.g., conclusions or decisions) that are either *true* or *false*. Each proposition has an opposite. This classical logic, therefore, deals with combinations of *variables* that represent propositions. As each variable stands for a hypothetical proposition, any combination of them eventually assumes a truth value (either true or false), but never is in between or both (i.e., is not true and false at the same time).

The main content of classical logic is the study of *rules* that allow new logical variables to be produced as *functions* of certain existing variables. Suppose that n logical variables, $x_1, \ldots, x_n$, are given, say

x_1 is true;
x_2 is false;
⋮
x_n is false.

Then a new logical variable, y, can be defined by a rule as a function of $x_1, \ldots, x_n$ that has a particular truth value (again, either true or false). One example of a rule is the following:

Rule: IF x_1 is true AND x_2 is false AND ... AND x_n is false THEN y is false.

Because one, and only one truth value (either true or false) is assumed by a logical function of a (finite) number of logical variables (hypothetical propositions), the classical logic is also called a *two-valued logic*.

The fundamental assumption upon which the classical logic is based is that every proposition is either true or false. This principle has been questioned by many philosophers, ever since Aristotle (in his treatise *On Interpretation*).

It is now well understood and well accepted that many propositions are both partially true and partially false. To describe such partial truth values by some new rules, in a way to extend and generalize the two-valued logic, multi-valued logics were proposed and developed. As the first attempt, several three-valued logics have now been well established, with their own rationale. It is common in these logics to introduce a "neither" in between "true" and "false." It has turned out that three-valued logics are successful both logically and mathematically. Motivated by the useful three-valued logics, n-valued logics were developed in the 1930s. In particular, the n-valued logic of Lukasiewicz even allows $n = \infty$.

It has lately been understood that there exists an isomorphism between the two-valued logic and the crisp set theory, and, similarly, there is an isomorphism between the Lukasiewicz logic and the fuzzy set theory. In fact, the isomorphism for the former is the standard characteristic function

$$y = X_A(x) = \begin{cases} 1 & \text{if } x \in A, \\ 0 & \text{if } x \notin A, \end{cases}$$

which can be interpreted as

$$y = \begin{cases} \text{true} & \text{if } x \text{ is true}, \\ \text{false} & \text{if } x \text{ is false}; \end{cases}$$

and the latter is a fuzzy membership function, which has been thoroughly studied in Chapter 1.

In this chapter, we study in somewhat detail the classical two-valued logic and its operations (the Boolean algebra), a three-valued logic, an n-valued logic (the Zadeh-Lukasiewicz logic with $n = \infty$), and the fuzzy logic. Since the ultimate goal of developing the fuzzy logic is to provide foundations for approximate reasoning, approximate reasoning using fuzzy logic will also be discussed briefly in this chapter. Finally, two important topics, fuzzy relation and fuzzy logic rule base, are studied. In particular, fuzzy logic rule base is the core for fuzzy systems and fuzzy control – the main subjects to be studied in detail in the rest of the book.

I. CLASSICAL LOGIC THEORY

A. Fundamental Concepts

To introduce the basic concepts in the classical logic theory, we consider a typical example of a process (a system, or a plant) that is operated by a console consisting of a lock (with a key) and an ON-OFF switch as shown in Figure 2.1.

For this console, we use L to denote the situation of the lock:

$$L = \begin{cases} 1 & \text{if the console is unlocked}, \\ 0 & \text{if the console is locked}. \end{cases}$$

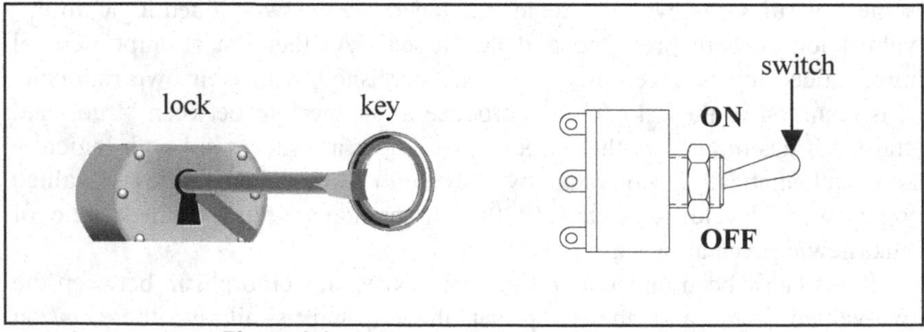

Figure 2.1 A console lock with a switch.

Table 2.1 Truth Table for the Console-Controlled Process

Inputs			Outputs
L	S	P	P
0	0	0	0
0	0	1	1
1	0	0	0
1	0	1	0
0	1	0	0
0	1	1	1
1	1	0	1
1	1	1	1

The two digits, 0 and 1, are used here to represent the two states of the logical variables. L, S, and P are logical functions, which only assume one of these two values, 0 or 1.

After the logical variables and functions are defined, it is common practice to establish a *truth table* for their relationships and possible outcomes of the relations. This truth table displays all *rules* of the two-valued logic for the particular problem in consideration. For this console example, we have the truth table shown as in Table 2.1, which is sometimes called a *look-up table*.

On the other hand, we use S to denote the state of the switch:

$$S = \begin{cases} 1 & \text{if the switch is ON,} \\ 0 & \text{if the switch is OFF.} \end{cases}$$

Moreover, we use P to denote the status of the process:

$$P = \begin{cases} 1 & \text{if the process is running,} \\ 0 & \text{if the process is shutdown.} \end{cases}$$

In Table 2.1, it is important to point out that the process P is an input (the current status) as well as an output (the resulting status), where its current status may affect its future status. For example, if $L = 0$ (the key is not in the lock, or the console is locked), then the process will remain its current status, regardless of the state of the ON-OFF switch.

A truth table is interpreted row by row. For instance, the fourth row of Table 2.1 is interpreted as follows:

IF the key is in the lock (the console is unlocked)
AND the switch is turned to OFF
AND the process is currently running
THEN the process will be shutdown.

This is a *rule*. In general, a rule may have any number of inputs and outputs like

IF (input 1) AND (input 2) AND ··· AND (input n)
THEN (output 1) AND (output 2) AND ··· AND (output m).

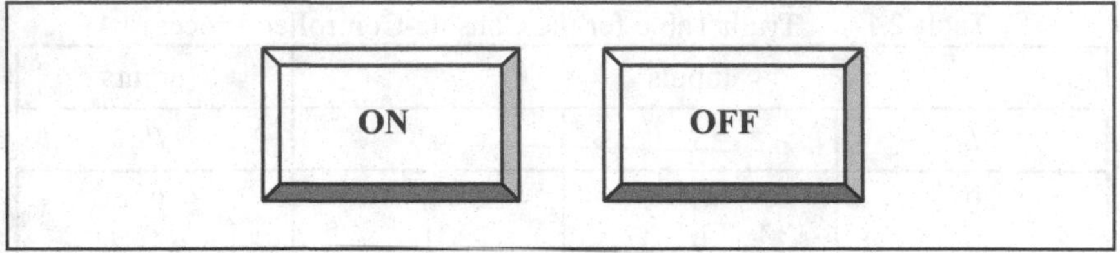

Figure 2.2 A push-button switch.

From this simple example, one can see that a classical logic string consists of several two-valued inputs, several rules, and several two-valued outputs. It will be seen later that fuzzy logic has the same structure except that both the inputs and outputs are n-valued, with $n \geq 2$ (probably, $n = \infty$).

As a second example, let us consider the push-button switch shown in Figure 2.2.

For this push-button switch, we use L to denote the situation of the ON button:

$$L = \begin{cases} 1 & \text{if the ON button is depressed,} \\ 0 & \text{if the ON button is released.} \end{cases}$$

On the other hand, we use S to denote the state of the OFF button:

$$S = \begin{cases} 1 & \text{if the OFF button is depressed,} \\ 0 & \text{if the OFF button is released.} \end{cases}$$

Finally, we use P to denote the status of the process under control:

$$P = \begin{cases} 1 & \text{if the current process is running,} \\ 0 & \text{if the current process is shutdown.} \end{cases}$$

Then, one can easily establish the truth table for this push-button control process, as shown in Table 2.2.

It is important and interesting to compare Tables 2.1 and 2.2. The most significant difference can be seen from the last rows of the two tables, in

Table 2.2 Truth Table for the Console-Controlled Process

Inputs			Outputs
L	S	P	P
0	0	0	0
0	0	1	1
0	1	0	0
0	1	1	0
1	0	0	1
1	0	1	1
1	1	0	0
1	1	1	0

Table 2.3 Logic Functions of Two Variables

x_1 / x_2	1 0 1 0 / 1 1 0 0	Name of Function	Symbol
y_1	0 0 0 0	zero function	0
y_2	0 0 0 1	nor (not-or) function	$x_1 \not\vee x_2$
y_3	0 0 1 0	proper inequality	$x_1 \neg x_2$
y_4	0 0 1 1	negation	$\bar{x}_2$
y_5	0 1 0 0	proper inequality	$x_1 \rightarrow x_2$
y_6	0 1 0 1	negation	$\bar{x}_1$
y_7	0 1 1 0	nonequivalence	$x_1 \not\Leftrightarrow x_2$
y_8	0 1 1 1	nand (not-and) function	$x_1 \not\wedge x_2$
y_9	1 0 0 0	and function	$x_1 \wedge x_2$
y_{10}	1 0 0 1	equivalence	$x_1 \Leftrightarrow x_2$
y_{11}	1 0 1 0	identity	x_1
y_{12}	1 0 1 1	implication	$x_1 \Leftarrow x_2$
y_{13}	1 1 0 0	identity	x_2
y_{14}	1 1 0 1	implication	$x_1 \Rightarrow x_2$
y_{15}	1 1 1 0	or function	$x_1 \vee x_2$
y_{16}	1 1 1 1	unity	1

which the three inputs (1,1,1) produce an output 1 in Table 2.1 but an output 0 in Table 2.2. The reason is that $S = 1$ means "ON" in Table 2.1 but it means "OFF" in Table 2.2, due to the nature of the two different switches.

B. Logical Functions of the Two-Valued Logic

Suppose that we are given n logical variables, $x_1, ..., x_n$. These n variables can have 2^n different combinations of truth values. Hence, there are a total of 2^{2^n} possible logical functions defining these variables. For example, for $n = 2$, there are $2^2 = 4$ combinations of logical variables: (1,1), (1,0), (0,1), and (0,0). The $2^{2^2} = 16$ logical functions are listed in Table 2.3, where the resulting logical variables are denoted by $y_1, ..., y_{16}$. Logical functions of one or two variables are usually called *logical operations*.

It can be seen from Table 2.3 that the number of logic functions of n variables grows extremely rapidly with increasing values of n. However, it is known that the logical functions listed in Table 2.3, and those created from them, can be obtained from a set of few primary logical functions. Two sets of primary logical functions are: (i) negation, and function, and or function; (ii) negation and the two implications. Using either set of primary logical functions, we can form any other logical functions via appropriate algebraic expressions, called *logical formulas*. For example, using the first set of primary logical functions, namely, *negation* $-$, *and function* $\wedge$, and *or function* $\vee$, logical formulas can be defined recursively as follows:

(a) The truth values 0 and 1 are logical formulas.

(b) If x is a logical variable, then both x and $\bar{x}$ are logical formulas.

(c) If x_1 and x_2 are logical formulas, then both $x_1 \wedge x_2$ and $x_1 \vee x_2$ are logical formulas.

(d) The only logical formulas are those defined by (a), (b), and (c).

Then, every logical formula of this type defines a logical function by composing it from the three primary logical functions $-$, $\wedge$, and $\vee$. To define a unique function, the composition order must be specified, say using parentheses. For example,

$$(\bar{x}_1 \vee \bar{x}_2) \wedge (x_1 \vee \bar{x}_3) \wedge (x_2 \vee x_3)$$

defines a unique logical function of three variables. When two logical formulas a and b are equivalent, we usually write $a = b$. For instance, one has

$$(\bar{x}_1 \wedge \bar{x}_2) \vee (x_1 \wedge \bar{x}_3) \vee (x_2 \wedge x_3)$$
$$= (\bar{x}_2 \wedge \bar{x}_3) \vee (\bar{x}_1 \wedge x_3) \vee (x_1 \wedge x_2).$$

This equivalence can be verified by evaluating each of the two formulas for all eight combinations of truth values of the logical variables x_1, x_2, and x_3. The Boolean algebra to be studied in the next section can help simplify such verifications.

II. THE BOOLEAN ALGEBRA

A. Basic Operations of the Boolean Algebra

An algebra of the two-valued logic is the *Boolean algebra*, named after the nineteenth-century English mathematician and logician George Boole. In this algebra, there are only three basic logic operations: *negation* $\neg$, *and* $\wedge$, and *or* $\vee$. For ease of algebraic operations, it is common to use symbols $^-$, $\times$ and $+$, as summarized in Table 2.4.

A Boolean algebraic formula can be completely described by a truth table, where all the variables appearing in the formula are listed as inputs and the value of the entire formula is the output. Conversely, one can write Boolean algebraic formulas for any truth table. For example, returning to the console lock example, the second, sixth, seventh, and eighth rows of Table 2.1 can be written together in Boolean algebra as

$$P_{output} = (\bar{L} \times \bar{S} \times P) + (\bar{L} \times S \times P) + (L \times S \times \bar{P}) + (L \times S \times P),$$

which is the same as the logic formula

$$P_{output} = (\bar{L} \wedge \bar{S} \wedge P) \vee (\bar{L} \wedge S \wedge P) \vee (L \wedge S \wedge \bar{P}) \vee (L \wedge S \wedge P),$$

Table 2.4 Boolean Operations

Name	Symbol	Example	Meaning
AND	•	$a \cdot b$	Both a and b must be true for the entire formula to be true.
OR	+	$a + b$	If either a or b, or both, is true then the entire formula is true.
NOT	–	$\overline{a}$	If a is true then the entire formula is false; if a is false then the entire formula is true.

and they both yield the same output $P_{\text{output}} = 1$. Indeed, we have

$$\begin{aligned} P_{\text{output}} &= (\overline{0} \times \overline{0} \times 1) + (\overline{0} \times 0 \times 1) + (0 \times 0 \times \overline{1}) + (0 \times 0 \times 0) \\ &= (1 \times 1 \times 1) + (1 \times 0 \times 1) + (0 \times 0 \times 0) + (0 \times 0 \times 0) \\ &= 1 + 0 + 0 + 0 \\ &= 1. \end{aligned}$$

We remark that this Boolean algebraic formula can be reduced to a simple, equivalent formula using some useful properties of the Boolean algebra to be discussed in the next subsection.

B. Basic Properties of the Boolean Algebra

Basic and useful properties of the Boolean algebra are summarized in Table 2.5, in which not all properties are necessary for an axiomatic characterization of the Boolean algebra.

It is important to note that some rules of the Boolean algebra are the same as those of the ordinary algebra, such as

$$a \cdot 0 = 0, \qquad a \cdot 1 = a, \qquad \text{and} \qquad a + 0 = a,$$

but some are quite different, e.g.,

$$a + 1 = 1.$$

Hence, one must be very careful not to fall into the trap of applying ordinary algebraic rules to logical formulas.

Now, to show an application of the Boolean algebraic laws to simplifying some complicated logic formulas, we return to the console lock example discussed in the last subsection:

$$\begin{aligned} P_{\text{output}} &= (\overline{L} \times \overline{S} \times P) + (\overline{L} \times S \times P) + (L \times S \times \overline{P}) + (L \times S \times P) \\ &= \overline{L} \times P \times (\overline{S} + S) + S \times L \times (\overline{P} + P) \\ &\qquad \text{[Commutative and Distributive Laws]} \\ &= \overline{L} \times P \times 1 + S \times L \times 1 \qquad \text{[Inclusion]} \\ &= \overline{L} \times P + S \times L \qquad \text{[Characteristics].} \end{aligned}$$

It can be easily seen that this final, simplified formula is equivalent to the original one: for the second row of Table 2.1, say, we have

$$P_{\text{output}} = \overline{0} \times 1 + 0 \times 0 = 1 \times 1 + 0 \times 0 = 1 + 0 = 1,$$

and for the sixth row, we have

$$P_{\text{output}} = \overline{0} \times 1 + 1 \times 0 = 1 \times 1 + 1 \times 0 = 1 + 0 = 1.$$

Table 2.5 Properties of the Boolean Algebra

Laws	Formulas
Characteristics	$a \cdot 0 = 0$ $a \cdot 1 = a$ $a + 0 = a$ $a + 1 = 1$
Commutative Law	$a + b = b + a$ $a \cdot b = b \cdot a$
Associative Law	$a + b + c = a + (b + c) = (a + b) + c$ $a \cdot b \cdot c = a(b \cdot c) = (a \cdot b) \cdot c$
Distributive Law	$a \cdot (b + c) = a \cdot b + a \cdot c$
Idempotence	$a \cdot a = a$ $a + a = a$
Negation	$\overline{\overline{a}} = a$
Inclusion	$a \times \overline{a} = 0$ $a + \overline{a} = 1$
Absorptive Law	$a + a \times b = a$ $a \times (a + b) = a$
Reflective Law	$a + \overline{a} \times b = a + b$ $a \times (\overline{a} + b) = a \cdot b$ $a \times b + \overline{a} \times b \times c = a \times b + b \times c$
Consistency	$a \times b + a \times \overline{b} = a$ $(a + b) \times (a + \overline{b}) = a$
DeMorgan's Laws	$\overline{a \times b} = \overline{a} + \overline{b}$ $\overline{a + b} = \overline{a} \times \overline{b}$

Table 2.6 Isomorphisms Among Classical Set Theory, Boolean Algebra, and Classical Logic

Classical Set Theory	Boolean Algebra	Two-Valued Logic
$\cup$	$+$	$\vee$
$\cap$	$\cdot$	$\wedge$
$-$	$-$	$-$
S	1	1
$\varnothing$	0	0
$\subset$	$<$	$\Rightarrow$
$=$	$=$	$\Leftrightarrow$

Table 2.7 A Three-Valued Logic

a	b	$\wedge$	$\vee$	$\Rightarrow$	$\Leftrightarrow$
0	0	0	0	1	1
0	1/2	0	1/2	1	1/2
0	1	0	1	1	0
1/2	0	0	1/2	1/2	1/2
1/2	1/2	1/2	1/2	1	1
1/2	1	1/2	1	1	1/2
1	0	0	1	0	0
1	1/2	1/2	1	1/2	1/2
1	1	1	1	1	1

We have to point out, however, that although the Boolean algebra is an effective method for reducing a logical formula to its simplest possible form, it can become quite tedious in real applications if many variables are involved in the formulas.

Finally, we note that there are interesting isomorphisms among the classical set theory, the Boolean algebra, and the two-valued logic. We list their correspondences in Table 2.6.

III. MULTI-VALUED LOGIC

A. Three-Valued Logic

The classical two-valued logic can be extended to a *three-valued logic* in various ways, each having its own rationale.

In a typical three-valued logic, we denote the truth, falsity, and indeterminacy by values of 1, 0, and 1/2, respectively. The negation $\overline{a}$ of a proposition a is usually defined to be $1 - a$. Thus, in this three-valued logic, we have

$$\overline{1} = 0, \quad \overline{0} = 1, \quad \text{and} \quad \overline{(1/2)} = 1/2.$$

Other primary logical operations, such as $\wedge$, $\vee$, $\Rightarrow$, and $\Leftrightarrow$, will differ from one logic to another. A commonly used three-valued logic is shown in Table 2.7.

Comparing the three-valued logic shown in Table 2.7 with the two-valued logic, one can see that the only new operations are those involving the new truth value 1/2. It should also be clear that the three-valued logic does not satisfy many basic operation laws of the two-valued logic, such as

$$a \wedge \overline{a} = 0 \quad \text{and} \quad a \vee \overline{a} = 1.$$

B. *n*-Valued Logic

Once a three-valued logic is accepted as a meaningful and useful tool for applications, it is not difficult to further extend it to an n-valued logic, for $3 < n < \infty$. For a given finite positive integer, $n > 3$, an n-valued logic assumes a rational truth value in the range [0,1], defined by the following equally spaced partition:

$$0 = \frac{0}{n-1}, \frac{1}{n-1}, \frac{2}{n-1}, \ldots, \frac{n-2}{n-1}, \frac{n-1}{n-1} = 1.$$

Each of these truth values describes a *degree of truth*.

The commonly used n-valued logic, particularly in fuzzy set and fuzzy systems theories, is the Lukasiewicz-Zadeh n-valued logic. Lukasiewicz developed an n-valued logic in the 1930s, using only the negation – and the implication $\Rightarrow$ logical operations. Based on that, one can define

$$a \vee b = (a \Rightarrow b) \Rightarrow b,$$

$$a \wedge b = \overline{\bar{a} \vee \bar{b}},$$

$$a \Leftrightarrow b = (a \Rightarrow b) \wedge (b \Rightarrow a).$$

The fuzzy set theory was developed by Zadeh in 1960s, which has been thoroughly studied in Chapter 1. To develop an n-valued logic, with $2 \le n \le \infty$, such that it is isomorphic to the fuzzy set theory in the same way as the two-valued logic is isomorphic to the classical set theory, Zadeh modified the Lukasiewicz logic and established an infinite-valued logic, by defining the following primary logic operations:

$$\begin{aligned}
\bar{a} &= 1 - a, \\
a \wedge b &= min\{ a, b \}, \\
a \vee b &= max\{ a, b \}, \\
a \Rightarrow b &= min\{ 1, 1 + \mathrm{b} - a \}, \\
a \Leftrightarrow b &= 1 - | a - b |.
\end{aligned}$$

It has been shown, in logic theory, that all these logical operations become the same as those for the two-valued logic when $n = 2$, and the same as those for the three-valued logic shown in Table 2.7 when $n = 3$. More importantly, when $n = \infty$, this logic does not restrict the truth values to be rational: they can be any real numbers in [0,1]. It has also been shown that this infinite-valued logic is isomorphic to the fuzzy set theory that employs the min, max, and $1-a$ operation for fuzzy set intersection, union, and complement, respectively, in the same way as the classical two-valued logic is isomorphic to the crisp set theory. These verifications are beyond the scope of this text and, hence, are omitted.

IV. FUZZY LOGIC AND APPROXIMATE REASONING

Fuzzy logic is a logic; its ultimate goal is to provide foundations for approximate reasoning using imprecise propositions based on fuzzy set theory, in a way similar to the classical reasoning using precise propositions based on the classical set theory.

To introduce this notion, we first recall how the classical reasoning works, using only precise propositions and the two-valued logic. The following syllogism is an example of such reasoning in linguistic terms:

(i) Everyone who is 40 years old or older is old.
(ii) David is 40 years old and Mary is 39 years old.
(iii) David is old but Mary is not.

This is a very precise deductive inference, correct in the sense of the two-valued logic.

In this classical (precise) reasoning using the two-valued logic, when the (output) logical variable represented by a logical formula is always true regardless of the truth values of the (input) logical variables, it is called a *tautology*. If, on the contrary, it is always false, then it is called a *contradiction*. Various tautologies can be used for making deductive inferences, which are referred to as *inference rules*. The four frequently used inference rules in classical reasoning are:

modus ponens: $(a \wedge (a \Rightarrow b)) \Rightarrow b$;
modus tollens: $(\bar{b} \wedge (a \Rightarrow b)) \Rightarrow \bar{a}$;
syllogism: $(a \Rightarrow b) \wedge (b \Rightarrow c) \Rightarrow (a \Rightarrow c)$;
contraposition: $(a \Rightarrow b) \Rightarrow (\bar{b} \Rightarrow \bar{a})$.

These inference rules are very easily understood and, indeed, have been commonly used in one's daily life. For example, the modus ponens is interpreted as follows:

IF "*a* is true" AND the statement "IF *a* is true THEN *b* is true" is true THEN "*b* is true."

Using this logic, we see that the deductive inference

IF "40 years old or older is old" AND "IF 40 years old or older is old THEN David is old" THEN "David is old"

is a modus ponens, and the deductive inference

IF "Mary is not old" AND "IF Mary is 40 years old or older then Mary is old" THEN "Mary is not 40 years old nor older"

is a modus tollens.

With the above discussion in mind, we now consider the following example of approximate reasoning in linguistic terms that cannot be handled by the classical (precise) reasoning using two-valued logic:

(i) Everyone who is 40 to 70 years old is old but is very old if he (she) is 71 years old or above; everyone who is 20 to 39 years old is young but is very young if he (she) is 19 years old or below.
(ii) David is 40 years old and Mary is 39 years old.
(iii) David is old but not very old; Mary is young but not very young.

This is of course a meaningful deductive inference, which indeed has been frequently used in one's daily life. This is an example of what is called *approximate reasoning*.

In order to deal with such imprecise inference, fuzzy logic can be employed. Briefly, fuzzy logic allows the imprecise linguistic terms such as:

- fuzzy predicates: old, rare, severe, expensive, high, fast
- fuzzy quantifiers: many, few, usually, almost, little, much
- fuzzy truth values: very true, true, unlikely true, mostly false, false, definitely false

Example 1.1 of Chapter 1 best explains the idea and approach of fuzzy logic for approximate reasoning.

To describe fuzzy logic mathematically, we introduce the following concepts and notation. Let **S** be a universe set and A a fuzzy set associated with a membership function, $\mu_A(x)$, $x \in \mathbf{S}$. If $y = \mu_A(x_0)$ is a point in [0,1], representing the truth value of the proposition "x_0 is a," or simply "a," then the truth value of "not a" is given by

$$\bar{y} = \mu_A(x_0 \text{ is not } a) = 1 - \mu_A(x_0 \text{ is } a) = 1 - \mu_A(x_0) = 1 - y.$$

Consequently, for n members $x_1, \ldots, x_n$ in **S** with n corresponding truth values $y_i = \mu_A(x_i)$ in [0,1], $i = 1,\ldots,n$, by applying the extension principle (Section IV.A of Chapter 1) the *truth values* of "not a" is defined as

$$\bar{y}_i = 1 - y_i, \qquad i=1,\ldots,n.$$

Here, we note that, actually, $n = \infty$ is allowed. With $n > 3$, following logic theory (which is beyond the scope of this book) we define the logical operations *and*, *or*, *not*, *implication*, and *equivalence* as follows: for any a, b $\in \mathbf{S}$,

$$\mu_A(a \wedge b) = \mu_A(a) \wedge \mu_A(b) = min\{\ \mu_A(a), \mu_A(b)\ \};$$
$$\mu_A(a \vee b) = \mu_A(a) \vee \mu_A(b) = max\{\ \mu_A(a), \mu_A(b)\ \};$$
$$\mu_A(\bar{a}) = 1 - \mu_A(a);$$
$$\mu_A(a \Rightarrow b) = \mu_A(a) \Rightarrow \mu_A(b) = min\{\ 1, 1 + \mu_A(\mathrm{b}) - \mu_A(\mathrm{a})\ \};$$
$$\mu_A(a \Leftrightarrow b) = \mu_A(a) \Leftrightarrow \mu_A(b) = 1 - |\ \mu_A(a) - \mu_A(b)\ |.$$

For multi-point cases, e.g., a_i, $b_j \in \mathbf{S}$, with $\mu_A(a_i)$, $\mu_A(b_j) \in [0,1]$, $i=1,\ldots,n$, $j=1,\ldots,m$, where $1 \le n$, $m \le \infty$, we can define

$$\mu_A(a_1,\ldots,a_n) \wedge \mu_A(b_1,\ldots,b_m) = \max_{1 \le i \le n, 1 \le j \le m} \{\ min\{\ \mu_A(a_i), \mu_A(b_j)\ \}\ \}.$$

This is equivalent to the minimum between two fuzzy numbers $a \wedge b$ (see the fuzzy minimum operation discussed in Section IV.B of Chapter 1). Other operations are defined accordingly.

In the classical two-valued logic, for instance in the modus ponens, the inference rule is

$$(\ a \wedge (\ a \Rightarrow b\ )\ ) \Rightarrow b.$$

In terms of membership values, this is equivalent to the following:

IF $\mu(a) = 1$

AND $\mu(a \Rightarrow b) = min\{\ 1, 1 + \mu(b) - \mu(a)\ \} = min\{\ 1, \mu(\mathrm{b})\} = 1$

THEN $\mu(b) = 1$.

Otherwise, it will contradict either $\mu(a) = 1$ or $\mu(a \Rightarrow b) = 1$.

In fuzzy logic, the inference rule reads the same: for the modus ponens, we have

$$(\ a \wedge (\ a \Rightarrow b\ )\ ) \Rightarrow b.$$

But in terms of membership values, we have

IF $\mu(a) > 0$

AND $\mu(a \Rightarrow b) = min\{\ 1, 1 + \mu(b) - \mu(a)\ \} > 0$

THEN $\mu(b) > 0$.

Here, of course, $\mu > 0$ is equivalent to $\mu \in (0,1]$ since, as usual, we have normalized all membership values. One can see that this reference is much more flexible. This fuzzy logic inference can be interpreted as follows: IF a is true with a certain degree of confidence AND "IF a is true with a certain degree of confidence THEN b is true with a certain degree of confidence" THEN b is true with a certain degree of confidence. All these "degrees of confidence" can be quantitatively evaluated by using the corresponding membership functions. This example is a *generalized modus ponens*, called *fuzzy modus ponens*.

It is important to note that in the extremal cases, fuzzy logic is consistent with the classical logic. This can be easily verified by replacing ">0" with "=1" in the above fuzzy logic inference.

Example 2.1. In the classical two-valued logic, the modus tollens is

$$(\bar{b} \wedge (a \Rightarrow b)) \Rightarrow \bar{a}\,.$$

The following is a simple case:

Premise	David cannot work
Implication	If David is young then he can work
Conclusion	David is not young

This example does not make much sense, of course, since it implies also that David cannot work at all. So we see a limitation of the two-valued logic in applications. But two-valued logic can only describe "young or old," "can or cannot," etc., and this is the best one can do for this example using two-valued logic. In contrast, the fuzzy modus tollens has the same rule:

$$(\bar{b} \wedge (a \Rightarrow b)) \Rightarrow \bar{a}\,,$$

but provides a much more meaningful inference as follows, which only implies that David cannot work so hard:

Premise	David cannot work much
Implication	If David is much younger then he can work more
Conclusion	David is not so young

In such examples, one only needs to select reasonable membership functions to describe "young, very young, old, very old, much, much more, hard, very hard," etc., such that they are meaningful and practical for the applications in consideration.

V. FUZZY RELATIONS

Let **S** be a universe set, and A and B be subsets of **S**. Similar to the Cartesian products, $A \times B$ denotes a product set in the universe set $\mathbf{S} \times \mathbf{S}$. A *fuzzy relation* is a relation between elements of A and elements of B, described by a membership function $\mu_{A \times B}(a,b)$, $a \in A$ and $b \in B$.

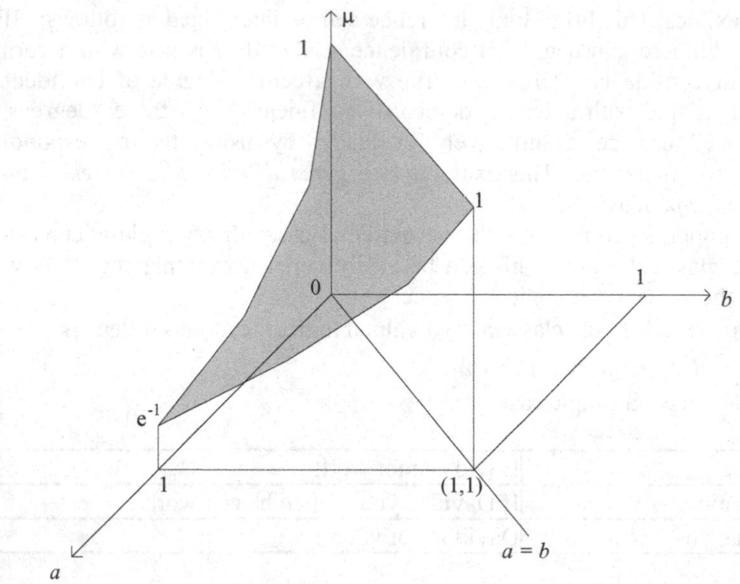

Figure 2.3 The fuzzy relation "*a* is slightly larger than *b*" on [0,1]×[0,1].

A simple example of a fuzzy relation is the following. Let **S** = R and $A = B = [0,1] \subset$ R. Define a membership function $\mu_{A\times B}(a,b)$ for the relation "*a* is slightly larger than *b*" by

$$\mu_{A\times B}(a,b) = \begin{cases} 0 & a \leq b, \\ e^{-(a-b)} & a > b, \end{cases}$$

which is shown in Figure 2.3. Then, *A*, *B*, and $\mu_{A\times B}$ together define a fuzzy relation between $a \in A$ and $b \in B$.

For discrete supports, fuzzy relations can be defined by tables or matrices. For example, let **S** = R, $A = \{ a_1, a_2, a_3, a_4 \} = \{ 1, 2, 3, 4 \}$ and $B = \{ b_1, b_2, b_3\} = \{ 0, 0.1, 2 \}$. Then the following table defines a fuzzy relation "*a* is considerably larger than *b*."

	b_1	b_2	b_3
a_1	0.6	0.6	0.0
a_2	0.8	0.7	0.0
a_3	0.9	0.8	0.4
a_4	1.0	0.9	0.5

Example 2.2. Let **S** = R with subsets A, $B \subseteq$ R. Consider two fuzzy subsets:

(i) A with the associate membership function $\mu_A(a)$, $a \in A$;
(ii) B with the associate membership function $\mu_B(b)$, $b \in B$.

Let $\mu_{A\times B}(a,b)$ be the membership function defined on $A \times B$ by

$$\mu_{A\times B}(a,b) = min\{ \mu_A(a), \mu_B(b) \}, \qquad (a,b) \in A \times B.$$

Then, A, B, and $\mu_{A\times B}$ together define a fuzzy relation.

For notational convenience, we denote by $\widetilde{R}_{(A,B)}$ a fuzzy relation defined on a product set $A \times B$ by two subsets A and B of a universe set, **S**, and an associate membership function, $\mu_{A\times B}$.

For two fuzzy relations $\widetilde{R}^1_{(A,B)}$ and $\widetilde{R}^2_{(A,B)}$, defined on the same product set $A \times B$, their *union* and *intersection*, $\widetilde{R}^1_{(A,B)} \cup \widetilde{R}^2_{(A,B)}$ and $\widetilde{R}^1_{(A,B)} \cap \widetilde{R}^2_{(A,B)}$, are defined by the membership functions

$$\mu_{\widetilde{R}^1 \cup \widetilde{R}^2}(a,b) = max\{ \mu_{\widetilde{R}^1}(a,b), \mu_{\widetilde{R}^2}(a,b) \}, \qquad (a,b) \in A\times B;$$

$$\mu_{\widetilde{R}^1 \cap \widetilde{R}^2}(a,b) = min\{ \mu_{\widetilde{R}^1}(a,b), \mu_{\widetilde{R}^2}(a,b) \}, \qquad (a,b) \in A\times B,$$

respectively.

Example 2.3. Let **S** = R, $A = \{ a_1, a_2, a_3, a_4 \} = \{ 1, 2, 3, 4 \}$, and $B = \{ b_1, b_2, b_3 \} = \{ 0, 0.1, 2 \}$. Let $\widetilde{R}^1_{(A,B)}$ be the fuzzy relation "a is considerably larger than b" defined by the following table:

$\widetilde{R}^1_{(A,B)}$:

	b_1	b_2	b_3
a_1	0.6	0.6	0.0
a_2	0.8	0.7	0.0
a_3	0.9	0.8	0.4
a_4	1.0	0.9	0.5

Let $\widetilde{R}^2_{(A,B)}$ be the fuzzy relation "a is considerably close to b" defined by the following table:

$\widetilde{R}^2_{(A,B)}$:

	b_1	b_2	b_3
a_1	0.2	0.2	0.5
a_2	0.1	0.1	1.0
a_3	0.0	0.0	0.3
a_4	0.0	0.0	0.5

Then, $\widetilde{R}^1_{(A,B)} \cup \widetilde{R}^2_{(A,B)}$ and $\widetilde{R}^1_{(A,B)} \cap \widetilde{R}^2_{(A,B)}$ are given by the following two tables, respectively:

$\widetilde{R}^1_{(A,B)} \cup \widetilde{R}^2_{(A,B)}$:

	b_1	b_2	b_3
a_1	0.6	0.6	0.5
a_2	0.8	0.7	1.0
a_3	0.9	0.8	0.4
a_4	1.0	0.9	0.5

$\widetilde{R}^1_{(A,B)} \cap \widetilde{R}^2_{(A,B)}$:

	b_1	b_2	b_3
a_1	0.2	0.2	0.0
a_2	0.1	0.1	0.0
a_3	0.0	0.0	0.3
a_4	0.0	0.0	0.5

Here, in this example, $\widetilde{R}^1_{(A,B)} \cup \widetilde{R}^2_{(A,B)}$ means that "*a* is either considerably larger than or considerably close to *b*," and $\widetilde{R}^1_{(A,B)} \cap \widetilde{R}^2_{(A,B)}$ means that "*a* is considerably larger than, as well as considerably close to, *b*." The latter statement is somewhat self-contradictory but is acceptable by fuzzy relations, and, however, has small membership values in general.

For two fuzzy relations, $\widetilde{R}^1_{(A,B)}$ and $\widetilde{R}^2_{(B,C)}$, we cannot perform their union and intersection by the same rule because they are defined on different product sets, $A \times B$ and $B \times C$, respectively. However, we can perform their *compositions*, since they have a common set *B* in between *A* and *C*. Among some other important compositions, the *max-min composition* is the most useful one in applications.

Let $\widetilde{R}^1_{(A,B)}$ and $\widetilde{R}^2_{(B,C)}$ be two fuzzy relations. Their *max-min* composition is defined to be the new relation

$$\widetilde{R}_{(A,B,C)} = \widetilde{R}^1_{(A,B)} \circ \widetilde{R}^2_{(B,C)}$$

with the membership function

$$\mu_{\widetilde{R}^1 \circ \widetilde{R}^2}(a,c) = \max_{b \in B} \{ \min\{ \mu_{\widetilde{R}^1}(a,b),\ \mu_{\widetilde{R}^2}(b,c) \}\}, \quad (a,c) \in A \times C.$$

Next, we provide some basic properties of the *max-min* composition of fuzzy relations for the case where $A = B = C$.

A fuzzy relation $\widetilde{R}_{(A,A)}$ is said to be *reflexive* if

$$\mu_{\widetilde{R}}(a,a) = 1 \qquad \text{for all } a \in A.$$

This fuzzy relation $\widetilde{R}_{(A,A)}$ is said to be *symmetric* if

$$\mu_{\widetilde{R}}(a,b) = \mu_{\widetilde{R}}(b,a) \qquad \text{for all } a, b \in A.$$

But, this fuzzy relation is said to be *antisymmetric* if

$$a \neq b \Rightarrow \begin{cases} \mu_{\widetilde{R}}(a,b) \neq \mu_{\widetilde{R}}(b,a) \\ or \\ \mu_{\widetilde{R}}(a,b) = \mu_{\widetilde{R}}(b,a) = 0 \end{cases} \qquad \text{for all } a, b \in A;$$

and *perfectly antisymmetric* if

$$a \neq b \Rightarrow (\mu_{\widetilde{R}}(a,b) > 0 \Rightarrow \mu_{\widetilde{R}}(b,a) = 0) \qquad \text{for all } a, b \in A.$$

Finally, this fuzzy relation $\widetilde{R}_{(A,A)}$ is said to be *transitive* if

$$\mu_{\widetilde{R}\circ\widetilde{R}}(a,b) \leq \mu_{\widetilde{R}}(a,b) \qquad \text{for all } a, b \in A.$$

Example 2.4. Consider the following three fuzzy relations, $\widetilde{R}^{i}(a,a)$, where $a \in A$ and $i = 1, 2, 3$.

$\widetilde{R}^{1}(a,a)$:

	a_1	a_2	a_3	a_4
a_1	0.4	0.0	0.1	0.8
a_2	0.8	1.0	0.0	0.0
a_3	0.0	0.6	0.7	0.0
a_4	0.0	0.2	0.0	0.0

$\widetilde{R}^{2}(a,a)$:

	a_1	a_2	a_3	a_4
a_1	0.4	0.8	0.7	0.0
a_2	0.0	1.0	0.9	0.6
a_3	0.8	0.4	0.7	0.4
a_4	0.0	0.1	0.0	0.0

$\widetilde{R}^{3}(a,a)$:

	a_1	a_2	a_3	a_4
a_1	0.4	0.8	0.1	0.8
a_2	0.8	1.0	0.0	0.2
a_3	1.0	0.6	0.7	0.1
a_4	0.0	0.2	0.0	0.0

It can be verified that $\widetilde{R}^1$ is perfectly antisymmetric while $\widetilde{R}^2$ is only antisymmetric but not perfectly. As to $\widetilde{R}^3$, it is a nonsymmetric fuzzy relation in the sense that there are $a, b \in A$ such that

$$\mu_{\widetilde{R}^3}(a,b) \neq \mu_{\widetilde{R}^3}(b,a),$$

which is not antisymmetric and so is not perfectly antisymmetric either.

Now, we can state the following basic properties of the max-min composition of fuzzy relations, which can be verified by some standard procedures discussed in Chapter 1.

Theorem 2.1. Let $\widetilde{R}$, $\widetilde{R}^1$, $\widetilde{R}^2$, $\widetilde{R}^3$, etc. be fuzzy relations defined on the same product set $A\times A$, and let "$\circ$" be the max-min composition operation for these fuzzy relations. Then

(1) The max-min composition is *associative*:

$$(\widetilde{R}^1 \circ \widetilde{R}^2) \circ \widetilde{R}^3 = \widetilde{R}^1 \circ (\widetilde{R}^2 \circ \widetilde{R}^3).$$

(2) If $\widetilde{R}^1$ is reflexive and $\widetilde{R}^2$ is arbitrary, then

$$\mu_{\widetilde{R}^2}(a,b) \leq \mu_{\widetilde{R}^1 \circ \widetilde{R}^2}(a,b), \qquad \text{for all } a, b \in A,$$

and

$$\mu_{\widetilde{R}^2}(a,b) \leq \mu_{\widetilde{R}^2 \circ \widetilde{R}^1}(a,b), \qquad \text{for all } a, b \in A.$$

(3) If $\widetilde{R}^1$ and $\widetilde{R}^2$ are reflexive, then so are $\widetilde{R}^1 \circ \widetilde{R}^2$ and $\widetilde{R}^2 \circ \widetilde{R}^1$.

(4) If $\widetilde{R}^1$ and $\widetilde{R}^2$ are symmetric and $\widetilde{R}^1 \circ \widetilde{R}^2 = \widetilde{R}^2 \circ \widetilde{R}^1$, then $\widetilde{R}^1 \circ \widetilde{R}^2$ is symmetric. In particular, if $\widetilde{R}$ is symmetric then so is $\widetilde{R} \circ \widetilde{R}$.

(5) If $\widetilde{R}$ is symmetric and transitive, then

$$\mu_{\widetilde{R}}(a,b) \leq \mu_{\widetilde{R}}(a,a), \qquad \text{for all } a, b \in A.$$

(6) If $\widetilde{R}$ is reflexive and transitive, then

$$\widetilde{R} \circ \widetilde{R} = \widetilde{R}.$$

(7) If $\widetilde{R}^1$ and $\widetilde{R}^2$ are transitive and $\widetilde{R}^1 \circ \widetilde{R}^2 = \widetilde{R}^2 \circ \widetilde{R}^1$, then $\widetilde{R}^1 \circ \widetilde{R}^2$ is transitive.

Finally in this section, we show by example an application of the fuzzy relations and their max-min compositions in fuzzy logic for approximate reasoning.

Example 2.5. Let **S** = { 1, 2, 3, 4 } and A = { 1, 2, 3, 4 } with the following membership function for the fuzzy description "small":

$$\mu_A(a) = \begin{cases} 1.0 & \text{if } a=1, \\ 0.7 & \text{if } a=2, \\ 0.3 & \text{if } a=3, \\ 0.0 & \text{if } a=4. \end{cases}$$

Let $\widetilde{R}$ be a fuzzy relation between two members in A, meaning "approximately equal," and be defined by the following table:

$\widetilde{R}$:

	1	2	3	4
1	1.0	0.5	0.0	0.0
2	0.5	1.0	0.5	0.0
3	0.0	0.5	1.0	0.5
4	0.0	0.0	0.5	1.0

Suppose that we want to perform the following fuzzy logic inference (approximate reasoning):

Premise	*a* is small
Implication	*a* and *b* are approximately equal
Conclusion	*b* is somewhat small

Then, we can apply the *max-min* composition of the fuzzy relations as follows:

(i) "*a* is small": $\mu_A(a)$ is available;

(ii) "*a* and *b* are approximately equal": $\mu_{\widetilde{R}}(a,b)$ is given by the table;

(iii) let $\mu_B(b)$ be the membership function for the conclusion (a fuzzy modus ponens for this example):

$$\mu_B(b) = \max_{a \in A} \{ \, min\{ \, \mu_A(a), \, \mu_{\widetilde{R}}(a,b) \, \} \, \}, \qquad b \in B = A.$$

The result, for $b = 2$, say, is

$$\begin{aligned} \mu_B(2) &= \max_{a \in A} \{ \, min\{ \, \mu_A(a), \, \mu_{\widetilde{R}}(a,2) \, \} \, \} \\ &= max\{ \, min\{1.0, 0.5\}, min\{0.7, 1.0\}, min\{0.3, 0.5\}, \\ &\qquad min\{0.0, 0.0\} \, \} \\ &= max\{ \, 0.5, 0.7, 0.3, 0.0 \, \} \\ &= 0.7. \end{aligned}$$

Similarly, one can evaluate $\mu_B(1)$, $\mu_B(3)$, and $\mu_B(4)$. The final result is

$$\mu_B(b) = \begin{cases} 1.0 & \text{if } b=1, \\ 0.7 & \text{if } b=2, \\ 0.5 & \text{if } b=3, \\ 0.3 & \text{if } b=4. \end{cases}$$

VI. FUZZY LOGIC RULE BASE

A. Fuzzy IF-THEN Rules

In Section IV, we discussed in somewhat detail the fuzzy logic operations *and*, *or*, *not*, *implication*, and *equivalence*:

$$a \wedge b, \qquad a \vee b, \qquad \overline{a}, \qquad a \Rightarrow b, \qquad a \Leftrightarrow b$$

and their evaluations on a fuzzy set A with the membership function $\mu_A(x)$:

$$\mu_A(a \wedge b) = \mu_A(a) \wedge \mu_A(b) = min\{ \mu_A(a), \mu_A(b) \};$$
$$\mu_A(a \vee b) = \mu_A(a) \vee \mu_A(b) = max\{ \mu_A(a), \mu_A(b) \};$$
$$\mu_A(\bar{a}) = \mu_{\bar{A}}(a) = 1 - \mu_A(a);$$
$$\mu_A(a \Rightarrow b) = \mu_A(a) \Rightarrow \mu_A(b) = \min\{ 1, 1 + \mu_A(b) - \mu_A(a) \};$$
$$\mu_A(a \Leftrightarrow b) = \mu_A(a) \Leftrightarrow \mu_A(b) = 1 - | \mu_A(a) - \mu_A(b) |.$$

In Section V, we also discussed fuzzy relations between elements of two subsets A and B, on which a membership function $\mu_{A \times B}(a,b)$ is defined, with $a \in A$ and $b \in B$. It is clear that one can consider the above fuzzy logic operations as some special fuzzy relations, with $A = B$ and $\mu_{A \times A} = \mu_A$.

In this section, we take a closer look at the implication relation $a \Rightarrow b$ and its application in fuzzy logic rules.

The implication relation $a \Rightarrow b$ can be interpreted, in linguistic terms, as "IF a is true THEN b is true." Of course, this is valid for both the classical (two-valued) logic and the fuzzy (multi-valued) logic. For fuzzy logic performed on a fuzzy subset A, we have a membership function μ_A describing the truth values of $a \in A$ and $b \in A$. In this case, a more complete linguistic statement would be

"(IF $a \in A$ is true with a truth value $\mu_A(a)$ THEN $b \in A$ is true with a truth value $\mu_A(b)$) has a truth value $\mu_A(a \Rightarrow b) = min\{ 1, 1 + \mu_A(b) - \mu_A(a) \}$."

In the above, both a and b belong to the same fuzzy subset A and share the same membership function μ_A. If they belong to different fuzzy subsets A and B with different membership functions μ_A and μ_B, then we have a nontrivial fuzzy relation, which can be quite complicated. In most cases, however, the implication relation $a \Rightarrow b$, performed on fuzzy subsets A and B, where $a \in A$ and $b \in B$, is simply defined in linguistic terms as

"IF $a \in A$ is true with a truth value $\mu_A(a)$ THEN $b \in B$ is true with a truth value $\mu_B(b)$."

Throughout this book, we often consider this kind of implication. Because such statements have a standard format and their meaning is clear, it is common to write them in the following simple form:

"IF a is A THEN b is B."

A fuzzy logic implication statement of this form is usually called a *fuzzy IF-THEN rule*.

To be more general, let $A_1,...,A_n$, and B be fuzzy subsets with membership functions $\mu_{A_1},...,\mu_{A_n}$, and μ_B, respectively.

Definition 2.1. A *General Fuzzy IF-THEN Rule* has the form

"IF a_1 is A_1 AND ... AND a_n is A_n THEN b is B."

Using the fuzzy logic AND operation, this rule is implemented by the following evaluation formula:

$$\mu_{A_1}(a_1) \wedge ... \wedge \mu_{A_n}(a_n) \Rightarrow \mu_B(b),$$

where

$$\mu_{A_i}(a_i) \wedge \mu_{A_j}(a_j) = min\{ \mu_{A_i}(a_i), \mu_{A_j}(a_j) \},$$

$1 \leq i, j \leq n$, and, therefore,

$$\mu_{A_1}(a_1) \wedge ... \wedge \mu_{A_n}(a_n) = min\{ \mu_{A_1}(a_1), ..., \mu_{A_n}(a_n) \}.$$

About this general fuzzy IF-THEN rule and its evaluation, a few issues have to be clarified:

(i) There is no fuzzy logic OR operation in a general fuzzy IF-THEN rule. What should we do if a fuzzy logic implication statement involves the OR operation?

(ii) There is no fuzzy logic NOT operation in a general fuzzy IF-THEN rule. What should we do if a fuzzy logic implication statement involves the NOT operation?

(iii) How do we interpret a fuzzy IF-THEN rule in a particular application? Is this interpretation unique?

We provide answers to these questions in the next two subsections.

B. Fuzzy Logic Rule Base

We first consider questions (i) and (ii). Let us first discuss, for example, the following fuzzy IF-THEN rule containing an OR operation:

"IF a_1 is A_1 AND a_2 is A_2 OR a_3 is A_3 AND a_4 is A_4 THEN b is B."

By convention, this is understood in logic as

"(IF a_1 is A_1 AND a_2 is A_2) OR (IF a_3 is A_3 AND a_4 is A_4) THEN (b is B)."

With this convention and understanding, it is clear that this statement is equivalent to the combination of the following two fuzzy IF-THEN rules:

(1) "IF a_1 is A_1 AND a_2 is A_2 THEN b is B."

(2) "IF a_3 is A_3 AND a_4 is A_4 THEN b is B."

Hence, the fuzzy logic OR operation is not necessary to use: it may shorten a statement of a fuzzy IF-THEN rule, but it increases the format complexity of the rules.

As to the fuzzy logic NOT operation, although we may have a negative statement like "IF a is not A," we can always interpret this negative statement by a positive one "IF $\overline{a}$ is A" or "IF a is $\overline{A}$," where $\overline{A}$ means "not A" in logic and "complement of A" in set theory. Moreover, the statement "$\overline{a}$ is A" or "a is $\overline{A}$" can be evaluated by

$$\mu_A(\overline{a}) = \mu_{\overline{A}}(a) = 1 - \mu_A(a).$$

Example 2.6. Given a fuzzy logic implication statement

"IF a_1 is A_1 AND a_2 is not A_2 OR a_3 is not A_3 THEN b is B,"

how can we rewrite it as a set of equivalent general fuzzy IF-THEN rules in the unified form?

We may first drop the fuzzy logic OR operation by rewriting the given statement as

(1) "IF a_1 is A_1 AND a_2 is not A_2 THEN b is B."

(2) "IF a_3 is not A_3 THEN b is B."

We may then drop the fuzzy logic NOT operation by rewriting them as

(1') "IF a_1 is A_1 AND $\overline{a}_2$ is A_2 THEN b is B."
(2') "IF $\overline{a}_3$ is A_3 THEN b is B."

Finally, these two general fuzzy IF-THEN rules can be evaluated as follows:

$$\mu_{A_1}(a_1) \wedge \mu_{A_2}(\overline{a}_2) \Rightarrow \mu_B(b),$$

where $\mu_{A_2}(\overline{a}_2) = 1 - \mu_{A_2}(a_2)$, and

$$\mu_{A_3}(\overline{a}_3) \Rightarrow \mu_B(b),$$

where $\mu_{A_3}(\overline{a}_3) = 1 - \mu_{A_3}(a_3)$. Therefore, we only need two general fuzzy IF-THEN rules (1') and (2') and three membership values $\mu_{A_2}(a_1)$, $\mu_{A_2}(a_2)$, and $\mu_{A_3}(a_3)$ to infer the conclusion "b is B," namely, $b \in B$ with the truth value $\mu_B(b)$.

All the other fuzzy logic operations can be simply defined and expressed by the AND and OR operations. They can be evaluated via the min and max operations as follows:

$$\mu_{A_1}(a_1) \wedge \mu_{A_2}(a_2) = min\{\, \mu_{A_1}(a_1), \mu_{A_2}(a_2) \,\};$$
$$\mu_{A_1}(a_1) \vee \mu_{A_2}(a_2) = max\{\, \mu_{A_1}(a_1), \mu_{A_2}(a_2) \,\};$$
$$\mu_A(\overline{a}) = \mu_{\overline{A}}(a) = 1 - \mu_A(a);$$
$$\mu_A(a \Rightarrow \tilde{a}) \;=\; \mu_A(a) \Rightarrow \mu_A(\tilde{a}) = min\{\, 1, 1+\mu_A(\tilde{a}) - \mu_A(a) \,\};$$
$$\mu_A(a \Leftrightarrow \tilde{a}) \;=\; \mu_A(a) \Leftrightarrow \mu_A(\tilde{a}) = 1 - |\, \mu_A(a) - \mu_A(\tilde{a}) \,|.$$

Thus, all finite combinations of these fuzzy logic operations can also be expressed by the AND and OR operations, so that in any finite fuzzy logic inference statement

IF ... THEN ...,

the condition part "IF ..." can be expressed only by the AND and OR operations. This eventually leads to using only the AND operation, as discussed above.

In summary, a finite fuzzy logic implication statement can always be described by a set of general fuzzy IF-THEN rules containing only the fuzzy logical AND operation, in the form

(1) "IF a_{11} is A_{11} AND ... AND a_{1n} is A_{1n} THEN b_1 is B_1."
(2) "IF a_{21} is A_{21} AND ... AND a_{2n} is A_{2n} THEN b_2 is B_2."
⋮
(m) "IF a_{m1} is A_{m1} AND ... AND a_{mn} is A_{mn} THEN b_m is B_m."

This family of general fuzzy IF-THEN rules is usually called a *fuzzy logic rule base*.

We remark that the number of components in each rule above needs not to be the same. If $n = 2$ but a rule has only one component in the condition part, say

"IF a_{11} is A_{11} THEN b_1 is B_1,"

we can formally rewrite it as

"IF a_{11} is A_{11} AND a_{12} is I_{12} THEN b_1 is B_1,"

where I_{12} is a fuzzy subset with $\mu_{I_{12}}(a) = 1$ for all $a \in I_{12}$. Here, we actually insert a "always true" (redundant) condition into the "IF ... AND ..." part to fill in the gap of the statement. In so doing, we can keep the format of a fuzzy logic rule base simple in all the general discussions throughout the book.

One can also verify that such a general form of a fuzzy logic rule base includes the nonfuzzy case and the unconditional (degenerate) case (with only "b is B") as special cases. Moreover, this general fuzzy logic rule base (with only the fuzzy logical AND operation in the condition part) also covers many unusual fuzzy logic implication statements, such as the one shown in the next example.

Example 2.7. Given a fuzzy logic implication statement

"b is B unless a_1 is A_1 AND ... AND a_n is A_n,"

which is understood in logic as

"(b is B) unless (a_1 is A_1 AND ... AND a_n is A_n),"

one can first convert it by using the fuzzy logic NOT and OR operations as follows:

"IF a_1 is $\overline{A}_1$ OR ... OR a_n is $\overline{A}_n$ THEN b is B,"

and then replace all the OR operations by a fuzzy logic rule base of the form:

(1) "IF a_1 is $\overline{A}_1$ THEN b is B,"

(2) "IF a_2 is $\overline{A}_2$ THEN b is B,"

⋮

(n) "IF a_n is $\overline{A}_n$ THEN b is B."

This rule base is in the general format, indeed.

In this example, however, we should note that the given statement is not equivalent to the following:

"IF a_1 is A_1 AND ... AND a_n is A_n THEN b is NOT B,"

since the conclusion can be "b has no relation with B."

Finally, we remark that a fuzzy rule base has to satisfy some properties or requirements. For example, a fuzzy rule base has to be *complete* in the sense that no other possible conditions are left out. The following rule base is incomplete:

(1) IF $a > 0$ THEN $b > 0$,

(2) IF $a = 0$ THEN $b < 0$,

because the case of $a < 0$ is left out. Also, a fuzzy rule base has to be *consistent* in the sense that no conclusions are contradictive. The following rule base is inconsistent:

(1) IF $a > 0$ THEN $b > 0$,

(2) IF $a > 0$ THEN $b = 0$,

(3) IF $a = 0$ THEN $b < 0$,

(4) IF $a < 0$ THEN $b = 0$,

since the first two rules contradict each other. Yet this rule base is complete. Note that the following two rules are consistent:

(1) IF $a > 0$ THEN $b > 0$,

(2) IF $a = 0$ THEN $b > 0$,

where two different conditions give the same conclusion, which is not a conflict, and the following two rules are consistent, too:

(1) IF $a > 0$ THEN $b > 0$,
(2) IF $a > 0$ THEN $c > 0$,
which are equivalent to "IF $a > 0$ THEN $b > 0$ AND $c > 0$."

Some other requirements may need to be imposed as well for a fuzzy rule base in a particular application. For example, a rule base should be *concise* with less or no redundancy.

C. Interpretation of Fuzzy IF-THEN Rules

Now, we return to Definition 2.1 and consider question (iii) thereafter: how do we interpret a fuzzy IF-THEN rule in a particular application, and is such an interpretation unique?

In the classical two-valued logic, the IF-THEN rule can be easily interpreted, namely,

"IF a is A THEN b is B"

is itself clear: the condition "a is A" infers the conclusion "b is B." For example, the statement

"IF a is positive THEN b is negative"

is crisp, nonvague, and absolute. In fuzzy multi-valued logic, however, both A and B are fuzzy subsets associated with fuzzy membership functions μ_A and μ_B. Depending on the actual membership values, $\mu_A(a)$ and $\mu_B(b)$ for the actual member values $a \in A$ and $b \in B$, respectively, both the condition "a is A" and the conclusion "b is B" can have various interpretations. We explain this in more detail by the following example.

Example 2.8. Let

$y = f(x)$

be a real-variable real-valued and invertible function defined on $X = [0,4]$ with range $Y = [-4,0]$ as shown in Figure 2.4 (a). If a crisp value x is given then $y = f(x)$ and if a crisp value y is given then $x = f^{-1}(y)$. Suppose that we don't actually know the exact formula of f. We let $\mu_S(\times)$, $\mu_M(\times)$, and $\mu_L(\times)$ be membership functions defined on X and Y, describing "small," "medium," and "large" in absolute values, respectively, as shown in Figure 2.4 (b). Thus, we may approximate the real function $y = f(x)$ by the following fuzzy rule base as shown in Figure 2.4 (c):

(1) "IF x is positive small THEN y is negative small."
(2) "IF x is positive medium THEN y is negative medium."
(3) "IF x is positive large THEN y is negative large."

Using the brief notation "a is A" to mean "$a \in A$ has a membership value $\mu_A(a)$" as we did before, one may now rewrite the above three implication statements as follows:

(1') "IF x is PS THEN y is NS."
(2') "IF x is PM THEN y is NM."
(3') "IF x is PL THEN y is NL."

Comparing it to the classical two-valued logic inference,

"IF x is positive THEN y is negative"

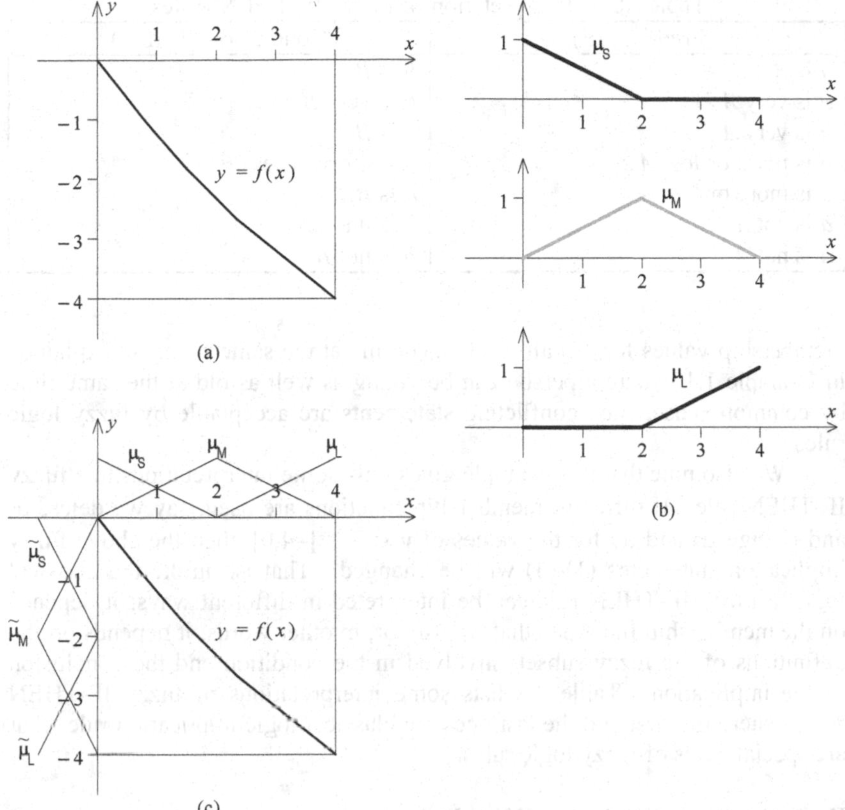

Figure 2.4 An example of approximating a real function by a fuzzy rule base.

or

"IF x is small THEN y is small,"

we see that we need the following long statement from the classical logic to express the same meaning of the fzzy logic inference (1):

"IF x is positive AND x is small THEN y is negative AND y is small."

Even if so, the classical logic can only be used to determine "x is small" or "x is not small," while the fuzzy membership function μ_S gives infinitely many different truth values to describe how small x is.

We note that in this example we only use one membership function for one evaluation, namely, if we are interested in the fuzzy implication statement (1), then we only apply the membership function μ_S to both x and y, but the other two membership functions μ_M and μ_L are not used. Later, we will discuss the cases where we apply more than one membership functions to one evaluation, to describe the situations like "x is small as well as medium" with different

Table 2.8 Interpretation of Fuzzy IF-THEN Rules

Premise (IF)	Conclusion (THEN)
a is *A*	*b* is *B*
a is very *A*	*b* is very *B*
a is very *A*	*b* is *B*
a is more or less *A*	*b* is more or less *B*
a is more or less *A*	*b* is *B*
a is not *A*	*b* is unknown
a is not *A*	*b* is not *B*

membership values for "small" and "medium" at the same time. As explained in Example 1.1, where a person can be young as well as old at the same time by common sense, such conflicting statements are acceptable by fuzzy logic rules.

We also note that this example gives only some interpretations to a fuzzy IF-THEN rule. If different membership functions are used, say we delete μ_L and change μ_S and μ_M for the values of $y \in Y = [-4,0]$, then the above fuzzy implication statements (1)-(3) will be changed. That is, unlike the classical logic, a fuzzy IF-THEN rule can be interpreted in different ways: it depends on the membership functions that we use, or, in other words, it depends on the definitions of the fuzzy subsets involved in the condition and the conclusion of the implication. Table 2.8 lists some interpretations of fuzzy IF-THEN rules, where the first and the last ones are classical logic implication rules that are special cases of fuzzy logic rules.

D. Evaluation of Fuzzy IF-THEN Rules

In this subsection, we discuss the problem of evaluating a fuzzy IF-THEN rule

$$\mu_{A\Rightarrow B}(a,b) = \mu_A(a) \Rightarrow \mu_B(b), \qquad a \in A,\ b \in B.$$

For the classical two-valued logic, this evaluation is simple:

$$\mu_B(b) = \begin{cases} 1 & \text{if}\,\mu_A(a)=1, \\ 0 & \text{if}\,\mu_A(a)=0, \end{cases}$$

namely, "$a \in A \Rightarrow b \in B$," and

$$\mu_{\bar{B}}(b) = \begin{cases} 1 & \text{if}\,\mu_{\bar{A}}(a)=1, \\ 0 & \text{if}\,\mu_{\bar{A}}(a)=0, \end{cases}$$

that is, "$a \notin A \Rightarrow b \notin B$." For fuzzy logic, we have the following options for the IF-THEN rule "$\mu_A(a) \Rightarrow \mu_B(b)$":

(a) $\mu_{A\Rightarrow B}(a,b) = min\{\ \mu_A(a),\ \mu_B(b)\ \}$;

(b) $\mu_{A\Rightarrow B}(a,b) = \mu_A(a) \times \mu_B(b)$;

(c) $\mu_{A\Rightarrow B}(a,b) = min\{\ 1,\ 1 + \mu_B(b) - \mu_A(\mathrm{a})\ \}$;

(d) $\mu_{A\Rightarrow B}(a,b) = max\{\ min\{\ \mu_A(a),\ \mu_B(b)\ \},\ 1 - \mu_A(a)\ \}$;

(e) $\mu_{A\Rightarrow B}(a,b) = max\{\ 1 - \mu_A(a),\ \mu_B(b)\ \}$;

(f) Goguen's formula:

$$\mu_{A\Rightarrow B}(a,b) = \begin{cases} 1 & \text{if } \mu_A(a) \le \mu_B(b), \\ \dfrac{\mu_B(b)}{\mu_A(a)} & \text{if } \mu_A(a) > \mu_B(b). \end{cases}$$

We remark that all these evaluation formulas are valid for the fuzzy logic inference purpose, provided that one uses consistently the same formula for the implication relation $\Rightarrow$. Of course, different formulas give different resulting values, which merely imply different degrees of inference based on different logical systems, but not the validness of the answers. Obviously, formulas (a) and (b) are very simple to use; but they are the same as the logical AND operation $\wedge$. The most common one in applications is formula (c), which we will use throughout the book.

We also remark that for the following general fuzzy IF-THEN rule:

"IF a_1 is A_1 AND ... AND a_n is A_n THEN b is B,"

we can first evaluate the condition part by either

$$\mu_A(a_1,\ldots,a_n) = min\{\ \mu_{A_1}(a_1),\ldots,\ \mu_{A_n}(a_n)\ \}$$

or, sometimes,

$$\mu_A(a_1,\ldots,a_n) = \mu_{A_1}(a_1) \ldots \mu_{A_n}(a_n),$$

and then evaluate

$$\mu_A(a_1,\ldots,a_n) \Rightarrow \mu_B(b) = \mu_{A\Rightarrow B}(a_1,\ldots,a_n,b) = min\{\ 1,\ 1+\mu_B(b)-\mu_A(a)\ \}.$$

PROBLEMS

P2.1 Construct a truth table for the output of the following expressions, in which A, B, and C are inputs:

(1) $A \times B \times \overline{C}$
(2) $(A + B) \times (B + C)$
(3) $(A + B) + (B + C)$
(4) $A + A \times B + A \times B \times C$
(5) $(A + \overline{A}) \times (B + \overline{B}) \times (C + \overline{C})$.

P2.2 Which of the following Boolean equations are true, and which are false?

(1) $A + \mathrm{B} = A + \overline{A} \times B$
(2) $\overline{A} \times B = \overline{A} \times (B \times A + B \times \overline{A})$
(3) $A = A + A \times B + A \times \overline{B}$
(4) $A = A \times (B + \overline{B}) + A \times (C + \overline{C})$
(5) $A = A \times (B + \overline{B}) + B \times C \times \overline{C}$.

P2.3 Verify:

(1) $A \vee (B \wedge \overline{A}) = A \vee B$.
(2) $\overline{(A \vee (B \wedge \overline{C})} \wedge B = \overline{A} \wedge (B \wedge C)$.

P2.4 Use truth tables to determine which of the following formulas are equivalent to each other:

(a) $(P \wedge Q) \vee (\overline{P} \wedge \overline{Q})$
(b) $\overline{P} \wedge Q$
(c) $(P \vee \overline{Q}) \wedge (Q \vee \overline{P})$
(d) $(\overline{P \vee Q})$
(e) $(Q \wedge P) \wedge \overline{P}$.

P2.5 Use truth tables to determine which of the following statements are tautologies, which are contradictions, and which are neither:

(a) $(P \vee Q) \wedge (\overline{P} \vee \overline{Q})$
(b) $(P \vee Q) \wedge (\overline{P} \wedge \overline{Q})$
(c) $(P \vee Q) \vee (\overline{P} \vee \overline{Q})$
(d) $(P \vee (Q \vee \overline{R})) \vee (\overline{P} \vee R)$.

P2.6
(a) Show that $P \Leftrightarrow Q$ is equivalent to $(P \wedge Q) \vee (\overline{P} \wedge \overline{Q})$.
(b) Show that $(P \rightarrow Q) \wedge (Q \rightarrow R)$ is equivalent to $(P \vee Q) \rightarrow R$.

P2.7 Simplify the following formula:

$$(P \wedge \overline{Q}) \vee (\overline{P} \wedge Q) \vee (\overline{P} \wedge \overline{Q}).$$

P2.8 Which of the following formulas are equivalent?

(a) $P \Rightarrow (Q \Rightarrow R)$
(b) $Q \Rightarrow (P \Rightarrow R)$
(c) $(P \Rightarrow Q) \wedge (P \Rightarrow R)$
(d) $(P \wedge Q) \Rightarrow R$
(e) $P \Rightarrow (Q \wedge R)$.

P2.9 Complete the following truth tables:

(a)

P	Q	$\overline{Q}$	$P \vee \overline{Q}$	$\overline{(P \vee \overline{Q})}$
F	F			
F	T			
T	F			
T	T			

(b)

P	Q	R	$P \wedge Q$	$\overline{(P \wedge Q)}$	$\overline{R}$	$\overline{(P \wedge Q)} \vee \overline{R}$
F	F	F				
F	F	T				
F	T	F				
F	T	T				
T	F	F				
T	F	T				
T	T	F				
T	T	T				

P2.10 Establish the logical form of the following statement:

(a) If it is raining, then it is windy and the sun is not shining.

Then establish the following statements. Also, for each statement, determine whether the statement is equivalent to either the above statement or its converse.

(b) It is windy and not sunny only if it is raining.
(c) Rain is a sufficient condition for wind with no sunshine.
(e) Rain is a necessary condition for wind with no sunshine.
(f) Wind is a necessary condition for it to be rainy, and so has no sunshine.

P2.11 Compute the following logical operation:

"IF $\mu(a) = 0.9$ AND $\mu(a \Rightarrow b) = 0.3$ THEN $\mu(b) = ?$"

P2.12 Let

$$\mu_1(x) = \begin{cases} \frac{x}{3} + \frac{5}{3}, & -5 \le x < -2, \\ -\frac{x}{3} + \frac{1}{3}, & -2 \le x < 1, \\ 0, & 1 \le x \le 12, \end{cases}$$

and

$$\mu_2(x) = \begin{cases} 0, & -5 \le x < -3, \\ \frac{x}{7} + \frac{3}{7}, & -3 \le x < 4, \\ -\frac{x}{8} + \frac{12}{8}, & 4 \le x \le 12. \end{cases}$$

Find

$$\max_{x \in [-5,12]} \ \min\{\mu_1(x), \mu_2(x)\} = ?$$

P2.13 Let the fuzzy relations $\widetilde{R}^1$, $\widetilde{R}^2$, and $\widetilde{R}^3$ be given by the three tables shown in Example 2.4. Find the composition results of $\widetilde{R}^1 \circ \widetilde{R}^2 \circ \widetilde{R}^3$. [Hint: Rename $\widetilde{R}^1$ (a,a) as $\widetilde{R}^1$ (a,b), ^{2} $\widetilde{R}^2$ (b,c), $\widetilde{R}^3$ (a,a) as $\widetilde{R}^3$ (c,a). Then perform ($\widetilde{R}^1$ (a,b) $\circ$ $\widetilde{R}^2$ (b,c)) $\circ$ $\widetilde{R}^3$ (c,a).]

P2.14 In Example 2.3, suppose that the first table gives the membership function μ_1 for the Premise "*a* is considerably larger than *b*" and the second table gives the membership function μ_2 for the Implication "*b* is considerably close to *a*." Find the membership value $\mu_3(b{=}2)$ for the Conclusion "a_3 is considerably larger than and considerably close to b_3."

P2.15 Verify that the logical statement "IF a_1 is A_1 AND ... AND a_n is A_n THEN *b* is NOT *B*" is not equivalent to the logical statement "*b* is *B* UNLESS a_1 is A_1 AND ... AND a_n is A_n."

REFERENCES

[1] C. R. Asfahl. *Robots and Manufacturing Automation*. New York, NY: Wiley (1985); and 2nd ed. (1992).

[2] G. J. Klir, U. H. St.Clair, and B. Yuan. *Fuzzy Set Theory: Foundations and Applications*. Upper Saddle River, NJ: Prentice-Hall PTR (1997).

[3] G. J. Klir and T. A. Folger. *Fuzzy Sets, Uncertainty, and Information*. Englewood Cliffs, NJ: Prentice-Hall (1988).

[4] L. X. Wang. *Adaptive Fuzzy Systems and Control: Design and Stability Analysis*. Englewood Cliffs, NJ: Prentice-Hall (1994).

[5] H. J. Zimmermann. *Fuzzy Set Theory and Its Applications* (*2nd Edition*). Boston, MA: Kluwer Academic Publishers (1991).

[6] K. Tanaka. *An Introduction to Fuzzy Logic for Practical Applications*. New York, NY: Springer (1996).

CHAPTER 3

Fuzzy System Modeling

Modeling, in a general sense, refers to the establishment of a description of a system (a plant, a process, etc.) in mathematical terms, which characterizes the input-output behavior of the underlying system.

To describe a physical system, such as a circuit or a microprocessor, we have to use a mathematical formula or equation that can represent the system both qualitatively and quantitatively. Such a formulation is a mathematical representation, called a *mathematical model*, of the physical system. Most physical systems, particularly those complex ones, are extremely difficult to model by an accurate and precise mathematical formula or equation due to the complexity of the system structure, nonlinearity, uncertainty, randomness, etc. Therefore, approximate modeling is often necessary and practical in real-world applications. Intuitively, approximate modeling is always possible. However, the key questions are what kind of approximation is good, where the sense of "goodness" has to be first defined, of course, and how to formulate such a good approximation in modeling a system such that it is mathematically rigorous and can produce satisfactory results in both theory and applications.

From the detailed studies in the last two chapters, it is clear that interval mathematics and fuzzy logic together can provide a promising alternative to mathematical modeling for many physical systems that are too vague or too complicated to be described by simple and crisp mathematical formulas or equations. When interval mathematics and fuzzy logic are employed, the interval of confidence and the fuzzy membership functions are used as approximation measures, leading to the so-called *fuzzy systems modeling*.

Following the traditional classification in the field of control systems, a system that describes the input-output behavior in a way similar to a mathematical mapping without involving a differential operator or equation is called a *static system*. In contrast, a system described by a differential operator or equation is called a *dynamic system*. In this chapter, static fuzzy systems modeling is first discussed, including its stability analysis, and the dynamic fuzzy systems modeling will then follow, along with its stability and controllability analyses.

The subject of fuzzy systems modeling is very important and useful in its own right, not necessarily related to controls. They are, of course, essential for the investigation of fuzzy systems control in later chapters of the book.

The fundamental concept of systems modeling can be illustrated as follows. Suppose that we have an unknown system ("black box"), for which only a set of its inputs x_1, ..., x_n and outputs y_1, ..., y_m can be measured (or observed) and so these data are available. Here, both inputs and outputs can be either discrete-time series or continuous signals. We want to find a

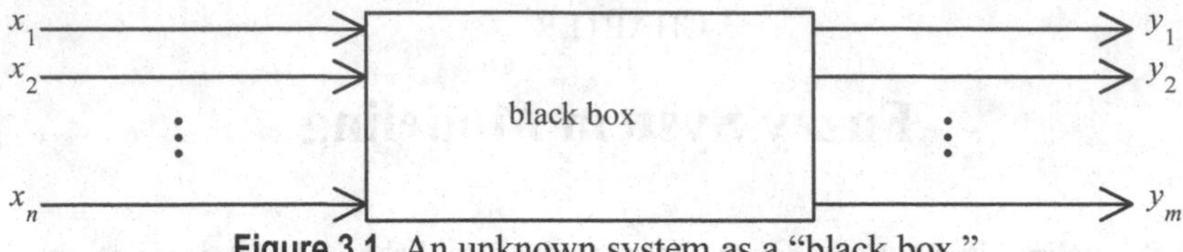

Figure 3.1 An unknown system as a "black box."

mathematical description to qualitatively and quantitatively characterize this unknown system, in the sense that by inputting $x_1, ..., x_n$ into the mathematical description we can always obtain the corresponding outputs $y_1, ..., y_m$. Establishing such a mathematical description is called *mathematical modeling* for the unknown system (Figure 3.1). As usual, the mathematical description can be a mathematical formula, such as a mapping or a functional that relates the inputs to the outputs in the form

$$\begin{cases} y_1 = f_1(x_1, ..., x_n), \\ \vdots \\ y_m = f_m(x_1, ..., x_n); \end{cases} \tag{3.1}$$

or a set of differential equations (assuming proper conditions) in the form

$$\begin{cases} y_1 = g_1(x_1, ..., x_n, \dot{x}_1, ..., \dot{x}_n), \\ \vdots \\ y_m = g_m(x_1, ..., x_n, \dot{x}_1, ..., \dot{x}_n); \end{cases} \tag{3.2}$$

or a logical linguistic statement, which can be quantified mathematically, in the form

IF (input x_1) AND ... AND (input x_n)
THEN (output y_1) AND ... AND (output y_m). (3.3)

Fuzzy systems modeling is to quantify the logical linguistic form (3.3) by using fuzzy logic and the mathematical functional model (3.1), or by using fuzzy logic together with the differential equation model (3.2). The result of the former is called a *static fuzzy system* and the latter, a *dynamic fuzzy system*.

In this chapter, we study both static and dynamic fuzzy systems modeling. We first study static fuzzy systems in Sections I and II, and then discuss dynamic fuzzy systems in the rest of the chapter.

I. MODELING OF STATIC FUZZY SYSTEMS

A. Fuzzy Logic Description of Input-Output Relations

Return to the "black box" shown in Figure 3.1. Since all the inputs $x_1, ..., x_n$ and outputs $y_1, ..., y_m$ are assumed to be available, the logical linguistic statement (3.3), namely,

IF (input x_1) AND ... AND (input x_n)
THEN (output y_1) AND ... AND (output y_m) (3.4)

has actually described the unknown system, on the basis of the available data. Yet this is not the ultimate purpose of mathematical modeling, since if a new input x_{n+1} comes in, we don't know what the corresponding output should be. The main purpose of mathematical modeling, therefore, is not only to correctly describe the existing input-output relations through the unknown system but also to enable the established model to approximately describe other possible hidden input-output relations of the system.

Thus, a general approach is to first quantify the linguistic statement (3.4) and then to relay the quantified logical input-output relations by using the mathematical functionals (3.1), or differential equations (3.2). In this section, we discuss this general approach using fuzzy logic quantization and the mathematical functionals (3.1), that is, the static fuzzy systems modeling.

Recall from Section VI.B of Chapter 2 that a finite fuzzy logic implication statement can always be described by a set of general fuzzy IF-THEN rules containing only the fuzzy logic AND operation, in the following multi-input single-output form:

(1) "IF (x_1 is X_{11}) AND ... AND (x_n is X_{1n}) THEN (y is Y_1)."
(2) "IF (x_1 is X_{21}) AND ... AND (x_n is X_{2n}) THEN (y is Y_2)."
⋮
(N) "IF (x_1 is X_{N1}) AND ... AND (x_n is X_{Nn}) THEN (y is Y_N)."

Here, we should recall that the phrase "x is X" is an abbreviation of the complete statement "x belongs to the fuzzy subset X with a corresponding membership value $\mu_X(x)$." We should now restrict our discussion on closed intervals for the fuzzy subsets X_{11}, ..., X_{Nn} and Y_1, ..., Y_N, so that interval arithmetic can be applied. Furthermore, we suppose that all these are fuzzy numbers in the sense that their membership functions are convex and normalized (Section IV.B of Chapter 1).

For simplicity of discussion, we first consider the simplest case where $N = 1$ with only one fuzzy IF-THEN rule:

"IF (x_1 is X_1) AND ... AND (x_n is X_n) THEN (y is Y)."

An example is the following:

R^1: IF (x_1 is X_1) AND ... AND (x_n is X_n)
THEN $y = a_0 + a_1x_1 + ... + a_nx_n$,

where $\{a_0,a_1,...,a_n\}$ are constants. When we are given a set of particular inputs,

$$x_1 = x_1^0 \in X_1, \ ..., \ x_n = x_n^0 \in X_n,$$

together with their available membership values $\mu_{X_1}(x_1^0)$, ..., $\mu_{X_n}(x_n^0)$, the output y will assume the value

$$y^0 = a_0 + a_1 x_1^0 + ... + a_n x_n^0, \tag{3.5}$$

with membership value given by the following *general rule* (Section IV.B of Chapter 1):

$$\mu_Y(y^0) = \sup_{y^0=a_0+...+a_nx_n^0} \{\mu_{X_1}(x_1^0)\wedge...\wedge\mu_{X_n}(x_n^0)\}. \tag{3.6}$$

If there are more than one fuzzy IF-THEN rule (N>1):

R^i: IF (x_1 is X_{i1}) AND ... AND (x_n is X_{in})
THEN $y_i = a_{i0} + a_{i1}x_1 + ... + a_{in}x_n$, $i=1,...,N$,

then, with the given set of inputs

$$x_1 = x_1^0 \in X_1, ..., x_n = x_n^0 \in X_n,$$

namely, the same inputs applied to all the different rules, we will have

$$\begin{aligned} y_1^0 &= a_{10} + a_{11}x_1^0 + ... + a_{1n}x_n^0, \\ y_2^0 &= a_{20} + a_{21}x_1^0 + ... + a_{2n}x_n^0, \\ &\vdots \\ y_N^0 &= a_{N0} + a_{N1}x_1^0 + ... + a_{Nn}x_n^0. \end{aligned} \tag{3.7}$$

The corresponding membership values for these outputs are given, again by the general rule, as

$$\mu_Y(y_i^0) = \sup_{y_i^0 = a_{i0} + ... + a_{in}x_n^0} \{\mu_{X_1}(x_1^0) \wedge ... \wedge \mu_{X_n}(x_n^0)\}, \tag{3.8}$$

where $i = 1, ..., N$. In a typical modeling approach, the final single output, y, is usually obtained via the following weighted average formula, using all y_i^0 with the weights $\mu_Y(y_i^0)$, usually called the center-of-gravity formula:

$$y = \frac{\sum_{i=1}^{N} \mu_Y(y_i^0) \times y_i^0}{\sum_{i=1}^{N} \mu_Y(y_i^0)}, \tag{3.9}$$

where "×" is the ordinary algebraic multiplication (since all quantities here are real numbers). This operation is a convex combination of all outputs.

It should be noted that we are talking about modeling here, so the weighted average formula (3.9) is merely a natural choice for the purpose of generating a sensible system output which, although has proven to be successful in many applications, is by no means the only possible formula to use (and may not be optimal in general).

For the most general situation, we assume that all the coefficients $\{a_{i0}, a_{i1}, ..., a_{in} \mid i=1,...,N\}$ are uncertain and belong to certain intervals:

$$a_{i0} \in A_0, ..., a_{in} \in A_n, \qquad i=1,...,N,$$

where, for example,

$$\begin{aligned} A_0 &= [\, min\{a_{10}, a_{20}, ..., a_{N0}\}, max\{a_{10}, a_{20}, ..., a_{N0}\} \,], \\ A_1 &= [\, min\{a_{11}, a_{21}, ..., a_{N1}\}, max\{a_{11}, a_{21}, ..., a_{N1}\} \,], \\ &\vdots \\ A_n &= [\, min\{a_{1n}, a_{2n}, ..., a_{Nn}\}, max\{a_{1n}, a_{2n}, ..., a_{Nn}\} \,]. \end{aligned}$$

Thus, with the given inputs

$$x_1 \in X_1, x_2 \in X_2, ..., x_n \in X_n,$$

the output becomes

$$Y = A_0 + A_1 \times X_1 + ... + A_n \times X_n, \tag{3.10}$$

which yields the fuzzy subset (interval) for y, with the membership functions given by the general rule as

$$\mu_{Y,i}(y_i) = \sup_{y_i = a_{i0} + \ldots + a_{in}x_n} \{\mu_{X_1}(x_1^0) \wedge \ldots \wedge \mu_{X_n}(x_n^0)\}, \tag{3.11}$$

where $i = 1, \ldots, N$. In the α-cut notation, this algorithm can be represented by

$$(S_y)_{\overline{\alpha}} = \{\, y_i \in Y_i \mid y_i = a_{i0} + a_{i1}x_1 + \ldots + a_{in}x_n,$$

$$x_1 \in (S_{x_1})_{\overline{\alpha}}, \ldots, x_n \in (S_{x_n})_{\overline{\alpha}} \,\},$$

in which, since all a_{ij}'s are real numbers, their membership functions are set identically to be 1. Finally, the output y is computed by

$$y = \frac{\sum_{i=1}^{N} \mu_{Y,i}(y_i) \times y_i}{\sum_{i=1}^{N} \mu_{Y,i}(y_i)} = \sum_{i=1}^{N} \beta_i \times y_i \tag{3.12}$$

where

$$\beta_i = \frac{\mu_{Y,i}(y_i)}{\sum_{i=1}^{N} \mu_{Y,i}(y_i)}.$$

This is a convex combination of the outputs y_i, $i=1,\ldots,N$.

The three formulas (3.7)-(3.9) or (3.10)-(3.12) are sometimes called the *input-output algorithm* for static fuzzy system modeling under the fuzzy IF-THEN rules R^i, $i = 1, \ldots, N$, where the former has constant coefficients while the latter has interval coefficients.

Before we demonstrate the static fuzzy system modeling, some remarks are in order. First, the coefficients a_{ij} do not explicitly contribute to the computation of $\mu_{Y,i}$ in (3.11) because they only rescale the values of x_i but do not change their confidence degrees. Second, the center-of-gravity formula (3.12) is just a natural and commonly used one, which is by no means the only choice for the purpose. A designer can, indeed, use any formula that is meaningful for the application, since mathematical modeling usually does not lead to a unique choice of its mathematical framework. For instance, one may use

$$y = \frac{\sum_{i=1}^{N} (\mu_{Y,i}(y_i))^{\alpha} \times y_i}{\sum_{i=1}^{N} (\mu_{Y,i}(y_i))^{\alpha}}, \qquad 0 \le \alpha < \infty,$$

which produces formula (3.12) when $\alpha = 1$, and yields a non-weighted average formula when $\alpha = 0$. Finally, for $n > 2$, the calculation

$$\mu_{X_1}(\cdot) \wedge \ldots \wedge \mu_{X_n}(\cdot)$$

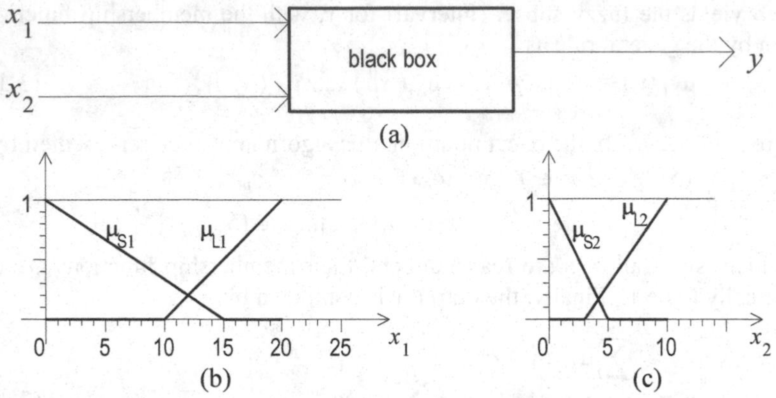

Figure 3.2 The unknown system and its two inputs in Example 3.1.

in formula (3.11) can be performed pairwise by repeating the general rule studied in Section IV.B of Chapter 1 (see, also, Example 1.12) since the logical AND operation satisfies the associative law (Table 2.5).

The next example explains the above discussion and computation, and particularly shows how to apply the input-output algorithm for static fuzzy modeling.

Example 3.1. Consider an unknown system with two inputs, x_1, x_2 , and one output, y, as shown in Figure 3.2 (a). Let inputs x_1 be within the range $X_1 = [0,20]$ and x_2 in $X_2 = [0,10]$. Suppose that X_1 has two associate membership functions, $\mu_{S_1}(x)$ and $\mu_{L_1}(x)$, describing "small" and "large," respectively, and similarly X_2 has $\mu_{S_2}(x)$ and $\mu_{L_2}(x)$, as shown in Figure 3.2 (b)-(c).

Now, suppose that we know, from experiments, the following input-output relations (fuzzy IF-THEN implication rules):

R^1: IF x_1 is small AND x_2 is small THEN $y = x_1 + x_2$

R^2: IF x_1 is large THEN $y = \frac{1}{2}x_1$

R^3: IF x_2 is large THEN $y = \frac{1}{3}x_2$.

Otherwise, $y = 0$. Here, we assume that x_1 and x_2 will not both be large in this example.

Let $x_1^0 = 13$ and $x_2^0 = 4$ be given, for instance. It can be found from Figures 3.2 (b)-(c) that

$$\mu_{S_1}(x_1^0) = 2/15, \qquad \mu_{L_1}(x_1^0) = 3/10,$$

$$\mu_{S_2}(x_2^0) = 1/5, \qquad \mu_{L_2}(x_2^0) = 1/7.$$

Then, following the input-output algorithm, we compute the following:

(i) The corresponding outputs:

$$y_1^0 = x_1^0 + x_2^0 = 13 + 4 = 17,$$

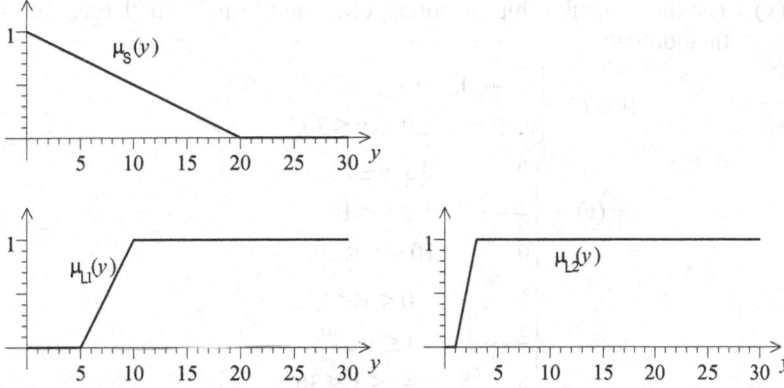

Figure 3.3 The three output membership functions in Example 3.1.

$$y_2^0 = \frac{1}{2}\, x_1^0 = \frac{13}{2},$$
$$y_3^0 = \frac{1}{3}\, x_2^0 = \frac{4}{3}.$$

(ii) The fuzzy subset Y: Since $a_{11} = 1$, $a_{12} = 1$, $a_{21} = \frac{1}{2} a_{22} = 0$, $a_{31} = 0$, $a_{32} = 1/3$, we have

$$A_1 = [\, min\{1, \tfrac{1}{2}, 0\},\ max\{1, \tfrac{1}{2}, 0\}\,] = [0,1],$$
$$A_2 = [\, min\{1, 0, \tfrac{1}{3}\},\ \max\{1, 0, \tfrac{1}{3}\}\,] = [0,1],$$
$$Y = A_1 \times X_1 + A_2 \times X_2 = [0,1] \times [0,20] + [0,1] \times [0,10] = [0,30].$$

(iii) The fuzzy membership values of the outputs:

$$\begin{aligned}\mu_{Y,1}(y_1^0) &= \sup_{y_1^0 = x_1^0 + x_2^0} \{ \mu_{S_1}(x_1^0) \wedge \mu_{S_2}(x_2^0) \} \\ &= \{ \mu_{S_1}(x_1^0) \wedge \mu_{S_2}(x_2^0) \} \\ &= min\, \{ \tfrac{2}{15}, \tfrac{1}{5} \} \\ &= \tfrac{2}{15},\end{aligned}$$

$$\mu_{Y,2}(y_2^0) = min\, \{ \mu_{L_1}(x_1^0) \} = \frac{3}{10},$$

$$\mu_{Y,3}(y_3^0) = min\, \{ \mu_{L_2}(x_2^0) \} = \frac{1}{7}.$$

(iv) The average value of the final output, corresponding to the input $x_1^0 = 13$ and $x_2^0 = 4$:

$$y^0 = \frac{\sum_{i=1}^{3} \mu_{Y,i}(y_i^0) \times y_i^0}{\sum_{i=1}^{3} \mu_{Y,i}(y_i^0)} = \frac{\frac{2}{15} \times 17 + \frac{3}{10} \times \frac{13}{2} + \frac{1}{7} \times \frac{4}{3}}{\frac{2}{15} + \frac{3}{10} + \frac{1}{7}} \approx 7.649.$$

(v) The three membership functions, one "small" and two "large," for the final output:

$$\mu_S(y) = \begin{cases} -\frac{y}{20}+1, & 0 \le y \le 20, \\ 0, & 20 \le y \le 30, \end{cases}$$

$$\mu_{L_1}(y) = \begin{cases} 0, & 0 \le y \le 5, \\ \frac{y}{5}-1, & 5 \le y \le 10, \\ 0 & 10 \le y \le 30, \end{cases}$$

$$\mu_{L_2}(y) = \begin{cases} 0, & 0 \le y \le 1, \\ \frac{3y}{7}-\frac{3}{7}, & 1 \le y \le \frac{10}{3}, \\ 0 & \frac{10}{3} \le y \le 30, \end{cases}$$

in which $\mu_S(y)$ can be computed from μ_{S_1} and μ_{S_2} by the α-cut operations (see Example 1.12), and, similarly, $\mu_{L_1}(y)$ from $\mu_{L_1}(x_1^0)$ and $\mu_{L_2}(y)$ from $\mu_{L_2}(x_2^0)$, respectively. Here, to find $\mu_{L_1}(y)$ from $\mu_{L_1}(x_1^0)$, for instance, we use α-cut on the membership function curve $\mu_{L_1}(x_1)$: $\alpha = (x - 10)$, and obtain $x_{11} = 10\alpha + 10$, while $x_{12} = 20$ is obtained from Figure 3.2 (b). Thus, we have $[x_{11},x_{12}] = [10\alpha+10,20]$. It then follows from the implication rule R^2 ($y_2 = \frac{1}{2}x_1$) that the output interval is $[y_{21},y_{22}] = [5\alpha+5,10]$. Then, by setting $y_{21} = 5\alpha + 5$, we have $\alpha = y_{21}/5 - 1$ on [5,10] (since $\mu_{L_1}(y=5) = 0$ and $\mu_{L_1}(y=10) = 1$). These membership functions are finally extended to the interval $Y = [0,30]$. The resulting three output membership functions are shown in Figure 3.3 (see Problem *P3.1*).

B. Parameters Identification in Static Fuzzy Modeling

It should be pointed out that the approach discussed in the last subsection, including the input-output algorithm and the example given therein, is about how to describe the input-output relations there for an unknown system. Note that the input-output relations are partially known, namely, the constant coefficients in the relations

$$y_i = a_{i0} + a_{i1}x_1 + \dots + a_{in}x_n, \qquad x_i \in X_i, \qquad i=1,\dots,N,$$

are all given (see Example 3.1). Hence, this is not a complete issue of modeling.

Very often, we only have the knowledge of the *structure* of the unknown system, say in the form of the above linear relations (implications), where the constant coefficients are unknown. We note that if the structure is also unknown, then the modeling problem becomes extremely difficult, which will not be further discussed in this book. Nevertheless, we can use the above relations (implications) to approximate the unknown system, regardless of its hidden structure, by finding out good (if not the best) coefficients such that these relations can well represent (approximate) the unknown system. In general, we need to identify the unknown coefficients (system parameters) based on the available input-output data under a certain meaningful model-matching criterion such as least-squares. This is called the *system parameter identification* problem.

In order to solve the system modeling problem, we have to determine the following five items by using the available input-output data:

(i) $x_1,\ldots,x_n$: input variables, used as the premises of fuzzy logic implications.

(ii) $X_1,\ldots,X_n$: input variable intervals, used as fuzzy subsets.

(iii) $\mu_{X_1},\ldots,\mu_{X_n}$: membership functions of the input variables, used to measure the qualities and quantities of the inputs.

(iv) R^i: relations (implications), used as descriptions of system input-output behavior, which take the form

$$y_i = a_{i0} + a_{i1}x_1 + \ldots + a_{in}x_n, \qquad x_i \in X_i, \qquad i=1,\ldots,N.$$

(v) $a_{i0},\ldots,a_{in}$ ($i=1,\ldots,N$): constant parameters of the model, used for the overall mathematical model.

Once these steps have been carried out, the fuzzy system modeling can be completed by applying the input-output algorithm studied in the last subsection.

In Step (i), generally speaking, to choose input variables $x_1,\ldots,x_n$, we first select some important ones from all possible inputs based on knowledge and experience. Then we carry out the next few steps to come out with the corresponding system outputs. Finally, we compute the errors between the output values of the model and the actual (measured) output data of the unknown system, and then improve the choice of the input variables by minimizing these errors.

In Step (ii), all the input variable intervals are determined by estimating the ranges of the input values, or taking the minimum and maximum values of input variables directly from the data.

In Step (iii), all the membership functions of input variables are selected, more or less subjectively, by the designer based on his experience and the meaning of these functions to the real physical system. Again, this is a modeling (or design) problem which does not lead to a unique choice in general. An experienced designer usually comes up with a better mathematical model using the same set of available data. Uniform shape(s) of membership functions are usually desirable for computational efficiency, simple memory, easy analysis, etc. The most common choices of simple and efficient membership functions are triangular, trapezoidal, Gaussian functions, etc., as shown in Figure 1.4. Since all input variables assume real values in general, or can be identified by (or mapped to) real values, the uniform triangular membership functions describing negative large (NL), negative medium (NM), negative small (NS), zero (Z), positive small (PS), positive medium (PM), and positive large (PL), all in absolute values, as shown in Figure 3.4, are very effective for system input (and output) variables. One may also use

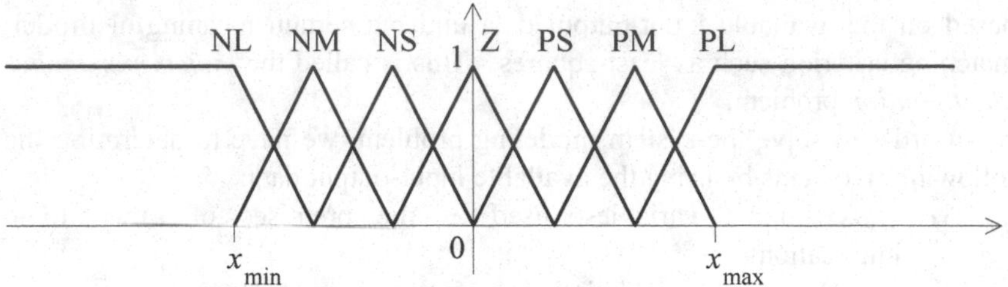

Figure 3.4 Uniform triangular membership functions over the value range $[x_{min}, x_{max}]$ of an input variable x.

very small, *very large*, *very very large*, etc., to extend the covering of the membership functions whenever necessary.

In Step (iv), for simplicity of design we have taken the linear input-output relations (implications) in this discussion, which will be fixed after being selected. In general, these relations should be obtained from the physics of the unknown system, reasonable mathematical description of the system, or human experience.

Finally, in Step (v), constant parameters are determined by minimizing the errors between the output values of the model and the actual (measured) output data of the unknown system. Here, the most convenient way to perform the minimization is to use the standard least-squares method. In the case that measurement noise is taken into account in the data, some successful optimal estimation techniques such as the Kalman filtering may also be employed.

We now show how the least-squares method can be applied to determine optimal constant parameters for the static fuzzy modeling problem discussed above.

The Least-Squares Parameter Identification Scheme: Start with the fuzzy IF-THEN rules that provide the input-output relations (implications):

R^1: IF x_1 is X_{11} AND ... AND x_n is X_{1n}
THEN $y_1 = a_{10} + a_{11}x_1 + \ldots + a_{1n}x_n$,

$\vdots$

R^N: IF x_1 is X_{N1} AND ... AND x_n is X_{Nn}
THEN $y_N = a_{N0} + a_{N1}x_1 + \ldots + a_{Nn}x_n$,

where, for each $i = 1,\ldots,N$, all X_{ij} share the same fuzzy subset X_j and the same membership function μ_{X_j}, $j = 1,\ldots,n$, and

$$a_{i0} \in A_0,\ a_{i1} \in A_1,\ \ldots,\ a_{in} \in A_n.$$

The final output y for the inputs $x_1,\ldots,x_n$ is given by formulas (3.11)-(3.12):

$$y = \frac{\sum_{i=1}^{N}\big(\mu_{X_{i1}}(x_1) \wedge \ldots \wedge \mu_{X_{in}}(x_n)\big) \times \big(a_{i0} + a_{i1}x_1 + \ldots + a_{in}x_n\big)}{\sum_{i=1}^{N}\big(\mu_{X_{i1}}(x_1) \wedge \ldots \wedge \mu_{X_{in}}(x_n)\big)}, \tag{3.13}$$

where $\mu_{X_{ij}}(x_j) = \mu_{X_j}(x_j)$ is the membership value of x_j obtained in the rule R^i, $i = 1,\ldots,N$, $j=1,\ldots,n$. Let

$$\beta_i = \frac{\mu_{X_{i1}}(x_1)\wedge\ldots\wedge\mu_{X_{in}}(x_n)}{\sum_{i=1}^{N}\left(\mu_{X_{i1}}(x_1)\wedge\ldots\wedge\mu_{X_{in}}(x_n)\right)}, \qquad i=1,\ldots,N, \tag{3.14}$$

and rewrite (3.13) as

$$\begin{aligned} y &= \sum_{i=1}^{N}\beta_i(a_{i0}+a_{i1}x_1+\ldots+a_{in}x_n) \\ &= \sum_{i=1}^{N}\begin{bmatrix}\beta_i & \beta_i x_1 & \ldots & \beta_i x_n\end{bmatrix}\begin{bmatrix}a_{i0}\\ a_{i1}\\ \vdots \\ a_{in}\end{bmatrix}. \end{aligned} \tag{3.15}$$

Suppose that a set of input-output data is given:

$$\begin{aligned} &x_1^1, \quad x_2^1, \quad \ldots, \quad x_n^1, \quad y^1; \\ &x_1^2, \quad x_2^2, \quad \ldots, \quad x_n^2, \quad y^2; \\ &\vdots \\ &x_1^M, \quad x_2^M, \quad \ldots, \quad x_n^M, \quad y^M. \end{aligned}$$

To determine the coefficient parameters $\{a_{i0},a_{i1},\ldots,a_{in} | \, i = 1,\ldots,N\}$ by using this set of data, we use the standard least-squares method and write, by substituting the data set into (3.15) successively, a system of algebraic equations,

$$\Lambda\theta = \boldsymbol{b}, \tag{3.16}$$

where

$$\begin{aligned} \Lambda &= \Lambda_{M\times(N\times(n+1))} \\ &= \begin{bmatrix} \beta_{11} & \cdots & \beta_{N1} & x_1^1\beta_{11} & \cdots & x_1^1\beta_{N1} & \cdots & x_n^1\beta_{11} & \cdots & x_n^1\beta_{N1} \\ \vdots & & \vdots & \vdots & & \vdots & & \vdots & & \vdots \\ \beta_{1M} & \cdots & \beta_{NM} & x_1^M\beta_{1M} & \cdots & x_1^M\beta_{NM} & \cdots & x_n^M\beta_{1M} & \cdots & x_n^M\beta_{NM} \end{bmatrix}, \\ \theta &= \begin{bmatrix} a_{10} & \cdots & a_{N0} & a_{11} & \cdots & a_{N1} & \cdots & a_{1n} & \cdots & a_{Nn} \end{bmatrix}^{\mathrm{T}}_{(N\times(n+1))\times 1}, \\ \boldsymbol{b} &= \begin{bmatrix} y^1 & y^2 & \cdots & y^M \end{bmatrix}^{\mathrm{T}}_{M\times 1}, \end{aligned}$$

$$\beta_{kj} = \frac{\mu_{X_{k1}}(x_1^j)\wedge\ldots\wedge\mu_{X_{kn}}(x_n^j)}{\sum_{k=1}^{N}\left(\mu_{X_{k1}}(x_1^j)\wedge\ldots\wedge\mu_{X_{kn}}(x_n^j)\right)}, \qquad k = 1,\ldots,N; \quad j = 1,\ldots,M.$$

It then follows from the standard least-squares method that the optimal coefficients θ^* are given by (i) if Λ has a full column-rank:

$$\theta^* = [\Lambda^{\mathrm{T}}\Lambda]^{-1}\Lambda^{\mathrm{T}}\boldsymbol{b}, \tag{3.17}$$

or (ii) if Λ does not have a full column-rank, then decompose $\Lambda = \boldsymbol{LC}$, where $\boldsymbol{L}$ has a full column-rank and $\boldsymbol{C}$ has a full row-rank; both need not be square, which is always possible, and

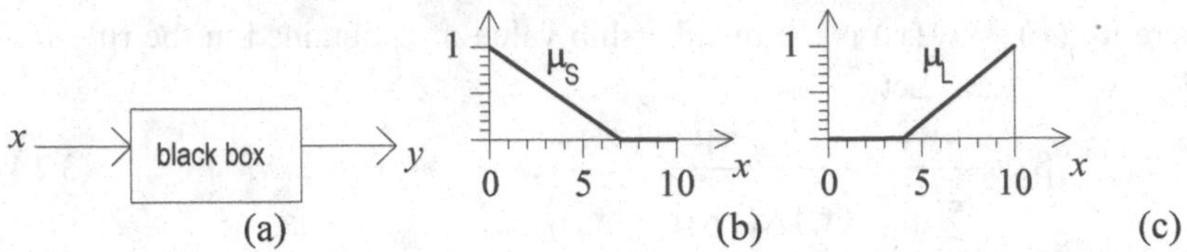

Figure 3.5 The unknown system and its two input membership functions.

$$\theta^* = C^{\mathrm{T}} [\, CC^{\mathrm{T}} \,]^{-1} [\, L^{\mathrm{T}} L \,]^{-1} L^{\mathrm{T}} b. \qquad (3.18)$$

This is the pseudo-inverse solution of the least-squares optimization.

Example 3.2. Consider an unknown system with one input x and one output y, as shown in Figure 3.5 (a). Let the input x be within the range X = [0,10], with the associate membership functions $\mu_S(x)$ and $\mu_L(x)$ describing "small" and "large," respectively, as shown in Figure 3.5 (b)-(c).

Suppose that the unknown system is the following precise fuzzy static system:

<u>Real system</u>:

R_0^1: IF x is small with a membership value $\mu_S(x)$
THEN $y = 2 + 0.5x$.

R_0^2: IF x is large with a membership value $\mu_L(x)$
THEN $y = 9 + 0.3x$.

Suppose, on the other hand, that we try to identify this system by the following estimated fuzzy static model:

R^1: IF x is small with a membership value $\tilde{\mu}_S(x)$
THEN $y = a_{10} + a_{11}x$.

R^2: IF x is large with a membership value $\tilde{\mu}_L(x)$
THEN $y = a_{20} + a_{21}x$.

Here, in general, $\tilde{\mu}_S$ and $\tilde{\mu}_L$ are likely different from μ_S and μ_L, respectively, since we don't exactly know the real system. For simplicity of discussion in this example, however, let us suppose that we set the membership functions in the modeling phase in such a way that $\tilde{\mu}_S$ = s and $\tilde{\mu}_L$ = μ , but we don't

a_{10} a_{11} a_{20} a_{21}

Now, suppose that we have carried out some experiments on the unknown

Input Output Data

	0.2	0.4		0.8	1.0		10.0
y		2.8	3.5	1.5		...	10.4

Suppose also that in this example, we created the data by using the true input-

R_0^1 and R_0^2 in the real system, namely, $y = 2 + 0.5x$ and $y = 9 + 0.3x$, respectively, by incorporating with pseudo Gaussian white noise of zero mean and a small covariance.

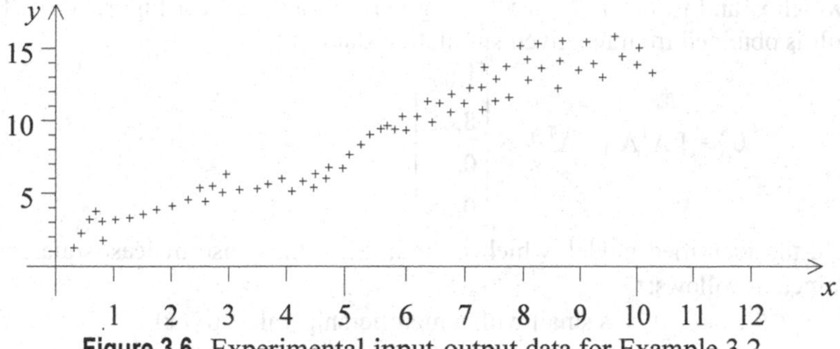

Figure 3.6 Experimental input-output data for Example 3.2.

To follow the least-squares parameter identification scheme, we first compute the following (see Figure 3.5 (b)-(c) and Table 3.1):

$$
\begin{aligned}
\mu_S(x_1) &= \mu_S(0.2) = 0.97,\\
\mu_S(x_2) &= \mu_S(0.4) = 0.94,\\
&\vdots\\
\mu_S(x_{50}) &= \mu_S(10.0) = 0.00,
\end{aligned}
$$

and

$$
\begin{aligned}
\mu_L(x_1) &= \mu_L(0.2) = 0.00,\\
&\vdots\\
\mu_L(x_{49}) &= \mu_L(9.8) = 0.97,\\
\mu_L(x_{50}) &= \mu_L(10.0) = 1.00.
\end{aligned}
$$

Then, we compute:

$$\beta_{1,1} = \frac{\mu_S(x_1)}{\mu_S(x_1)+\mu_L(x_1)}, \qquad \beta_{2,1} = \frac{\mu_L(x_1)}{\mu_S(x_1)+\mu_L(x_1)},$$

$$\beta_{1,2} = \frac{\mu_S(x_2)}{\mu_S(x_2)+\mu_L(x_2)}, \qquad \beta_{2,2} = \frac{\mu_L(x_2)}{\mu_S(x_2)+\mu_L(x_2)},$$

$$\vdots \qquad\qquad \vdots$$

$$\beta_{1,50} = \frac{\mu_S(x_{50})}{\mu_S(x_{50})+\mu_L(x_{50})}, \qquad \beta_{2,50} = \frac{\mu_L(x_{50})}{\mu_S(x_{50})+\mu_L(x_{50})}.$$

Finally, we solve the following equation by using the least-squares formula (3.17) or (3.18):

$$\Lambda\theta = \boldsymbol{b},$$

where, in this example,

$$\Lambda = \begin{bmatrix} \beta_{1,1} & \beta_{2,1} & x_1\beta_{1,1} & x_1\beta_{2,1} \\ \vdots & \vdots & \vdots & \vdots \\ \beta_{1,50} & \beta_{2,50} & x_{50}\beta_{1,50} & x_{50}\beta_{2,50} \end{bmatrix}_{50\times 4},$$

$$\theta = \begin{bmatrix} a_{10} & a_{20} & a_{11} & a_{21} \end{bmatrix}^T_{4\times 1},$$

$$\boldsymbol{b} = \begin{bmatrix} y_1 & y_2 & \ldots & y_{50} \end{bmatrix}^T_{50\times 1},$$

in which x_k and y_k, $k = 1, 2, ..., 50$, are given in Table 3.1 (or Figure 3.6). The result is obtained from the given simulation data as

$$\theta^* = [\,\Lambda^T\Lambda\,]^{-1}\,\Lambda^T\,\boldsymbol{b} = \begin{bmatrix} 1.87 \\ 8.68 \\ 0.56 \\ 0.38 \end{bmatrix}.$$

Thus, the identified model, which is optimal in the sense of least-squares, is obtained as follows:

R^1: IF x is small with a membership value $\tilde{\mu}_S(x)$
THEN $y = 1.87 + 0.56x$.

R^2: IF x is large with a membership value $\tilde{\mu}_L(x)$
THEN $y = 8.68 + 0.38x$.

The associate membership functions are $\tilde{\mu}_S = \mu_S$ and $\tilde{\mu}_L = \mu_L$, as shown in Figure 3.5 (b)-(c).

II. DISCRETE-TIME DYNAMIC FUZZY SYSTEMS AND THEIR STABILITY ANALYSIS

In the last section, we discussed how to establish a static fuzzy model for an unknown system that has a linear input-output relationship. The result is a static fuzzy model given by a set of fuzzy IF-THEN rules in the form

R^i: IF x_1 is X_{i1} AND ... AND x_n is X_{in}
THEN $y_i = a_{i0} + a_{i1}x_1 + ... + a_{in}x_n$, $\quad i=1,...,N,$ (3.19)

and the final output is

$$y = \frac{\sum_{i=1}^{N} \mu_{Y,i}(y_i) \times y_i}{\sum_{i=1}^{N} \mu_{Y,i}(y_i)}, \tag{3.20}$$

where $i=1,...,N$.

This fuzzy modeling by its nature is *static*. Yet if the input-output relations are such that

$$x_1 = x(k),$$
$$x_2 = x(k-1),$$
$$\vdots$$
$$x_n = x(k-(n-1)),$$

and

$$y_i = y_i(k), \qquad i=1,...,N,$$

so that we have a system of input-output relations of the form

$$y_i(k) = a_{i0} + a_{i1}x(k) + a_{i2}x(k-1) + ... + a_{in}x(k-(n-1)), \tag{3.21}$$

$i=1,...,N$, then this new system is considered to be *dynamic*, since it is described by a set of difference equations rather than a simple input-output mapping. This case is special in that all the inputs $x_1,\ldots, x_n$ are related (each state is the delay of its previous one) and the output is given at the last step of the process. Equation (3.21) is a typical discrete-time dynamic system in the classical systems theory if all quantities involved are crisp.

A. Dynamic Fuzzy Systems without Control

We are interested in the stability of a discrete-time dynamic (fuzzy) system described by (3.21). The reason is that when we are doing system modeling, we would like that the resulting model works well in the sense that it approximates the unknown system closely (preferably, optimal in some sense) and performs its actions stably, particularly for dynamic processes. In other words, stability is one criterion, or one of the few most important issues, in dynamic system modeling, as is in common practice of system engineering.

Definition 3.1. A typical *single-input/single-output* (SISO), *discrete-time, dynamic fuzzy system* is a fuzzy model described by a set of fuzzy IF-THEN rules of the form

$$R^i: \quad \text{IF } x(k) \text{ is } X_{i1} \text{ AND} \ldots \text{AND } x(k-(n-1)) \text{ is } X_{in}$$
$$\text{THEN } y_i(k) = a_{i0} + a_{i1}x(k) + a_{i2}x(k-1) + \ldots + a_{in}x(k-(n-1)), \quad i=1,\ldots,N,$$

with $x(k+1) = c_k y(k+1)$, $k = 0, 1, 2, \ldots$, in which

$$y(k+1) = \frac{\sum_{i=1}^{N} w_i \times y_i(k+1)}{\sum_{i=1}^{N} w_i}, \qquad k=0,1,2,\ldots.$$

Here, the fuzzy sets consist of intervals $\{X_j | j=1,\ldots,n\}$ with the associate fuzzy membership functions $\{\mu_{X_j} | j=1,\ldots,n\}$, and $\{w_i | i=1,\ldots,N\}$ is a set of weights satisfying $w_i \geq 0$, $i=1,\ldots,N$, and $\sum_{i=1}^{N} w_i > 0$.

In this definition, in all the rules R^i, $i=1,\ldots,N$, X_{ij} share the same fuzzy subset X_j and the same membership function μ_{X_j}, for each $j=1,\ldots,n$.

We also note that, similar to formula (3.20), the weights $\{w_i \mid i=1,\ldots,N\}$ usually are chosen to be equal to $\mu_{Y_i}(y_i)$. Moreover, for simplicity of notation and discussion in the following, we will always let $c_k = 1$ for all $k=0,1,2,\ldots$, and $a_{i0} = 0$ for all $i=1,\ldots,N$, although this is not necessary in general.

Now, we start with a given SISO, discrete-time, dynamic fuzzy system, where the fuzzy sets (both intervals and membership functions) and constant coefficients are all known (given, or have been identified). We want to study under what conditions on those constant coefficients the dynamic fuzzy system is asymptotically stable. This criterion is important and useful for a designer to verify if the established fuzzy model is in good working conditions.

We first recall some standard concepts and stability results in classical systems theory.

Definition 3.2. A multi-input/multi-output (MIMO), nonlinear, discrete-time, dynamic system of the form

$$x(k+1) = f(x(k)), \qquad x(k) \in R^m, \qquad k=0,1,2,\cdots,$$

is said to be *asymptotically stable* about an equilibrium point x_e, or x_e is an *asymptotically stable equilibrium point* of the system, if

$$x_e = f(x_e)$$

and, starting from any $x(0) \in R^m$, all $x(k)$ are bounded and

$$x(k) \to x_e \qquad (k \to \infty).$$

In this definition, the convergence $x(k) \to x_e$ $(k \to \infty)$ is usually measured by the l_2-norm (the "length" of a vector), namely,

$$\lim_{k \to \infty} \| x(k) - x_e \|_2 = 0,$$

where

$$\| [\, x_1 \cdots x_m \,]^T \|_2 = \sqrt{x_1^2 + \ldots + x_m^2}\,.$$

Also, in this definition, since we allow the initial state $x(0)$ to be arbitrary, this asymptotic stability is sometimes said to be "in the large," meaning "globally" in R^m. Since we only discuss this global asymptotic behavior of the system trajectories in this section, we will not distinguish it from any other local stabilities.

The following stability criterion is well known in classical systems theory.

Theorem 3.1. Suppose that the MIMO, nonlinear, discrete-time, dynamic system

$$x(k+1) = f(x(k)), \qquad x(k) \in R^m, \qquad k=0,1,2,\ldots,$$

has an equilibrium point $x_e = 0$, and that there exists a scalar-valued function $V(x(k))$ satisfying

(i) $V(0) = 0$;

(ii) $V(x(k)) > 0$ for all $x(k) \neq 0$;

(iii) $V(x(k)) \to \infty$ as $\| x(k) \|_2 \to \infty$; and

(iv) $V(x(k+1)) - V(x(k)) < 0$ for all $x(k) \neq 0$, and for all $k=0,1,2,\cdots$.

Then this system is asymptotically stable about the equilibrium point 0.

In this theorem, the function $V(x(k))$, if it exists, is called a *Lyapunov function*.

We note that this theorem particularly applies to all linear time-invariant state-space systems of the form

$$x(k+1) = A\,x(k), \qquad x(k) \in R^m, \qquad k=0,1,2,\cdots. \tag{3.22}$$

For this linear time-invariant system, however, there is a very simple criterion for the determination of its asymptotic stability.

Theorem 3.2. Let λ_j, $j = 1, \ldots, m$, be eigenvalues, counting multiple ones, of the constant matrix A in the linear time-invariant system (3.22). Then the system is asymptotically stable about the equilibrium point 0 if and only if

$$| \lambda_j | < 1 \qquad \text{for all } j=1,\ldots,m.$$

Now, return to the single-input/single-output, linear, discrete-time dynamic fuzzy system described in Definition 3.1. To apply the classical stability theorem 3.1 or 3.2 to it, we first reformulate it in the state-space setting as follows. Let $c_k = 1$, $k = 0, 1, 2, \ldots$, and $a_{i0} = 0$, $i = 1, \ldots, N$, in the system (for simplicity, as mentioned above). Define

$$\boldsymbol{x}(k) = [\, x(k)\ \ x(k-1)\ \ x(k-2)\ \ \ldots\ \ x(k-(n-1))\,]^{\mathrm{T}},$$

$$A_i = \begin{bmatrix} a_{i1} & a_{i2} & \cdots & a_{i,n-1} & a_{in} \\ 1 & 0 & \cdots & 0 & 0 \\ 0 & 1 & \cdots & 0 & 0 \\ \vdots & \vdots & \ddots & \vdots & 0 \\ 0 & 0 & \cdots & 1 & 0 \end{bmatrix}.$$

Although this (canonical) formulation has some redundancy (only the first equation is essential), it is convenient to use for stability analysis. We can then rewrite the system as

$$\boldsymbol{x}(k+1) = \sum_{i=1}^{N} w_i A_i \boldsymbol{x}(k) \Big/ \sum_{i=1}^{N} w_i \qquad k=0,1,2,\cdots. \tag{3.23}$$

Corollary 3.1. In the dynamic fuzzy system (3.23), let

$$A = \sum_{i=1}^{n} \frac{w_i}{\sum_{i=1}^{n} wi} A_i ,$$

and assume that $\{w_i\}$ are constants independent of k. Then, the system is asymptotically stable if and only if all eigenvalues of A, λ_i, i=1,...,n, satisfying

$$|\,\lambda_i\,| < 1, \qquad i=1,\ldots,n.$$

Proof. This is an immediate consequence of Theorem 3.2.

We must point out an important fact that differing from the classical linear system (3.22), the linear fuzzy system (3.23) may not be asymptotically stable about its equilibrium point 0 even if each constant matrix A_i has eigenvalues satisfying

$$|\,\lambda_{ij}\,| < 1 \qquad i=1,\ldots,N; \qquad j=1,\ldots,m.$$

This is because (3.23) is generally a time-varying system due to the fact that membership functions are involved, so that its weights depend on the discrete-time variable, k, and the system actually switches between its subsystems from time to time, as explained by the following example.

Example 3.3. Consider the discrete-time, linear, dynamic fuzzy system described by

R^1: IF x(k−1) is X_a THEN $x_1(k+1) = x(k) - \frac{1}{2}x(k-1)$,

R^1: IF x(k−1) is X_b THEN $x_2(k+1) = -x(k) - \frac{1}{2}x(k-1)$,

where $X_a = X_b = [-1,1]$ with membership functions μ_a and μ_b shown in Figure 3.7, and with initial conditions $x(1) = 0.90$ and $x(0) = -0.70$. Here, $x_1(k+1)$ and $x_2(k+1)$ are the same $x(k+1)$ in which the subscripts 1 and 2 only indicate that they are the results of rules R^1 and R^2, respectively. In the next iteration, $x(k+1)$ comes into the right-hand sides of the two equations.

It can be easily verified that the two subsystem matrices

$$A_1 = \begin{bmatrix} 1 & -1/2 \\ 1 & 0 \end{bmatrix}, \qquad A_2 = \begin{bmatrix} -1 & -1/2 \\ 1 & 0 \end{bmatrix}$$

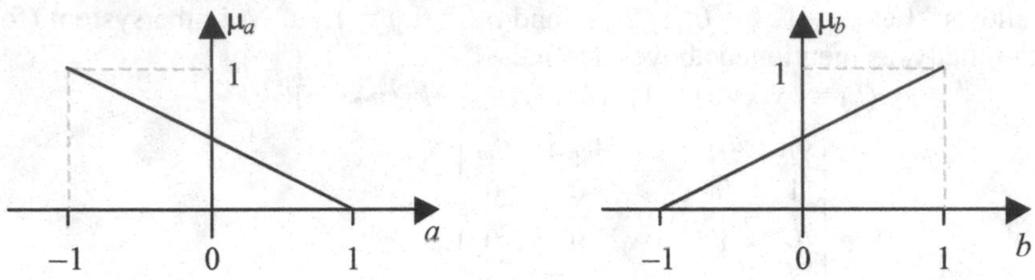

Figure 3.7 Two membership functions in Example 3.3.

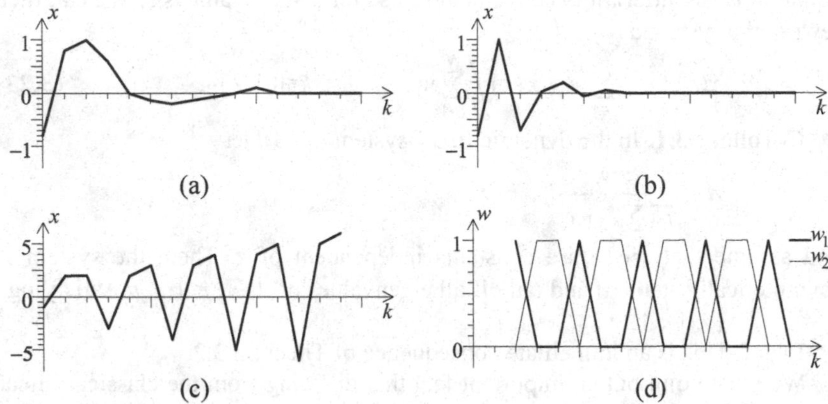

Figure 3.8 System outputs of Example 3.3:
(a) output of $x(k+1) = A_1 x(k)$;
(b) output of $x(k+1) = A_2 x(k)$;
(c) output of the entire fuzzy system;
(d) weights (output membership functions).

have all eigenvalues with absolute values strictly less than 1, so the two subsystems are asymptotically stable. This is shown in Figure 3.8 (a)-(b). However, the entire system is unstable, whose output plots are shown in Figure 3.8 (c). The reason is that the fuzzy system output is given by

$$y(k+1) = x(k+1) = \frac{w_1 A_1 x(k) + w_2 A_2 x(k)}{w_1 + w_2},$$

in which the weights w_1 and w_2 are time-varying, as shown in Figure 3.8 (d). For this type of linear time-varying system, Theorem 3.2 does not apply in general.

We now return to the discrete-time dynamic fuzzy system (3.23), which generally is a linear time-varying (or, oftentimes, nonlinear) system due to the use of the fuzzy membership functions in the weights even if all system matrices A_i are constant. We next establish a useful stability criterion (condition) for this kind of fuzzy system. To do so, we first need a preliminary result from the classical systems theory.

Lemma 3.1. Let P be a positive definite matrix such that

$$A^{\mathrm{T}}PA - P < 0 \qquad \text{and} \qquad B^{\mathrm{T}}PB - P < 0$$

for some matrices A and B of the same dimension. Then

$$A^{\mathrm{T}}PB + B^{\mathrm{T}}PA - 2P < 0.$$

Proof. Since P is positive definite, we have

$$-(A-B)^{\mathrm{T}}P(A-B) \le 0,$$

so that

$$\begin{aligned} A^{\mathrm{T}}PB + B^{\mathrm{T}}PA - 2P &= -(A-B)^{\mathrm{T}}P(A-B) + A^{\mathrm{T}}PA + B^{\mathrm{T}}PB - 2P \\ &= -(A-B)^{\mathrm{T}}P(A-B) + A^{\mathrm{T}}PA - P + B^{\mathrm{T}}PB - P \\ &< 0. \end{aligned}$$

Now, we are in a position to prove the following result.

Theorem 3.3. The discrete-time dynamic fuzzy system (3.23) is asymptotically stable about the equilibrium point 0 if there exists a common positive definite matrix P such that

$$A_i^{\mathrm{T}} PA_i - P < 0, \qquad \text{for all } i=1,2,\ldots,N.$$

Proof. We prove the theorem by using Theorem 3.1. In so doing, wc construct a Lyapunov function candidate

$$V(x(k)) = x^{\mathrm{T}}(k)\, P\, x(k), \quad k=0,1,2,\ldots,$$

where P is the common positive definite matrix stated in the theorem. Clearly, this function satisfies the following properties:

(i) $V(0) = 0$;

(ii) $V(x(k)) > 0$ for all $x(k) \neq 0$;

(iii) $V(x(k)) \to \infty$ as $\| x(k) \|_2 \to \infty$.

To verify that this function also satisfies the property

(iv) $V(x(k+1)) - V(x(k)) < 0$ for all $x(k) \neq 0$, $k=0,1,2,\ldots$,

we carry out the calculations as follows:

$$\begin{aligned} &V(x(k+1)) - V(x(k)) \\ &\quad= x^{\mathrm{T}}(k+1)P\,x(k+1) - x^{\mathrm{T}}(k)P\,x(k) \\ &\quad= [\sum_{i=1}^{N} w_i A_i x(k) / \sum_{i=1}^{N} w_i\,]^{\mathrm{T}} P\, [\sum_{i=1}^{N} w_i A_i x(k) / \sum_{i=1}^{N} w_i\,] - x^{\mathrm{T}}(k)P\,x(k) \\ &\quad= x^{\mathrm{T}}(k)\, \{\, [\sum_{i=1}^{N} w_i A_i^{\mathrm{T}} / \sum_{i=1}^{N} w_i\,]\, P\, [\sum_{i=1}^{N} w_i A_i / \sum_{i=1}^{N} w_i\,] - P \,\}\, x(k) \\ &\quad= \sum_{i=1}^{N}\sum_{j=1}^{N} w_i w_j x^{\mathrm{T}}(k)[A_i^{\mathrm{T}} PA_j - P]x(k) / \sum_{i=1}^{N}\sum_{j=1}^{N} w_i w_j \\ &\quad= \left\{ \sum_{i=1}^{N} w_i^2 x^{\mathrm{T}}(k)[A_i^{\mathrm{T}} PA_i - P]x(k) \right. \\ &\qquad \left. + \sum_{i<j}^{N} w_i w_j x^{\mathrm{T}}(k)[A_i^{\mathrm{T}} PA_j + A_j^{\mathrm{T}} PA_i - 2P]x(k) \right\} \Big/ \sum_{i=1}^{N}\sum_{j=1}^{N} w_i w_j \\ &\quad< 0, \end{aligned}$$

where the last inequality follows from Lemma 3.1 and the facts that $w_i \geq 0$ for all i=1,...,N, and $\sum_{i=1}^{N} w_i > 0$.

Therefore, the conditions of Theorem 3.1 are all satisfied, so that its conclusion follows.

We remark that the linear matrix inequality (LMI) problem stated in Theorem 3.3 is equivalent to the following convex programming problem:

$$\min_{P>0} \max_{0\leq i\leq N} \lambda_{\max}(A_i^{\mathrm{T}} P A_i - P) < 0$$

where $\lambda_{\max}(\times)$ is the maximum eigenvalue of a matrix. This convex programming problem can be solved efficiently by the "interior point algorithm" using the LMI Control Toolbox with Matlab.

Theorem 3.3 brings up a question: in general, how can we tell if such a common positive definite matrix P exists? The following theorem provides a necessary (but not sufficient) condition.

Theorem 3.4. Let A_i, i=1,...,N, be stable and nonsingular matrices, namely, with eigenvalues

$$0 < |\lambda_i| < 1, \qquad i=1,...,N.$$

If there exists a common positive definite matrix P such that

$$A_i^{\mathrm{T}} P A_i - P < 0, \qquad i=1,...,N,$$

then the product matrices $A_i A_j$ are all stable for i,j=1,...,N.

Proof. If such a matrix P exists, then we have

$$P < [A_i^{-1}]^{\mathrm{T}} P [A_i^{-1}],$$

since A_i are nonsingular and $[A_i^{-1}]^{\mathrm{T}} = [A_i^{\mathrm{T}}]^{-1}$, i=1,...,N. Thus, it follows, again from the given inequality, that

$$A_i^{\mathrm{T}} P A_i < P < [A_j^{-1}]^{\mathrm{T}} P [A_j^{-1}], \qquad i,j=1,...,N.$$

This yields

$$A_j^{\mathrm{T}} A_i^{\mathrm{T}} P A_i A_j - P < 0, \qquad i,j=1,...,N.$$

That is, the product matrix A_iA_j is a stable matrix (with all the eigenvalues having absolute values strictly less than one).

Returning to Example 3.3, we see that although the two subsystems are stable, the product matrix

$$A_1A_2 = \begin{bmatrix} 1 & -1/2 \\ 1 & 0 \end{bmatrix} \begin{bmatrix} -1 & -1/2 \\ 1 & 0 \end{bmatrix} = \begin{bmatrix} -3/2 & -1/2 \\ -1 & -1/2 \end{bmatrix}$$

has eigenvalues $\lambda_1 = -0.135$ and $\lambda_2 = -1.865$, and so is unstable. Hence, by Theorem 3.4 there will not be a common positive definite matrix P satisfying

$$A_i^{\mathrm{T}} P A_i - P < 0, \qquad i=1,...,N.$$

In other words, the asymptotic stability of the system in Example 3.3 cannot be ensured. Note, however, that this does not mean that the system must be unstable, since the existence of P is only a sufficient condition for the stability. The instability (not the stability) has to be determined by other methods. Plotting the system outputs, as shown in Figure 3.8 (c), is one practical test, particularly for complex systems.

B. Dynamic Fuzzy Systems with Control

Consider, again, the typical single-input/single-output (SISO), linear, discrete-time, dynamic fuzzy system discussed in the last subsection (Definition 3.1). We now take into consideration also control inputs.

Definition 3.3. An SISO, *discrete-time, dynamic fuzzy control system* is a fuzzy control model described by a set of fuzzy IF-THEN rules of the form

$$
\begin{aligned}
R^i: \quad & (\text{IF } x(k) \text{ is } X_{i1} \text{ AND } \ldots \text{ AND } x(k-(n-1)) \text{ is } X_{in}) \text{ AND} \\
& (\text{IF } u(k) \text{ is } U_{i1} \text{ AND } \ldots \text{ AND } u(k-(m-1)) \text{ is } U_{im}) \\
& \text{THEN } y_i(k) = a_{i0} + a_{i1}x(k) + \ldots + a_{in}x(k-(n-1)) \\
& \qquad\qquad + b_{i0} + b_{i1}u(k) + \ldots + b_{im}u(k-(m-1)), \\
& i=1,\ldots,N,
\end{aligned}
$$

with $m \le n$ and

$$x(k+1) = c_k\, y(k+1), \qquad k=0,1,2,\ldots,$$

in which

$$y(k+1) = \frac{\sum_{i=1}^{N} w_i y_i(k+1)}{\sum_{i=1}^{N} w_i}, \qquad k=0,1,2,\ldots,$$

where the fuzzy sets consist of intervals $\{X_j | j=1,\ldots,n\}$ and $\{U_j | j=1,\ldots,m\}$, and their associate fuzzy membership functions $\{\mu_{X_j} | j=1,\ldots,n\}$ and $\{\mu_{U_j} | j=1,\ldots,m\}$, respectively, and $\{w_i | i=1,\ldots,N\}$ is a set of weights satisfying $w_i \ge 0$, $i=1,\ldots,N$, and $\sum_{i=1}^{N} w_i > 0$.

Here, we note again that for each rule R^i, $i=1,\ldots,N$, all the X_{ij} share the same fuzzy subset X_j and the same membership function μ_{X_j}, for each $j=1,\ldots,n$. Also, the weights are usually quantified by the output membership functions in the same way as in (3.20).

As before, for simplicity of notation and discussion, we will always assume $c_k = 1$ for all $k=0,1,2,\ldots$, and $a_{i0} = b_{i0} = 0$ for all $i=1,\ldots,N$, in this model.

It is clear that Definitions 3.1 and 3.3 have no essential difference if the control inputs $\{u(k)\}$ in Definition 3.3 are independent of the states $\{x(k)\}$. In engineering control systems, however, most of the time we would like to have negative state-feedback controllers of the form

$$
\begin{aligned}
u(k) &= -K_{11}x(k) - K_{12}x(k-1) - \ldots - K_{1n}x(k-(n-1)), \\
u(k-1) &= \qquad\qquad\;\; - K_{22}x(k-1) - \ldots - K_{2n}x(k-(n-1)), \\
&\;\;\vdots \\
u(k-m) &= \qquad\qquad\;\; - K_{mm}x(k-m) - \ldots - K_{mn}x(k-(n-1)),
\end{aligned}
$$

where K_{ij} are constant control gains to be determined.

Again, if these control inputs have crisp relations with the states, then substituting them into the fuzzy control system described in Definition 3.3 results in a dynamic fuzzy system described by Definition 3.1. However, if these controllers are given by fuzzy descriptions, then the situation becomes much more involved, both conceptually and mathematically. In addition, a control input may also depend on control inputs at the previous steps, which will further complicate the computation.

To facilitate our discussion, let us consider a simple example. Suppose that a fuzzy control system is given by

$$R_S^1: \text{ IF } x(k) \text{ is } X_1 \text{ AND } u(k) \text{ is } U_1$$
$$\text{THEN } x_1(k+1) = a_1x(k) + b_1u(k).$$
$$R_S^2: \text{ IF } x(k) \text{ is } X_2 \text{ AND } u(k) \text{ is } U_2$$
$$\text{THEN } x_2(k+1) = a_2x(k) + b_2u(k).$$

Assume also that a state-feedback fuzzy controller is designed to be one described by

$$R_C^1: \text{ IF } x(k) \text{ is } X_1 \text{ AND } x(k-1) \text{ is } X_1$$
$$\text{THEN } u_1(k) = -K_{11}x(k) - K_{12}x(k-1).$$
$$R_C^2: \text{ IF } x(k) \text{ is } X_2 \text{ AND } x(k-1) \text{ is } X_2$$
$$\text{THEN } u_2(k) = -K_{21}x(k) - K_{22}x(k-1).$$

Then, we can combine all possibilities, where it is important to note that for "$x(k)$ is X_1 and $u(k)$ is U_1" there are two possibilities for the previous control input $u(k-1)$, either "is U_1" or "is U_2." And the same is true for "$x(k)$ is X_2 and $u(k)$ is U_1." We thus obtain

$$R^{11}: \text{ (IF } x(k) \text{ is } X_1) \text{ AND (IF } u(k) \text{ is } U_1 \text{ AND } u(k-1) \text{ is } U_1)$$
$$\text{THEN } x_{11}(k+1) := a_1x(k) + b_1u_1(k)$$
$$= (a_1 - b_1K_{11})\, x(k) - b_1K_{12}x(k-1).$$
$$R^{12}: \text{ (IF } x(k) \text{ is } X_1) \text{ AND (IF } u(k) \text{ is } U_1 \text{ AND } u(k-1) \text{ is } U_2)$$
$$\text{THEN } x_{12}(k+1) := a_1x(k) + b_1u_2(k)$$
$$= (a_1 - b_1K_{21})\, x(k) - b_1K_{22}x(k-1).$$
$$R^{21}: \text{ (IF } x(k) \text{ is } X_2) \text{ AND (IF } u(k) \text{ is } U_2 \text{ AND } u(k-1) \text{ is } U_1)$$
$$\text{THEN } x_{21}(k+1) := a_2x(k) + b_2u_1(k)$$
$$= (a_2 - b_2K_{11})\, x(k) - b_2K_{12}x(k-1).$$
$$R^{22}: \text{ (IF } x(k) \text{ is } X_2) \text{ AND (IF } u(k) \text{ is } U_2 \text{ AND } u(k-1) \text{ is } U_2)$$
$$\text{THEN } x_{22}(k+1) := a_2x(k) + b_2u_2(k)$$
$$= (a_2 - b_2K_{21})\, x(k) - b_2K_{22}x(k-1).$$

After all these state-feedback control inputs have been used, this fuzzy system, described by Definition 3.3, reduces to a new closed-loop dynamic fuzzy system, given by the rules R^{11}, R^{12}, R^{21}, and R^{22}, in which there are no more control variables $u(k)$. Thus, the stability conditions obtained in Theorem 3.3 can be applied, namely, the designer can choose the feedback control gains K_{11}, K_{12}, K_{21}, and K_{22} to ensure the asymptotic stability of the overall controlled system.

Example 3.4. Consider a fuzzy control system described by the following rule base:

$$R_S^1: \text{ IF } x(k) \text{ is } X_1 \text{ AND } x(k-1) \text{ is } X_1$$
$$\text{THEN } x_1(k+1) = 2.178x(k) - 0.588x(k-1) + 0.603u(k).$$

R_S^2: IF $x(k)$ is X_2 AND $x(k-1)$ is X_2

THEN $x_2(k+1) = 2.256x(k) - 0.361x(k-1) + 1.120u(k)$.

Let the fuzzy state-feedback controller be described by

R_C^1: IF $x(k)$ is X_1 AND $x(k-1)$ is X_1

THEN $u_1(k) = r(k) - K_{11}x(k) - K_{12}x(k-1)$.

R_C^2: IF $x(k)$ is X_2 AND $x(k-1)$ is X_2

THEN $u_2(k) = r(k) - K_{21}x(k) - K_{22}x(k-1)$.

Here, $\{r(k)\}$ is a reference signal (set-points).

Since the control input $u(k)$ does not appear in the condition parts, we consider its membership values to be identically equal to 1 therein. The resulting closed-loop dynamic fuzzy system is then obtained as follows:

R^{11}: IF $x(k)$ is X_1 AND $x(k-1)$ is X_1

THEN $x_{11}(k+1) = (2.178-0.603K_{11})\, x(k)$
$+ (-0.588-0.603K_{12})\, x(k-1) + 0.603\, r(k)$.

R^{12}: IF $x(k)$ is X_1 AND $x(k-1)$ is X_2

THEN $x_{12}(k+1) = (2.178-0.603K_{21})\, x(k)$
$+ (-0.588-0.603K_{22})\, x(k-1) + 0.603\, r(k)$.

R^{21}: IF $x(k)$ is X_2 AND $x(k-1)$ is X_1

THEN $x_{21}(k+1) = (2.256-1.120K_{11})\, x(k)$
$+ (-0.361-1.120K_{12})\, x(k-1) + 1.120\, r(k)$.

R^{22}: IF $x(k)$ is X_2 AND $x(k-1)$ is X_2

THEN $x_{22}(k+1) = (2.256-1.120K_{21})\, x(k)$
$+ (-0.361-1.120K_{22})\, x(k-1) + 1.120\, r(k)$.

The overall state output $x(k+1)$ is then computed by

$$x(k+1) = \frac{w_1 x_{11}(k+1) + w_2 x_{12}(k+1) + w_3 x_{21}(k+1) + w_4 x_{22}(k+1)}{w_1 + w_2 + w_3 + w_4}$$

for some weights $w_i \geq 0$, $i=1,2,3,4$, and $\sum_{i=1}^{4} w_i > 0$.

Suppose now that the membership functions μ_{X_1} and μ_{X_2} are given as shown in Figure 3.9. Then, the weights can be determined by using these membership functions, as discussed before. Also, for simplicity, assume that the reference signal $r(k) = 0$ for all $k = 0,1,2,\ldots$, and that the controller in both R_C^1 and R_C^2 are the same:

$$u_1(k) = u_2(k) = -K\, x(k), \quad k=0,1,2,\ldots.$$

Then, the closed-loop fuzzy system reduces to the following simple one:

R^1: IF $x(k)$ is X_1 AND $x(k-1)$ is X_1

THEN $x_1(k+1) = (2.178-0.603K)\, x(k) - 0.588\, x(k-1)$.

R^2: IF $x(k)$ is X_2 AND $x(k-1)$ is X_2

THEN $x_2(k+1) = (2.256-1.120K)\, x(k) - 0.361\, x(k-1)$.

To analyze the stability of this closed-loop fuzzy system, Theorem 3.3 can be applied. Here, for this example of a linear time-invariant system, we show another method of classical root locus analysis.

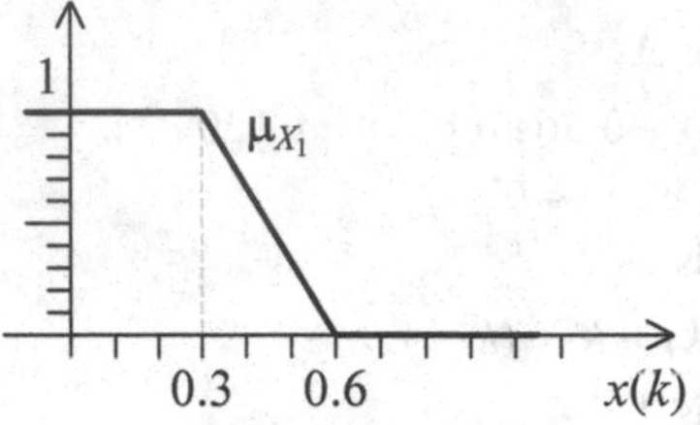

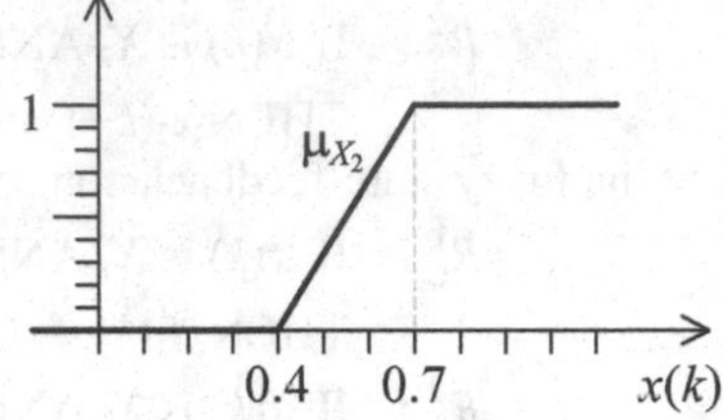

Figure 3.9 The two membership functions in Example 3.4.

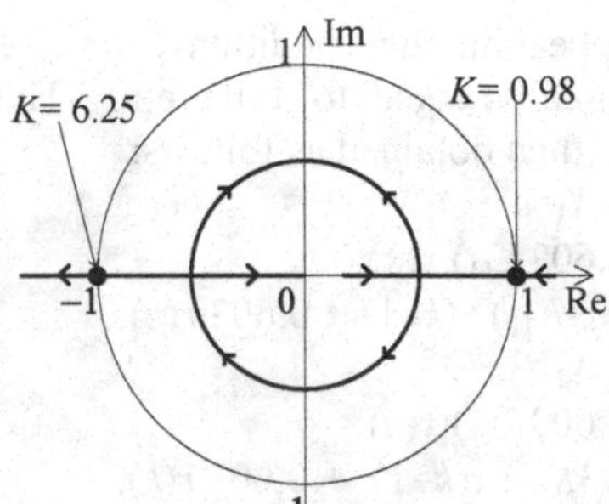

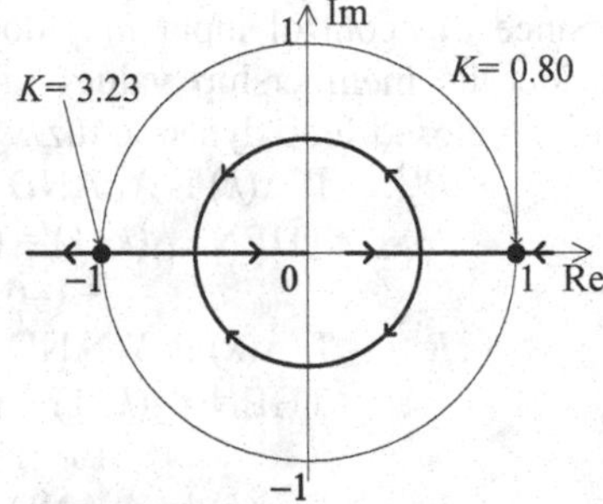

Figure 3.10 The root locus plots for Example 3.4.

Figure 3.10 shows the root locus plots for the linear subsystems described by R^1 and R^2, respectively, where $0 < K < \infty$. Recall that for discrete-time systems, the stability boundary is the unit circle $|z| = 1$ on the eigenplane. It is clear from Figure 3.10 that the two subsystems can be stabilized by choosing $0.98 < K < 6.25$ and $0.80 < K < 3.23$, respectively, which together yield the common stabilizing control gain range $0.98 < K < 3.23$.

To verify this, we can also reformulate the closed-loop fuzzy control system to the state-space setting, with

$$A_1 = \begin{bmatrix} 2.178 - 0.603K & -0.588 \\ 1 & 0 \end{bmatrix},$$

$$A_2 = \begin{bmatrix} 2.256 - 1.120K & -0.361 \\ 1 & 0 \end{bmatrix}.$$

Pick a value for K, say $K = 1.12$, from the range $0.98 < K < 3.23$, and pick a positive definite matrix

$$P = \begin{bmatrix} 2.0 & -1.3 \\ -1.3 & 1.0 \end{bmatrix}.$$

Then, one can verify that the conditions

$$A_i^{\mathrm{T}} P A_i - P < 0\,, \quad i=1,2,$$

are satisfied, so Theorem 3.3 ensures that the overall fuzzy control system is asymptotically stable about its equilibrium point 0.

Finally, we note that if we combine the two subsystems together by the weighted average formula

$$x(k+1) = \frac{w_1 x_1(k+1) + w_2 x_2(k+1)}{w_1 + w_2}$$

for some constant weights $w_1 \geq 0$ and $w_2 \geq 0$, with $w_1 + w_2 > 0$, independent of k, then the stability of the overall closed-loop control system can be analyzed by Corollary 3.1. In this case, the root locus method may not be convenient to use. Instead, by Corollary 3.1 one may simply choose a control gain K such that the matrix

$$A = \frac{1}{w_1 + w_2}(w_1 A_1 + w_2 A_2)$$

has all the eigenvalues with absolute value strictly less than 1, which is quite easy in the case of this example.

Finally in this section, we return to system (3.23), now with control inputs, namely,

$$x(k+1) = \boldsymbol{A}\, x(k) + \boldsymbol{B}\, u(k), \qquad k=0,1,2,... \tag{3.24}$$

where, in a general form,

$$\boldsymbol{A} = \sum_{i=1}^{N} \beta_i \boldsymbol{A}_i, \qquad \boldsymbol{B} = \sum_{i=1}^{N} \gamma_i \boldsymbol{B}_i,$$

with constant matrices $\boldsymbol{A}_i$, $\boldsymbol{B}_i$ and coefficients β_i, γ_i satisfying

$$0 \leq \beta_i \leq 1, \qquad 0 \leq \gamma_i \leq 1, \qquad i=1,2,...,N,$$

$$\sum_{i=1}^{N} \beta_i = 1, \qquad \sum_{i=1}^{N} \gamma_i = 1.$$

Note, again, that these coefficients usually depend on the discrete-time variable k, and, in fact, depend on the state vector $x(k)$ due to the involvement of the membership functions.

Suppose that a negative linear state-feedback controller is used:

$$u(k) = -\boldsymbol{K}\, x(k), \qquad k=0,1,2,..., \tag{3.25}$$

where $\boldsymbol{K}$ is a constant gain matrix to be determined. The controlled system is

$$x(k+1) = \overline{\boldsymbol{A}}\, x(k) \tag{3.26}$$

where

$$\overline{\boldsymbol{A}} = \boldsymbol{A} - \boldsymbol{B}\,\boldsymbol{K} = \sum_{i=1}^{N} (\beta_i \boldsymbol{A}_i - \gamma_i \boldsymbol{B}_i \boldsymbol{K}).$$

Let $\|\times\|_s$ be the spectral norm of a constant matrix, which is equal to the largest singular value (namely, the largest absolute value of all the eigenvalues) of the matrix. Then, we have the following stability condition:

Theorem 3.5. If the feedback control gain matrix $\boldsymbol{K}$ in (3.25) is designed such that the spectral norm of the matrix $\overline{\boldsymbol{A}} = \boldsymbol{A} - \boldsymbol{B}\,\boldsymbol{K}$ is strictly less than 1, uniformly with respect to $x(k)$ and k, then the dynamic fuzzy controlled system (3.24) is asymptotically stable about its equilibrium point 0.

Proof. It is easily seen that if $\|\overline{\boldsymbol{A}}\|_s < 1$ uniformly with respect to $\underline{x}(k)$ and k, then

$$x(k+1) = \overline{\boldsymbol{A}}\, x(k) = \overline{\boldsymbol{A}}^2\, x(k-1) = ... = \overline{\boldsymbol{A}}^{k+1}\, x(0),$$

so that

$$\| x(k+1) \| \leq \| \overline{A} \|_s^{k+1} \| x(0) \| \to 0 \qquad (k \to \infty).$$

An easier (but rather conservative) condition is the following:

Corollary 3.2. Let

$$\alpha_i = max\{ | \beta_i |, | \gamma_i | \},$$

$$\| \boldsymbol{A}_i \|_s \leq a_i, \qquad \| \boldsymbol{B}_i \|_s \leq b_i, \qquad i=1,2,...,N.$$

If

$$\sum_{i=1}^{N} \alpha_i (a_i + b_i) < 1 \tag{3.27}$$

uniformly with respect to $x(k)$ and k, then the dynamic fuzzy control system (3.24) is asymptotically stable about its equilibrium point 0.

Proof. It suffices to note that

$$\| \overline{A} \|_s \leq \sum_{i=1}^{N} (| \beta_i | \times \| \boldsymbol{A}_i \|_s + | \gamma_i | \times \| \boldsymbol{B}_i \|_s) \leq \sum_{i=1}^{N} \alpha_i (a_i + b_i).$$

III*. MODELING OF CONTINUOUS-TIME DYNAMIC FUZZY CONTROL SYSTEMS

Modeling of a continuous-time dynamic system in the conventional approach is to describe the physical system by, for example, a nonlinear autonomous ordinary differential equation of the form

$$\begin{cases} \dot{x}(t) = f(x(t), u(t)), & t_0 \leq t < \infty, \\ x(t_0) = x_0, \end{cases} \tag{3.28}$$

where $x(t) \in \mathrm{R}^n$ is the state-vector, $u(t) \in \mathrm{R}^m$ $(m \leq n)$ is the control vector, f:$\mathrm{R}^n \times \mathrm{R}^m \to \mathrm{R}^n$ is a nonlinear, usually integrable, function, and x_0 is the initial state of the system. We will always assume that this mathematical model is well defined on $[t_0,\infty)$ in the sense that for each initial state x_0 and each control input $u(t)$ it has a unique solution $x(t)$. When certain control objectives are taken into account, the model has to satisfy some additional conditions in general.

The mathematical model (3.28), once established and accepted as an appropriate description of the underlying physical system, represents a time-evolutionary (dynamic) process via its explicit or implicit solution function $x(t) = x(t;u(t),x_0)$.

We may consider this differential equation from a mathematical logic point of view: at any instant $t = t^* \in [t_0,\infty)$, the system is described by the relation that if the state vector x is equal to $x(t^*)$ and the control vector u is equal to $u(t^*)$ then the derivative of the state vector, $\dot{x}$, is equal to $\left.\frac{dx(t)}{dt}\right|_{t=t^*}$. As mentioned above, since the differential equation (3.28) represents a time-evolutionary process, we should view the system in a dynamic way. However, this logic-based interpretation of the relation embedded in the differential equation is also reasonable and, in fact, can be used for fuzzy modeling of continuous-time dynamic systems.

Our purpose here is to establish a fuzzy version of the conventional mathematical model (3.28), by using fuzzy logic and fuzzy mathematics.

To prepare for the development, we first briefly describe the fuzzy contents that we need. In the continuous-time fuzzy modeling to be discussed in this section, only triangular membership functions are used for simplicity. Both the state and the control vectors, x and u, are assumed to have "fuzzy values," instead of having crisp values as in the classical setting, in the sense that their values at any instant are located within certain subsets (usually, intervals) with certain membership values, respectively. Consequently, the derivative of the state, $\dot{x}$, has a fuzzy value at each instant, as a logic-inference result of both the state and the control. Using the standard terminology, a function $y = y(t)$ assumes a fuzzy value at instant $t = \tilde{t}$ is described as follows: If I is an interval with μ_I being the membership function defined on it, then the fuzzy value $y(\tilde{t})$ is located inside I with a corresponding membership value measured by μ_I, namely, with the membership value equal to $\mu_I(y(\tilde{t}))$. As usual, this is said "y is I."

A. Fuzzy Interval Partitioning

To develop a continuous-time fuzzy modeling technique, we first need to clearly describe a fuzzy set operation, which serves as a key in this approach.

Let $\sigma > 0$ be a chosen and fixed real number, and let

$$I_n^\sigma = [\,(n-1)\sigma\,,\,(n+1)\sigma\,), \qquad n = 0,\pm1,\pm2,\ldots, \tag{3.29}$$

be a semi-closed interval in R, where the right-hand side of the interval can actually be closed when interval arithmetic (Chapter 1) is performed, with the associate normalized piecewise continuous membership function $\mu_{I_n^\sigma}(x)$ defined on R, assuming nonzero values only on the interval I_n^σ. A simple triangular membership function is used in this section for simplicity, which is defined by

$$\mu_{I_n^\sigma}(x) = \begin{cases} (x/\sigma)-(n-1) & \text{if}\,(n-1)\sigma < x \le (n)\sigma, \\ (n+1)-(x/\sigma) & \text{if}\,(n)\sigma < x \le (n+1)\sigma, \\ 0 & \text{else.} \end{cases} \tag{3.30}$$

As usual, such an interval and its associated membership function together define a fuzzy subset, denoted by S_n in the following. We first observe that when n runs over all possible integers: $n = 0,\pm1,\pm2,\ldots$, these intervals together cover the whole real line:

$$R = (-\infty\,,\,\infty) = \bigcup_{n=-\infty}^{\infty} I_n^\sigma\,,$$

which is called a *fuzzy interval covering*.

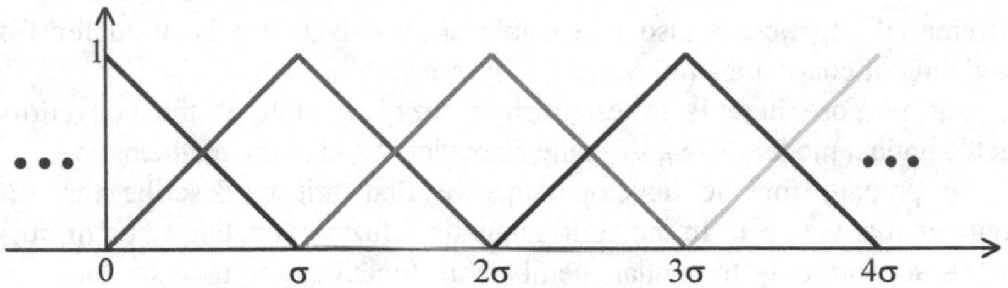

Figure 3.11 Fuzzy interval covering of the real line.

A fuzzy interval covering is visualized in Figure 3.11, where such a partition of R is called a *fuzzy interval partitioning* because for $x \in$ R, $\sum_{n=-\infty}^{\infty} S_n^\sigma(x) = 1$. While a nonfuzzy partition specifies $S_n^\sigma(x) = 1$ for exactly one n, the fuzzy partition defined here allows nonzero values for two adjacent intervals, with zero elsewhere, as is clear from Figure 3.11.

We note that the choice of equal-width intervals entails no loss of generality, particularly in applications. We also note that Figure 3.11 serves as a fuzzification algorithm, which is to be discussed in more detail below. The corresponding membership values can be associated with linguistic terms such as positive small (PS), positive medium (PM), and positive large (PL). These linguistic labels have no impact on the modeling and analysis, but make fuzzy rules easy to understand and explain, as seen previously in discrete-time fuzzy modeling. Note that the real line can be covered simultaneously by several fuzzy interval coverings with (probably different) scaling parameters, σ_1, σ_2,....

We next introduce an interval operation into the coverings of the real line. Suppose that the real line has been covered by two interval coverings with (probably different) scaling parameters σ_1 and σ_2:

$$\text{R} = \bigcup_{n=-\infty}^{\infty} I_n^{\sigma_1} \quad \text{and} \quad \text{R} = \bigcup_{m=-\infty}^{\infty} I_m^{\sigma_2},$$

with the associate membership functions $\mu_{I_n^{\sigma_1}}(x)$ and $\mu_{I_m^{\sigma_2}}(x)$ defined on them, respectively. We first note that, in the covering R = $\bigcup_{n=-\infty}^{\infty} I_n^{\sigma_1}$, for any particular real-value x there always exists a unique integer n such that

$$n\sigma_1 \leq x < (n+1)\sigma_1;$$

namely, x belongs to the intersection of two adjoining intervals, as shown in Figure 3.12. In this case, we simply say that "x is $I_n^{\sigma_1} \cap I_{n+1}^{\sigma_1}$."

Similarly, for any particular value of u in the second covering of R, there is a unique integer m satisfying

$$m\sigma_2 \leq u < (m+1)\sigma_2;$$

and in this case, we say that "u is $I_m^{\sigma_2} \cap I_{m+1}^{\sigma_2}$."

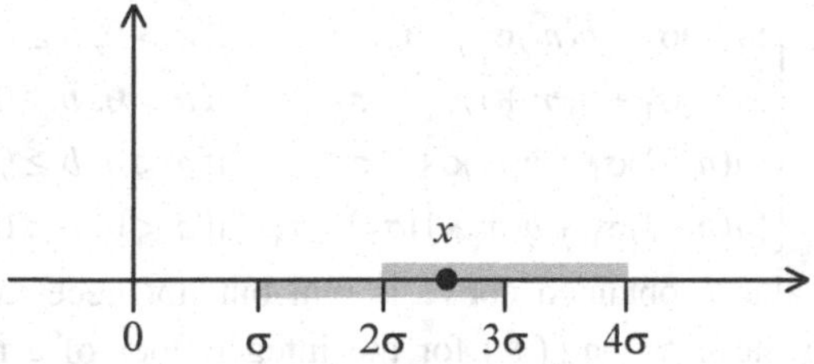

Figure 3.12 Position of a particular value of x.

Given these definitions, we now want to see IF x is $I_n^{\sigma_1} \cap I_{n+1}^{\sigma_1}$ AND u is $I_m^{\sigma_2} \cap I_{m+1}^{\sigma_2}$, THEN $f(x, u)$. We are mainly concerned with the linear function $f(x,u) = ax + bu$ for two arbitrary (yet crisp) real numbers a and b (not both zero), which will eventually be extended to the case of a nonlinear function $f(x,u)$ in the last part of the section. In order for this interval operation to be closed, we expect to have the result in the form "$ax + bu$ is $I_k^{\sigma_3} \cap I_{k+1}^{\sigma_3}$ " for some real number $\sigma_3 > 0$ and some integer k, calculated from the given σ_1, σ_2, n, m, a, and b. Here, we must have $\sigma_3 > 0$ but the integer k can be either positive, zero, or negative.

If this can be done, then we should have

$$n\sigma_1 \le x < (n+1)\sigma_1,$$
$$m\sigma_2 \le u < (m+1)\sigma_2,$$
$$k\sigma_3 \le ax + bu < (k+1)\sigma_3.$$

For a nontrivial linear combination $ax + bu$, which describes a single-input/single-output (SISO) control system, we require $a \neq 0$. Thus, we have only four cases to consider: (1) $a > 0, b \ge 0$; (2) $a > 0, b < 0$; (3) $a < 0, b \ge 0$; (4) $a < 0, b < 0$. We only discuss the first case here, since the other cases are similar. In the first case, by matching both the lower and upper bounds of the above three inequalities, we have

$$k\sigma_3 = a(n)\sigma_1 + b(m)\sigma_2,$$
$$(k+1)\sigma_3 = a(n+1)\sigma_1 + b(m+1)\sigma_2,$$

which yields

$$\begin{cases} \sigma_3 = a\sigma_1 + b\sigma_2, \\ k = \dfrac{a(n)\sigma_1 + b(m)\sigma_2}{a\sigma_1 + b\sigma_2}, \end{cases}$$

guaranteeing $k\sigma_3 \le ax + bu < (k+1)\sigma_3$. All cases considered, where $a \neq 0$ always, we arrive at

$$\sigma_3 = |a|\sigma_1 + |b|\sigma_2 \tag{3.31a}$$

and

$$k = \begin{cases} \{a(n)\sigma_1 + b(m)\sigma_2\}/\sigma_3 & \text{if } a > 0,\ b \geq 0, \\ \{a(n)\sigma_1 + b(m+1)\sigma_2\}/\sigma_3 & \text{if } a > 0,\ b < 0, \\ \{a(n+1)\sigma_1 + b(m)\sigma_2\}/\sigma_3 & \text{if } a < 0,\ b \geq 0, \\ \{a(n+1)\sigma_1 + b(m+1)\sigma_2\}/\sigma_3 & \text{if } a < 0,\ b < 0. \end{cases} \tag{3.31b}$$

Note that the value k obtained above is real but not necessarily an integer. However, using the notation $\lfloor(\ell)$ for the integer-floor of a real number ℓ, namely:

$$\lfloor(\ell) \leq \ell < \lfloor(\ell) + 1,$$

we always have

$$\lfloor(k)\sigma_3 \leq ax + bu \leq (\lfloor(k) + 2)\sigma_3,$$

so that, with $p = \lfloor(k)$,

$$\text{IF } x \text{ is } I_n^{\sigma_1} \cap I_{n+1}^{\sigma_1} \text{ AND } u \text{ is } I_m^{\sigma_2} \cap I_{m+1}^{\sigma_2}$$

$$\text{THEN } ax + bu \text{ is } I_p^{\sigma_3} \cap I_{p+1}^{\sigma_3},$$

in which σ_3 and k are calculated from (3.31a) and (3.31b), respectively.

We remark that in making this operation closed, we have actually enlarged the range of the values of $ax+bu$ whenever k is not an integer. It is however important to point out that the enlargement makes the conclusion "THEN" sufficient, although it is not necessary, and the situation is actually more robust in the sense that if $ax + bu$ satisfies

$$k\sigma_3 \leq ax + bu < (k+1)\sigma_3,$$

then it is always in $I^{\sigma_3} \cap I_{p+1}^{\sigma_3}$. This is consistent with the interval arithmetic studied in Chapter 1. Moreover, using $\lfloor(k)$ (or $\lfloor(k)+1$) to replace (or, approximate) k, we obtain the same partition of R:

$$\mathrm{R} = \bigcup_{(k} {}^{3}_{k)} \qquad \bigcup_{\lfloor(k)=-\infty}^{\infty} I_{\lfloor(k)+1}^{\sigma_3}.$$

This implies that using the integer $\lfloor(k)$ to replace (approximate) the real number k only shifts the partition, and this, as will be seen in the sequel, has no effect on the stability, controllability, and observability analyses as well as optimal control computations discussed later in the book.

We next establish a mathematical formulation for fuzzy logic inference rules, using the fuzzy interval operation introduced above. To simplify the statement of a rule, and to take into account the associate membership functions, we will simply state as before that "x is S_n^σ " instead of saying "x is $I_n^\sigma \cap I_{n+1}^\sigma$ with the associate membership function value $\mu_{,\sigma}(x)$." For example, if x is located as shown in Figure 3.12, then we will say "x is S_2^σ," meaning that x belongs to the set $I_2^\sigma \cap I_3^\sigma = [2\sigma, 3\sigma)$ on the x-axis.

Formally, we define the following *fuzzy logic inference rule*:

$$\text{IF } x \text{ is } S_n^{\sigma_1} \text{ AND } u \text{ is } S_m^{\sigma_2}$$

$$\text{THEN } y = f(x,u) = ax + bu \text{ is } S_p^{\sigma_3}, \tag{3.32}$$

where

$$\begin{cases} \sigma_3 = |a|\sigma_1 + |b|\sigma_2, \\ p = \lfloor(k), \end{cases}$$

with k being defined in (3.31b).

This fuzzy rule formulation is a very natural logic-based description for the one-dimensional SISO system, conventionally described by $y = ax + bu$ if all functional values are crisp, which now has fuzzy means of both the state x and the controller u. Although at this moment what we have formulated is a simple SISO setting, it will shortly be extended to general multiple-input/multiple-output (MIMO) systems and, eventually, generalized to some nonlinear systems of the form $y = f(x,u)$. This simple rule (3.32) is key in all the developments to be carried out below.

Because the S_i^{σ}'s overlap, a single rule for each of x or u may trigger more than one rule. To determine a unique value for y, we use the following weighted average defuzzification formula (see (3.12)):

$$y = \frac{\sum (\textit{value of member} \times \textit{membership value})}{\sum (\textit{membership value})} \tag{3.33}$$

B. Dynamic Fuzzy System Modeling

In this subsection, we describe continuous-time fuzzy modeling of a linear MIMO control system, containing the SISO systems as a special case.

To motivate, we start with an SISO control system, with state $x = x(t)$ and control input $u = u(\mathrm{t})$ defined on the time domain $[0,\infty)$. Suppose that the system dynamics are described by the derivative of the state, $\dot{x}$, which is a function of both x and u. If the relation between $\dot{x}$ and both (x,u) is known exactly, then classical mathematical modeling techniques and the systems control theory can be employed. The exact modeling $\dot{x} = ax + bu$ simply means, from a mathematical logic point of view, that at any instant $t = t^* \in [0,\infty)$, if x is equal to $x(t^*)$ and u is equal to $u(t^*)$ then $\dot{x}$ is equal to $\left.\frac{\mathrm{d}x(t)}{\mathrm{d}t}\right|_{t=t^*}$. However, if this relation is not precisely given, then it is not well-defined and, hence, fuzzy modeling becomes necessary. In this case, our knowledge of the plant (e.g., from experimental data) consists of statements like if x and u belong to certain fuzzy sets, respectively, then $\dot{x}$ belongs to some other fuzzy set. Each of such correspondence is called a *realization* of the hidden relation between $\dot{x}$ and (x,u). We next apply the fuzzy rule formulation (3.32) to model such realizations.

We first consider the hidden relation to be *linear* in the sense that the system dynamics $\dot{x}$ depends only on a linear combination of x and u; namely, there are two real (crisp) numbers a and b, not both zero, such that $\dot{x}$ is determined by $ax + bu$, where both x and u are given only in fuzzy terms (fuzzy sets to which they belong, and the values of membership functions that they assume). We model this linear realization in the following way: Let $R^{(q)}$ be the qth realization of the system dynamics $\dot{x}$, described by the fuzzy rule (3.32), and suppose that there are a total of ℓ realizations: $\dot{x}^q$, $q = 1, ..., \ell$:

Fuzzy Rules Base:

$$R^{(q)}:\quad \text{IF } x \in S_{n_q}^{\sigma_1} \text{ AND } u \in S_{m_q}^{\sigma_2} \text{ THEN } \dot{x} \in S_{p_q}^{\sigma_3}, \tag{3.34}$$

where $\in$ is used to mean "is" or "is in," and

$$\begin{cases} \sigma_3 = |a|\sigma_1 + |b|\sigma_2, \\ p_q = \lfloor (k_q), \end{cases}$$

in which k_q is defined in (3.31b), with $n = n_q$ and $m = m_q$ therein, $q = 1, ..., \ell$, and $\lfloor(\cdot)$ is the integer-floor of a real number.

We should point out that since normalized membership functions are used, this fuzzy model becomes nonfuzzy as all the scaling parameters σ_i approach 0, $i = 1, 2, 3$. That is, in this case the following logic-based description for the nonfuzzy model is obtained: at any instant $t = t^*$, if x is equal to $x(t^*)$ and u is equal to $u(t^*)$ then $\dot{x}$ is equal to $\dot{x}(t^*)$, consistent with the classical theory.

Also, we note that for any $\dot{x}$ so determined, $\dot{x}$ belongs to two subsets, $I_{p_q}^{\sigma_3}$ and $I_{p_q+1}^{\sigma_3}$, assuming two (probably different) membership values, say μ_{q1} and μ_{q2}, respectively, for each realization $q = 1, ..., \ell$. For example, if triangular membership functions are used, then the situation can be visualized by Figure 3.11. Using the defuzzification formula (3.33), the resulting system dynamics, $\dot{x}(t)$, is calculated via the weighted-average defuzzification formula

$$\dot{x}(t) = \frac{\sum_{q=1}^{\ell} \{\mu_{q_1} \dot{x}^{q_1}(t) + \mu_{q_2} \dot{x}^{q_2}(t)\}}{\sum_{q=1}^{\ell} \{\mu_{q_1} + \mu_{q_2}\}}, \tag{3.35}$$

where all the multiplications are the ordinary arithmetic multiplications.

In this way, we have modeled the system dynamics as response to the system states and control inputs that take fuzzy values. It is very important to point out that the unknown system dynamics under modeling can indeed be nonlinear or uncertain. As long as each of the fuzzy rules approximates the system dynamics as a linear combination of the system state and control input, the unknown dynamics are modeled (and covered) by the above family of fuzzy rules. The performance of this fuzzy modeling of an unknown system dynamics will be evaluated via the analysis of stability and controllability given later in this chapter.

The fuzzy model of an SISO control system works as follows: starting with a given initial state x_0 and an initial control input u_0, we first have

$$x_0 \in I_n^{\sigma_1} \cap I_{n+1}^{\sigma_1} \quad \text{and} \quad u_0 \in I_m^{\sigma_2} \cap I_{m+1}^{\sigma_2}$$

for some integers n and m. Then, the dynamics of the control system, $\dot{x}(t)$, can be described by fuzzy IF-THEN rules (3.34), with the corresponding scaling parameter σ_3 and integer p determined for a small duration of time, $[t_0,t_1]$. When the time goes on, until $t > t_1$, so that either $x(t) \in I_n^{\sigma_1} \cap I_{n+1}^{\sigma_1}$ or $u(t) \in I_m^{\sigma_2} \cap I_{m+1}^{\sigma_2}$ is not satisfied, a new fuzzy rule, with a new n and/or a new m, will be applied. As the time goes on and on, the fuzzy rules will be switched again and again, and the system dynamics are then specified in this piecewise manner throughout the entire process.

With the above motivation from the linear SISO system case, we now extend the basic concepts and modeling techniques to MIMO control systems.

Consider a general MIMO linear control system of the form

$$\begin{cases} \dot{x}(t)=Ax(t)+Bu(t), \\ x(t_0)=x_0, \end{cases} \tag{3.36}$$

where

$$x(t) = [\, x_1(t) \cdots x_r(t)\,]^{\mathrm{T}},$$
$$u(t) = [\, u_1(t) \cdots u_s(t)\,]^{\mathrm{T}},$$
$$A = \begin{bmatrix} a_{11} & \cdots & a_{1r} \\ \vdots & \ddots & \vdots \\ a_{r1} & \cdots & a_{rr} \end{bmatrix},$$
$$B = \begin{bmatrix} b_{11} & \cdots & b_{1s} \\ \vdots & \ddots & \vdots \\ b_{r1} & \cdots & b_{rs} \end{bmatrix},$$

in which, usually, $1 \le s \le r$. We first investigate the free system

$$\begin{cases} \dot{x}(t)=Ax(t), \\ x(t_0)=x_0, \end{cases} \tag{3.37}$$

with the fuzzy rules:

$$R_i^q: \quad \text{IF } x \in S_{n_i}^{\sigma_{1i}} \text{ THEN } \dot{x} \in S_{p_i}^{\sigma_{3i}}, \tag{3.38}$$

where

$$\begin{cases} \sigma_{3i}=\sum_{v=1}^{r} | a_{iv} | \sigma_{1v}, \\ p_i=\lfloor(\sum_{v=1}^{r} \alpha_{iv}\sigma_{1v}/\sigma_{3i}), \end{cases}$$

in which

$$\sigma_{iv} = \begin{cases} a_{iv} n_v & \text{if } a_{iv} \ge 0, \\ a_{iv}(n_v+1) & \text{if } a_{iv} < 0, \end{cases}$$

with i=1,...,r; q=1,..., ℓ ; for the special case of $\sigma_{3i} = 0$, we define $p_i = 0$.

The free system (3.37), together with the fuzzy IF-THEN rules (3.38), is called a *fuzzy free system* below.

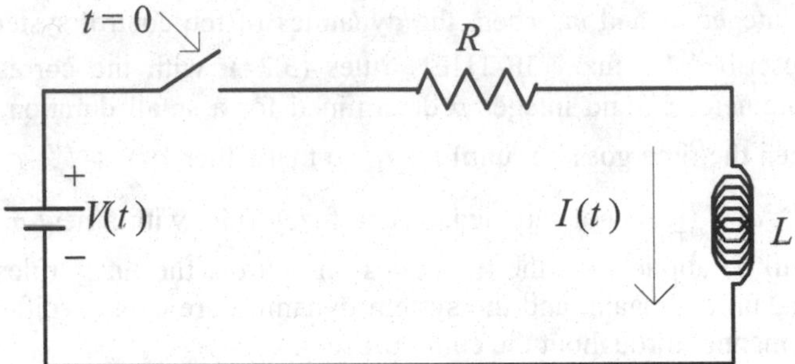

Figure 3.13 A serial RL circuit for Example 3.1.

In the above rules, σ_{3i} and p_i are determined as follows. First, it is clear that $\sigma_{3i} = 0$ if and only if all $a_{i\nu} = 0$, ν=1,...,r, namely, the ith row of the matrix A is identically zero, which means that the ith component of $\dot{x}$ has to be zero. Hence, we define $p_i = 0$ for this component of $\dot{x}$. Next, for a fixed implication q (hereafter, the fixed q is omitted), if $x_i(t)$ is $S_{n_i}^{\sigma_{1i}}$, then we have

$$n_i\,\sigma_{1i} \le x_i(t) < (n_i + 1)\,\sigma_{1i}\,, \qquad i=1,\ldots,r,$$

so that, in the case where all $a_{ij} \ge 0$,

$$\sum_{i=1}^{r} a_{ji} n_i \sigma_{1i} \le \dot{x}_j(t) = \sum_{i=1}^{r} a_{ji} x_i(t) < \sum_{i=1}^{r} a_{ji}(n_i + 1)\sigma_{1i}\,,$$

where j=1,...,r. All other cases can be verified similarly (see the discussion on (3.32) above), where the algebra is somewhat tedious but straightforward. Now, to continue the case with all $a_{ij} \ge 0$, we compare the above r inequalities with

$$k_j\,\sigma_{3j} \le \dot{x}_j < (k_j + 1)\,\sigma_{3j}\,, \qquad j=1,\ldots,r,$$

and obtain

$$k_j\,\sigma_{3j} = \sum_{i=1}^{r} a_{ji} n_i \sigma_{1i}\,,$$

$$(k_j + 1)\,\sigma_{3j} = \sum_{i=1}^{r} a_{ji}(n_i + 1)\sigma_{1i}\,,$$

where j=1,...,r. The final rules (3.38) are then obtained by summarizing all possible cases.

It is clear that in the special case where $r = 1$, this formulation reduces to the one for SISO systems discussed in the last subsection.

Example 3.5. Consider a simple serial RL circuit shown in Figure 3.13, where the input is the voltage source $V(t)$, and the output is the current $I(t)$ observed at the ampmeter. Assume that the initial condition for $I(t)$ at $t = 0$ is $I(0) = 0$.

This typical RL circuit can be modeled in crisp mathematical formulation using the Kirchoff law as

$$\frac{dI(t)}{dt} = -\frac{R}{L}\,I(t) + \frac{1}{L}\,V(t).$$

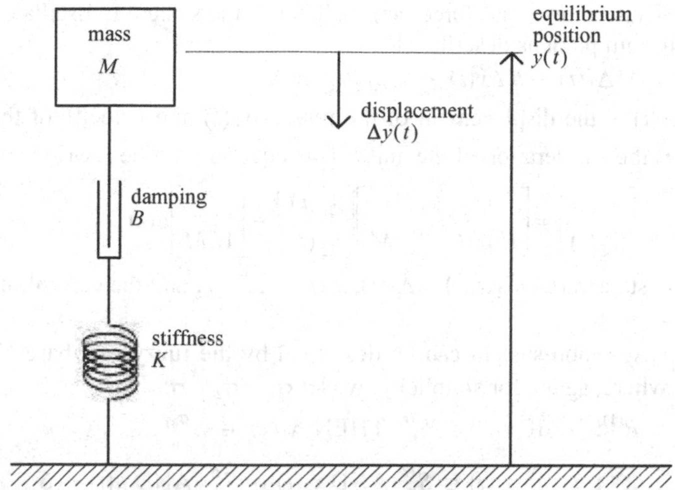

Figure 3.14 A mechanical suspension system.

This LC circuit can also be modeled with the fuzzy rules base (3.34) as follows, if it has uncertainty, where for simplicity we let $\sigma_1 = \sigma_2 = \sigma$:

$$R^{(1)}: \quad \text{IF } I(t) \in S_n^{\sigma} \text{ AND } V(t) \in S_m^{\sigma} \text{ THEN } dI(t)/dt \in S_p^{\sigma_3}.$$

In this example,

$$\sigma_3 = (-\frac{R}{L} + \frac{1}{L})\sigma,$$

$$k = [-\frac{R}{L}(n+1) + \frac{1}{L}(m)]\sigma,$$

$$p = \lfloor(k).$$

Note that k is derived using equation (3.31b), with the knowledge that $R > 0$ and $L > 0$, which implies $-R/L < 0$ and $1/L > 0$. As mentioned earlier, as $\sigma \to 0$, this model reduces to the crisp equation.

In applications, one can also use the parameter identification technique described in Section I.B and the observed data in a real circuit to determine p and σ.

Example 3.6. Consider a simple mechanical system shown in Figure 3.14, which is used in an automobile as a shock absorber for balancing (to provide comfort to the driver on a road). This system consists of three basic mechanical components: (i) a spring, (ii) a hydraulic, and (iii) a mass. The spring is characterized by its stiffness K which provides a force proportional to the displacement (position) of being stretched or compressed. The hydraulic is characterized by its damping B which provides a force proportional to the rate of displacement (velocity) of being stretched or compressed. The mass is characterized by the material of M which provides a force proportional to the rate of the changing rate of displacement (acceleration) of being stretched or compressed. This typical system is sometimes referred to as a *suspension* *system*. Thus, an external force, $f(t)$, will cause the system to oscillate around an equilibrium point as described by

$$M\,\Delta\ddot{y}(t)\; - B\,\Delta\dot{y}(t)\; - K\,\Delta y(t) = f(t),$$

where $\Delta y(t)$ is the displacement of the mass, $\Delta\dot{y}(t)$ the velocity of the mass, and $\Delta\ddot{y}(t)$ the acceleration of the mass. This equation can be rewritten as

$$\begin{bmatrix} \dot{x}_1(t) \\ \dot{x}_2(t) \end{bmatrix} = \begin{bmatrix} 0 & 1 \\ K/M & B/M \end{bmatrix} \begin{bmatrix} x_1(t) \\ x_2(t) \end{bmatrix} + \begin{bmatrix} 0 \\ 1/M \end{bmatrix} u(t),$$

where the state variables $x_1(t) = \Delta y(t)$, $x_2(t) = \Delta\dot{y}(t)$, and the control input $u(t) = f(t)$.

This suspension system can be described by the fuzzy rule base (3.34) as follows, where, again, for simplicity, we let $\sigma_1 = \sigma_2 = \sigma$:

$R^{(1)}$: IF $x_2(t) \in S^{\sigma}_{n_2}$ THEN $\dot{x}_1(t) \in S^{\sigma_{31}}_{p_1}$,

$R^{(2)}$: IF $x_1(t) \in S^{\sigma}_{n_1}$ AND $x_2(t) \in S^{\sigma}_{n_2}$ AND $u(t) \in S^{\sigma}_{m}$ THEN $\dot{x}_2(t) \in S^{\sigma_{32}}_{p_2}$,

where

$$\sigma_{31} = \sigma,$$

$$\sigma_{32} = (K + B + 1)\frac{\sigma}{M},$$

$$p_1 = \lfloor(n_2) = n_2,$$

$$p_2 = \lfloor(\,K\,n_1 + B\,n_2 + m\,)\;\frac{1}{M}.$$

IV*. STABILITY ANALYSIS OF CONTINUOUS-TIME DYNAMIC FUZZY SYSTEMS

In this section, we first discuss the stability issue of the fuzzy modeling established above, and then seek an answer to the question: If the nonfuzzy linear MIMO free system

$$\begin{cases} \dot{x} = Ax, \\ x(t_0) = x_0, \end{cases} \tag{3.39}$$

is asymptotically stable about its equilibrium point 0, then is it also true for its corresponding fuzzy model? Many other general and significant results can then be derived based on the positive answer to this question.

For this purpose, some basic concepts first need to be clarified. Recall that the concept of *asymptotic stability* given in Definition 3.2 is in the classical sense. The linear free system (3.39) is said to be globally asymptotically stable about 0, if starting from any bounded initial state the system trajectory remains to be bounded and eventually $\|x(t)\|_2 \to 0$ as $t \to \infty$.

Now, we consider the fuzzy modeling of this free system, namely, the free system (3.37) with the fuzzy IF-THEN rules (3.38) described in the last

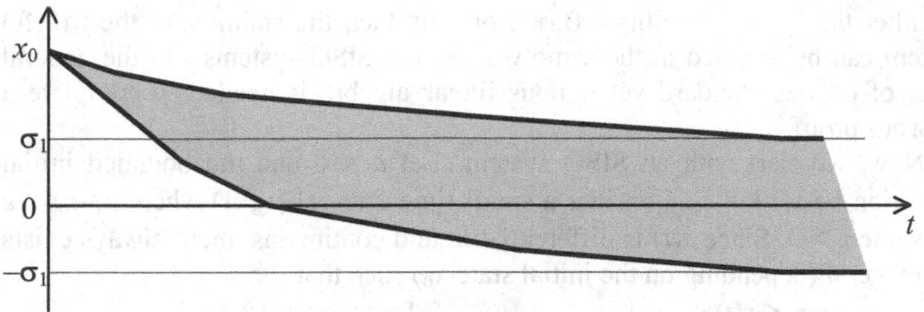

Figure 3.15 Fuzzy σ-asymptotic stability for a 1-D free system (with $a < 0$).

subsection. For this fuzzy version of the free system, the conventional asymptotic stability definition cannot be directly applied. Yet, the fuzzy model has the following stability, called the *fuzzy σ-asymptotic stability*, which is consistent with the classical one when the model changes from fuzzy to nonfuzzy as $\sigma \to 0$.

Definition 3.4. The fuzzy system (3.37) is said to be (globally) *fuzzy σ-asymptotically stable* about its equilibrium point 0, if starting from any bounded initial state the system trajectory remains to be bounded and eventually satisfies $\|x(t)\|_2 \leq \sigma$ as $t \to \infty$, where $\sigma = \max\{\sigma_{11},\ldots,\sigma_{1r}\}$, which are given in the fuzzy rules (3.38).

Figure 3.15 illustrates this concept for a simple one-dimensional free system with $A = a < 0$.

Obviously, a fuzzy σ-asymptotically stable system is asymptotically stable in the classical sense if $\sigma_{1i} \to 0$ for all i=1,...,r, namely, if the fuzzy model becomes nonfuzzy.

When the triangular membership functions (see Figure 3.11) are used, we have the following result.

Theorem 3.6. The free fuzzy system described by (3.37) is fuzzy σ-asymptotically stable about 0 if and only if all the eigenvalues of the system matrix A have a negative real part.

Proof. The condition of negative real part for all eigenvalues of the system matrix A is necessary for the nonfuzzy case, even for the boundedness of the solution trajectory alone, and is therefore necessary for the fuzzy setting. Therefore, we only need to establish the sufficiency, namely, that if all the eigenvalues of A have a negative real part then the fuzzy free system is σ-asymptotically stable about 0.

To facilitate the arguments, we organize the proof as follow. We first give a complete proof for the SISO systems. Then, we notice that a unitary transform $y = Ux$, with any unitary matrix U, can change the free system $\dot{x} = Ax$ to $\dot{y} = \mathrm{A}y$, such that A is a (block) diagonal matrix consisting of all the complex eigenvalues of A, counting multiplicities. More importantly, observe that this transform is stability-preserving, in the sense that the two systems have the same stability, because the real parts of their eigenvalues are on the diagonal and the imaginary parts and the multiplicities of the eigenvalues do not alter the system stability. Based on this fact, the stability of the MIMO system can be justified in the same way as the SISO systems. In the general case, of course, standard yet tedious linear algebra is needed to complete a rigorous proof.

Now, we start with an SISO system. Let $a < 0$ and the bounded initial condition be $x(t_0) = x_0$. Consider a small time interval $[t_0, t_1)$ where $|t_1 - t_0| < \varepsilon$ for some $\varepsilon > 0$. Since $x(t)$ is differentiable and continuous, there always exists an integer n (depending on the initial state x_0) such that

$$n\sigma_1 \leq x(t) < (n+1)\sigma_1, \qquad \text{for all } t \in [t_0, t_1),$$

provided that $\varepsilon > 0$ is small enough. Note that in so doing both $x(t)$ and $\dot{x}(t)$ are well defined by the fuzzy rules on the interval $[t_0, t_1)$. We next calculate $x(t)$ and $\dot{x}(t)$. First, $x(t)$ belongs to the two subsets $I_n^{\sigma_1}$ and $I_{n+1}^{\sigma_1}$ with the triangular membership function values

$$\begin{aligned} \mu_{11} &= \mu_{I_n^{\sigma_1}}(x) = [(n+1)\sigma_1 - x(t)]/\sigma_1, \\ \mu_{12} &= \mu_{I_{n+1}^{\sigma_1}}(x) = [x(t) - n\sigma_1]/\sigma_1, \end{aligned} \tag{3.40}$$

for the first implication with $q = 1$, where $t \in [t_0, t_1)$. For simplicity of notation, we only consider the case $\ell = 1$ in this proof, since for the general case where $\ell > 1$, the following procedure can be carried out in exactly the same way (which can be easily verified by noticing that the sum of the two triangular membership function values at each point is always equal to one).

Now, the corresponding fuzzy rule, namely,

$$R: \text{ IF } x \in S_n^{\sigma_1} \cap S_{n+1}^{\sigma_1} \text{ THEN } \dot{x} \in S_p^{\sigma_3} \cap S_{p+1}^{\sigma_3},$$

where

$$\sigma_3 = |a|\sigma_1 \qquad \text{and} \qquad p = -(n+1),$$

(since $a < 0$), gives the following possible solutions for $\dot{x}(t)$:

$$a(n+1)\sigma_1 \leq \dot{x}^{11}(t) < a(n-1)\sigma_1$$

and

$$a(n+2)\sigma_1 \leq \dot{x}^{12}(t) < a(n)\sigma_1,$$

with μ_{11} and μ_{12} given by (3.40). On the other hand, the weighted-average formula (3.35) yields

$$\dot{x}(t) = \frac{\mu_{11}\dot{x}^{11}(t) + \mu_{12}\dot{x}^{12}(t)}{\mu_{11} + \mu_{12}}.$$

It then follows that

$$\frac{\mu_{11}}{\mu_{11}+\mu_{12}} a(n+1)\sigma_1 + \frac{\mu_{12}}{\mu_{11}+\mu_{12}} a(n+2)\sigma_1$$

$$\leq \dot{x}(t) < \frac{\mu_{11}}{\mu_{11}+\mu_{12}} a(n-1)\sigma_1 + \frac{\mu_{12}}{\mu_{11}+\mu_{12}} a(n)\sigma_1$$

or, after simplification,

$$a(n+1)\sigma_1 + \frac{\mu_{12}}{\mu_{11}+\mu_{12}}\, a\sigma_1 \le \dot{x}(t) < a(n)\sigma_1 - \frac{\mu_{11}}{\mu_{11}+\mu_{12}}\, a\sigma_1.$$

Substituting μ_{11} and μ_{12}, which are given by (3.40), into the above yields

$$a\,x(t) + a\,\sigma_1 \le \dot{x}(t) < a\,x(t) - a\,\sigma_1.$$

Consequently, by integrating the two systems obtained from the two inequalities in the above, with the given initial condition, namely:

$$\dot{x}(t) = a\,x(t) \pm a\,\sigma_1, \qquad x(t_0) = x_0,$$

we arrive at

$$(x_0+\sigma_1)e^{a(t-t_0)} - \sigma_1 \le x(t) < (x_0-\sigma_1)e^{a(t-t_0)} + \sigma_1, \tag{3.41}$$

where $t_0 \le t < t_1$, regardless where the x_0 is located as long as it is bounded.

Note that this result is independent of the integer n. Hence, when the time variable t goes beyond the interval $[t_0,t_1)$, we can find another (probably different) integer, $\tilde{n}$ say, such that

$$(\tilde{n})\sigma_1 \le x(t) < (\tilde{n}+1)\sigma_1 \quad \text{for } t\in[t_1,t_2),$$

provided that we choose $|\, t_2 - t_1 \,| \le \tilde{\varepsilon}$ for another small (probably different) $\tilde{\varepsilon} > 0$. The same procedure can be repeated and leads to the relation (3.41) for all $t \in [t_1,t_2)$.

This process can then be continued, and eventually yields the following sequential results:

$$(x_0+\sigma_1)e^{a(t-t_0)} - \sigma_1 \le x(t) < (x_0-\sigma_1)e^{a(t-t_0)} + \sigma_1, \qquad (t_0 \le t < t_1)$$

$$(x_1+\sigma_1)e^{a(t-t_1)} - \sigma_1 \le x(t) < (x_1-\sigma_1)e^{a(t-t_1)} + \sigma_1, \qquad (t_1 \le t < t_2)$$

$$(x_2+\sigma_1)e^{a(t-t_2)} - \sigma_1 \le x(t) < (x_2-\sigma_1)e^{a(t-t_2)} + \sigma_1, \qquad (t_2 \le t < t_3)$$

$$\vdots$$

where x_1 in the second equation is given by the first equation with $t = t_1$, namely: $x_1 = x(t_1)$ with

$$(x_0+\sigma_1)e^{a(t_1-t_0)} - \sigma_1 \le x(t_1) < (x_0-\sigma_1)e^{a(t_1-t_0)} + \sigma_1.$$

Substituting it into the second equation gives

$$(x_0+\sigma_1)e^{a(t-t_0)} - \sigma_1 \le x(t) < (x_0-\sigma_1)e^{a(t-t_0)} + \sigma_1,$$

where $t \in [t_1,t_2)$. Then, repeating this procedure and observing that all the resulting inequalities have the same formulation, we obtain generally

$$(x_0+\sigma_1)e^{a(t-t_0)} - \sigma_1 \le x(t) < (x_0-\sigma_1)e^{a(t-t_0)} + \sigma_1, \tag{3.42}$$

where $t \in [t_0,t_m)$ for any integer $m \ge 1$.

To this end, we further argue that the sequence of time intervals, $[t_1,t_2)$, $[t_2,t_3)$, ..., can be so chosen that $t_m \to \infty$ as $m \to \infty$. Actually, since

$$a(x_0+\sigma_1)e^{a(t-t_0)} \le \dot{x}(t) < a(x_0-\sigma_1)e^{a(t-t_0)},$$

where $t \in [t_0,t_m)$ for any $m \ge 1$, we can show that the magnitude of the slope of $x(t)$, namely $|\dot{x}(t)|$, becomes smaller as t tends to be larger: at t_1 and t_2, for instance, we have

$$|\dot{x}(t_1)| \le max\{\, |a(x_0+\sigma_1)|e^{a(t_1-t_0)},\ |a(x_0-\sigma_1)|e^{a(t_1-t_0)} \,\}$$

and

$$|\dot{x}(t_2)| \le max\{\, |a(x_0+\sigma_1)|e^{a(t_2-t_0)},\ |a(x_0-\sigma_1)|e^{a(t_2-t_0)} \,\},$$

respectively, which implies that

$$max\,|\dot{x}(t_2)| \le max\,|\dot{x}(t_1)|.$$

Indeed, it amounts to noting that if

$$|a(x_0+\sigma_1)|e^{a(t_1-t_0)} \ge |a(x_0-\sigma_1)|e^{a(t_1-t_0)},$$

then multiplying to both sides by $e^{a(t_2-t_1)} \ge 0$ gives

$$|a(x_0+\sigma_1)|e^{a(t_2-t_0)} \ge |a(x_0-\sigma_1)|e^{a(t_2-t_0)},$$

which implies that we only need to compare $|a(x_0+\sigma_1)|e^{a(t_1-t_0)}$ with $|a(x_0+\sigma_1)|e^{a(t_2-t_0)}$, where the later is smaller. Another case,

$$|a(x_0+\sigma_1)|e^{a(t_1-t_0)} \le |a(x_0-\sigma_1)|e^{a(t_1-t_0)}$$

yields the same result. Hence, we always have $\max\,|\dot{x}(t_2)| \le \max\,|\dot{x}(t_1)|$. This means that the curve $x(t)$ becomes flat on the interval $[t_2,t_3)$ more than that on $[t_1,t_2)$, so that the length of the interval $[t_2,t_3)$ can be chosen to be at least as large as that of $[t_1,t_2)$. The same is true for the consequent intervals $[t_3,t_4)$, $[t_4,t_5)$, $\cdots$, which implies that $t \to \infty$ as $m \to \infty$.

Thus, it follows from the general formulation (3.42), namely,

$$(x_0+\sigma_1)\, e^{a(t-t_0)} - \sigma_1 \le x(t) < (x_0-\sigma_1)\, e^{a(t-t_0)} + \sigma_1,$$

where $t_0 \le t < \infty$, that

$$-\sigma_1 \le \lim_{t\to\infty} x(t) \le \sigma_1.$$

We next return to the MIMO systems. Recall that by a unitary transform we can assume without loss of any generality that the system matrix A is (block) diagonal and consists of all its eigenvalues with negative real part on the diagonal. The reason is, again, that the imaginary parts and multiplicities of the eigenvalues do not alter the system stability.

Similarly, we start with the initial state $x(t_0) = x_0$ and consider a small time interval $[t_0,t_1)$. Since $x(t)$ is differentiable, and is hence continuous, we can always find some integers n_i, i=1,...,r, such that

$$(n_i)\sigma_{1i} \le x_i(t) < (n_i+1)\sigma_{1i},$$

where $t \in [t_0,t_1)$ and i=1,...,r. With the corresponding membership function values

$$\mu_{i11} = \mu_{I_{n_i}^{\sigma_{i1}}}(x_i) = [(n_i+1)\sigma_{1i} - x_i(t)]/\sigma_{1i},$$

$$\mu_{i12} = \mu_{I_{n_i+1}^{\sigma_{i1}}}(x_i) = [x_i(t) - n_i\sigma_{1i}]/\sigma_{1i},$$

i=1,...,r, $t \in [t_0,t_1]$, we apply the MIMO version of the weighted average formula (3.35) to obtain

$$A\,x(t) \mp A\,\underline{\sigma}_1 \le \dot{x}(t) < A\,x(t) \pm A\,\underline{\sigma}_1,$$

where $\underline{\sigma}_1 = [\,\sigma_{11}\,...\,\sigma_{1r}\,]^T$ and the inequality is held componentwise.

By solving the two linear systems

$$\dot{x}(t) = A\,x(t) \pm A\,\underline{\sigma}_1$$

with the initial condition $x(t_0) = x_0$, we obtain

$$e^{A(t-t_0)}[x_0+\underline{\sigma}_1] - \underline{\sigma}_1 \le x(t) < e^{A(t-t_0)}[x_0-\underline{\sigma}_1] + \underline{\sigma}_1,$$

which is an MIMO analogue of (3.41) over $[t_0,\infty)$, in which the inequalities are held componentwise. Since A is (block) diagonal having all the diagonal elements with a negative real part, the above arguments can be similarly repeated for each component, where the imaginary part of the eigenvalues can be ignored since they do not affect the stability. Thus, eventually, we have

$$-\underline{\sigma}_1 \leq \lim_{t\to\infty} x(t) \leq \underline{\sigma}_1$$

componentwise. This completes the proof of Theorem 3.6.

V*. CONTROLLABILITY ANALYSIS OF CONTINUOUS-TIME DYNAMIC FUZZY SYSTEMS

In this section, we further study the important controllability property of the fuzzy modeling discussed above.

Consider a general (nonfuzzy) linear MIMO control system described by (3.36), namely,

$$\begin{cases} \dot{x}(t) = Ax(t) + Bu(t), \\ x(t_0) = x_0, \end{cases} \tag{3.43}$$

where

$$x(t) = [\, x_1(t) \cdots x_r(t) \,]^{\mathrm{T}},$$
$$u(t) = [\, u_1(t) \cdots u_s(t) \,]^{\mathrm{T}},$$

with $1 \leq s \leq r$. The concept of (complete) controllability, for both fuzzy and nonfuzzy systems, is defined as follows.

Definition 3.5. The linear MIMO control system (3.43), fuzzy or not, is said to be *(completely) controllable* if, starting from any bounded initial state, the system trajectory can be brought to the target point 0 (the origin of the state space), by a bounded control input within a finite period of time.

Recall from the classical control theory that a nonfuzzy linear MIMO control system (3.43) is (completely) controllable if and only if

$$rank\{\, [\, B \;\; AB \;\; \ldots \;\; A^{r-1}B \,] \,\} = r,$$

where r is the dimension of the system state vector $x(t)$, whereas the control vector $u(t)$ is s-dimensional as indicated above, with $1 \leq s \leq r$ in general. In particular, for one-dimensional systems the condition reduces to $B = b \neq 0$.

For the fuzzy version of the linear MIMO control system, described also by (3.43), we use the fuzzy rules:

$$R_i^q: \;\; \text{IF } x_i \in S_{n_i}^{\sigma_{1i}} \text{ AND } u_i \in S_{m_i}^{\sigma_{2i}} \text{ THEN } \dot{x}_i \in S_{p_i}^{\sigma_{3i}},$$

where

$$\begin{cases} \sigma_{3i} = \sum_{v=1}^{r} |\, a_{iv} \,|\, \sigma_{1v} + \sum_{v=1}^{s} |\, b_{iv} \,|\, \sigma_{2v}, \\ p_i = \lfloor \left(\sum_{v=1}^{r} \dfrac{\alpha_{iv}\sigma_{1v}}{\sigma_{3i}} + \sum_{v=1}^{s} \dfrac{\beta_{iv}\sigma_{2v}}{\sigma_{3i}} \right), \end{cases} \tag{3.44}$$

in which

$$\alpha_{iv} = \begin{cases} a_{iv} n_v & \text{if } a_{iv} \geq 0, \\ a_{iv}(n_v + 1) & \text{if } a_{iv} < 0, \end{cases}$$

$$\beta_{iv} = \begin{cases} b_{iv} m_v & \text{if } b_{iv} \ge 0, \\ b_{iv}(m_v + 1) & \text{if } b_{iv} < 0, \end{cases}$$

with i=1, ...,r, q=1,...,ℓ; and for the case of $\sigma_{3i} = 0$, we define $p_i = 0$, i=1,...,r.

Here, the superscript q has also been omitted, particularly for the case where q is fixed, for the simplicity of notation.

For the fuzzy model (3.43) with (3.44), we have the following result.

Theorem 3.7. *The linear MIMO fuzzy control system* (3.43) *is (completely) controllable if and only if*

$$rank\{ [\, B \ AB \ \ldots \ A^{r-1}B \,] \} = r.$$

Proof. Again, we first establish the theorem for the SISO systems, and then explain how the arguments can be modified for the MIMO setting.

For an SISO system, it is clear that the condition $b \neq 0$ is necessary for the controllability (otherwise there would not be any control input). We therefore only verify the sufficiency.

Start with the given initial state $x(t_0)$=x_0. Suppose that we first apply a control input $u(t)$ (to be determined later) and obtain the state $x(t)$ as the corresponding output response of the system. Consider an arbitrary (need not be small) instant $t > t_0$ after the control has been applied. For this t, there exist integers n and m such that

$$(n)\sigma_1 \le x(t) < (n+1)\sigma_1$$

and

$$(m)\sigma_2 \le u(t) < (m+1)\sigma_2,$$

respectively. Hence, $x(t)$ and $u(t)$ belong to the two subsets $I_n^{\sigma_1} \cap I_{n+1}^{\sigma_1}$ and $I_m^{\sigma_2} \cap I_{m+1}^{\sigma_2}$, respectively, with the membership function values

$$\begin{aligned} \mu_{11} &= \mu_{I_n^{\sigma_1}}(x) = [(n+1)\sigma_1 - x(t)]/\sigma_1, \\ \mu_{12} &= \mu_{I_{n+1}^{\sigma_1}}(x) = [x(t) - n\sigma_1]/\sigma_1, \end{aligned} \tag{3.45}$$

and

$$\begin{aligned} \mu_{13} &= \mu_{I_m^{\sigma_2}}(u) = [(m+1)\sigma_2 - u(t)]/\sigma_2, \\ \mu_{14} &= \mu_{I_{m+1}^{\sigma_2}}(u) = [u(t) - m\sigma_2]/\sigma_2, \end{aligned} \tag{3.46}$$

for the first implication with q = 1. Again, for simplicity we only consider the case where there is only one implication, $\ell = 1$, since the general case can be carried out in exactly the same manner.

Now, the fuzzy rule:

$$R^{(1)}: \quad \text{IF } x^1(t) \text{ is } S_n^{\sigma_1} \text{ AND } u^1(t) \text{ is } S_m^{\sigma_2} \text{ THEN } \dot{x}^1(t) \text{ is } S_p^{\sigma_3},$$

where

$$\sigma_3 = a\,\sigma_1 + b\,\sigma_2,$$
$$p = \lfloor(k),$$

in which k is defined in (3.31b), yields the following possible solutions:

(i) $a > 0$ and $b \ge -a\,\dfrac{\sigma_1}{\sigma_2}$; or $a < 0$ and $b \ge -a\,\dfrac{\sigma_1}{\sigma_2}$:

$$a(n-1)\sigma_1+b(m-1)\sigma_2 \le \dot{x}^{11}(t) < a(n+1)\sigma_1+b(m+1)\sigma_2$$

and

$$a(n)\sigma_1+b(m)\sigma_2 \le \dot{x}^{12}(t) < a(n+2)\sigma_1+b(m+2)\sigma_2\,; \tag{3.47}$$

(ii) $a < 0$ and $b \le -a\,\dfrac{\sigma_1}{\sigma_2}$; or $a > 0$ and $b \le -a\,\dfrac{\sigma_1}{\sigma_2}$:

$$a(n+1)\sigma_1+b(m+1)\sigma_2 \le \dot{x}^{13}(t) < a(n-1)\sigma_1+b(m-1)\sigma_2$$

and

$$a(n+2)\sigma_1+b(m+1)\sigma_1 \le \dot{x}^{14}(t) < a(n)\sigma_1+b(m)\sigma_2\,, \tag{3.48}$$

with μ_{11}, μ_{12} and μ_{13}, μ_{14} given respectively by (3.45) and (3.46).

Next, let us first consider Case (i), where $a > 0$ and $b \ge -a\,\sigma_1/\sigma_2$. In this case, we have the solutions given by (3.47). Observe that the result of $\dot{x}^1(t)$ contains two parts: One is contributed by $x^1(t)$ and the other by $u^1(t)$. Hence, we take the weighted average over these two parts, in the manner of the defuzzification defined in (3.33), as shown below:

$$\left\{\frac{\mu_{11}}{\mu_{11}+\mu_{12}}a(n-1)\sigma_1+\frac{\mu_{12}}{\mu_{11}+\mu_{12}}a(n)\sigma_1\right\}+$$

$$\left\{\frac{\mu_{13}}{\mu_{13}+\mu_{14}}b(m-1)\sigma_2+\frac{\mu_{14}}{\mu_{13}+\mu_{14}}b(m)\sigma_2\right\}\le \dot{x}(t) <$$

$$\left\{\frac{\mu_{11}}{\mu_{11}+\mu_{12}}a(n+1)\sigma_1+\frac{\mu_{12}}{\mu_{11}+\mu_{12}}a(n+2)\sigma_1\right\}+$$

$$\left\{\frac{\mu_{13}}{\mu_{13}+\mu_{14}}b(m+1)\sigma_2+\frac{\mu_{14}}{\mu_{13}+\mu_{14}}b(m+2)\sigma_2\right\},$$

or, after simplifying,

$$a(n)\sigma_1-\frac{\mu_{11}}{\mu_{11}+\mu_{12}}a\sigma_1+b(m)\sigma_2-\frac{\mu_{13}}{\mu_{13}+\mu_{14}}b\sigma_2\le \dot{x}(t) <$$

$$a(n+1)\sigma_1+\frac{\mu_{12}}{\mu_{11}+\mu_{12}}a\sigma_1+b(m+1)\sigma_2+\frac{\mu_{14}}{\mu_{13}+\mu_{14}}b\sigma_2.$$

Substituting μ_{11}, μ_{12}, μ_{13}, and μ_{14} into the above yields

$$ax(t)-a\sigma_1+bu(t)-b\sigma_2\le\dot{x}(t)<ax(t)+a\sigma_1+bu(t)+b\sigma_1. \tag{3.49}$$

To this end, it is very important to notice that this relationship is independent of the integers n and m, which means that at the very beginning of the proof, we can actually consider any time instant (need not be small) $t \ge t_0$, and note also that this relationship, (3.49), holds over the entire time-interval $[t_0,\infty)$.

Next, observe that the general solution of the left-hand equality in (3.49), namely:

$$\dot{x}(t) = ax(t)-a\sigma_1+bu(t)-b\sigma_2,$$

with initial condition $x(t_0) = x_0$, is given by

$$x(t) = \sigma_1 + (x_0 - \sigma_1)e^{a(t-t_0)} + \int_{t_0}^{t} e^{a(t-\tau)} b[u(\tau) - \sigma_2]d\tau . \qquad (3.50)$$

To find a bounded control input $u(t)$ to drive $x(t)$ from x_0 to 0 over a finite duration of time, $[t_0,t^*]$ say, we may first fix $t = t^*$ and then set

$$0 = \sigma_1 + (x_0 - \sigma_1)e^{a(t^*-t_0)} + \int_{t_0}^{t^*} e^{a(t^*-\tau)} b[u(\tau) - \sigma_2]d\tau$$

to solve for a solution input $u(t)$, $t \in [t_0,t^*]$. The result is that $u(t)$ must satisfy the following simple integral equation:

$$\int_{t_0}^{t^*} e^{-a\tau} u(\tau)d\tau = -\frac{\sigma_1}{b} + \frac{\sigma_2}{a} - \left(\frac{\sigma_2}{a} + \frac{x_0 - \sigma_1}{b}\right) e^{a(t^*-t_0)} .$$

Obviously, an exact control function $u(t)$, $t \in [t_0,t^*]$, always exists. For example, a constant (or piecewise constant) control can be found, which can be created by a standard fuzzy controller, where the fixed terminal instant t^* can be chosen arbitrarily according to the application at hand. The solvability of equation (3.50) is equivalent to that the system trajectory $x(t)$ can be controlled from x_0 at $t = t_0$ to the target point 0 at $t = t^*$.

Similarly, the right-hand equality of (3.49), namely:

$$\dot{x}(t) = ax(t) + a\sigma_1 + bu(t) + b\sigma_2,$$

with initial condition $x(t_0)=x_0$, gives

$$x(t) = -\sigma_1 + (x_0 + \sigma_1)e^{a(t-t_0)} + \int_{t_0}^{t} e^{a(t-\tau)} b[u(\tau) + \sigma_2]d\tau \qquad (3.51)$$

and the control input $u(t)$, $t \in [t_0,t^*]$, which always exists and satisfies

$$\int_{t_0}^{t^*} e^{-a\tau} u(\tau)d\tau = \frac{\sigma_1}{b} - \frac{\sigma_2}{a} + \left(-\frac{\sigma_2}{a} + \frac{x_0 + \sigma_1}{b}\right) e^{a(t^*-t_0)}$$

can drive the state $x(t)$ from x_0 to 0 over the time interval $[t_0,t^*]$.

Finally, for other cases indicated in (i) and (ii), the above analysis and calculation can be carried out in exactly the same way, using (3.48) instead of (3.47). And it leads to the same result, namely: the state $x(t)$ of the system can be driven from x_0 to 0 by a control input $u(t)$ over a finite duration of time. This completes the proof of the theorem for the SISO setting.

It is then routine to imitate the above proof, by using the control rules (3.44). In the same way, we can derive an MIMO analogue of (3.49), namely:

$$Ax(t) + Bu(t) - [A\underline{\sigma}_1 + B\underline{\sigma}_2] \le \dot{x}(t) < Ax(t) + Bu(t) + [A\underline{\sigma}_1 + B\underline{\sigma}_2], \qquad (3.52)$$

where again the inequality is held componentwise. Then, by examining the two extremal cases (with equalities in (3.52)) and observing that the constant vector $A\underline{\sigma}_0 + B\underline{\sigma}_0$ can be considered as a constant control input, which, as is well known, does not affect the controllability, we arrive at two classical (nonfuzzy) MIMO control systems. Since these two classical linear MIMO control systems have the same system matrix A and control matrix B, the result of the theorem follows immediately.

VI*. ANALYSIS OF NONLINEAR CONTINUOUS-TIME DYNAMIC FUZZY SYSTEMS

In this section, we extend the results obtained in the previous sections to certain nonlinear control systems. We only study the SISO setting here for simplicity of notation and discussion.

Consider a nonlinear autonomous system of the form

$$\begin{cases} \dot{x}(t)=f(x(t),u(t)), \\ x(t_0)=x_0, \end{cases} \tag{3.53}$$

where we assume, as usual, that the nonlinear system has unique solutions for all feasible control inputs and bounded initial conditions. To apply the same fuzzy modeling technique, developed above, to this nonlinear control system, we use the following fuzzy rule:

$$\text{IF } x(t) \in S_n^{\sigma_1} \text{ AND } u(t) \in S_m^{\sigma_2} \text{ THEN } \dot{x}(t) \in S_k^{\sigma_3},$$

where σ_3 and k are determined as follows: If

$$n\,\sigma_1 \le x(t) < (n+1)\sigma_1,$$

$$m\,\sigma_2 \le u(t) < (m+1)\sigma_2,$$

then we expect to have

$$k\,\sigma_3 \le \dot{x}(t) = f(x(t),u(t)) < (k+1)\sigma_3,$$

and, under certain conditions on f (to be determined below),

$$f(n\sigma_1,m\sigma_2) \le f(x,u) \le f((n+1)\sigma_1,(m+1)\sigma_2),$$

so that

$$\begin{cases} \sigma_3 = | f((n+1)\sigma_1,(m+1)\sigma_2) - f(n\sigma_1,m\sigma_2) |, \\ k = f(n\sigma_1,m\sigma_2)/\sigma_3. \end{cases}$$

Thus, following the same idea and arguments, we model the nonlinear control system (3.53) in the same manner as before, namely, we have the fuzzy nonlinear control system:

$$\text{IF } x(t) \in S_n^{\sigma_1} \text{ AND } u(t) \in S_m^{\sigma_2} \text{ THEN } \dot{x}(t) \in S_p^{\sigma_3}, \tag{3.54}$$

where

$$\sigma_3 = | f((n+1)\sigma_1,(m+1)\sigma_2) - f(n\sigma_1,m\sigma_2) |,$$

and

$$p = \lfloor(f(n\sigma_1,m\sigma_2) / \sigma_3),$$

in which $\lfloor(\cdot)$ is the integer-floor of a real number.

In order to study the fuzzy nonlinear control system modeled by (3.54), we first recall from (3.50)-(3.51) that for a linear fuzzy control system we have

$$\sigma_1 + (x_0-\sigma_1)e^{a(t-t_0)} + \int_{t_0}^{t} e^{a(t-\tau)} b[u(\tau)-\sigma_2]d\tau \le x(t) <$$

$$-\sigma_1 + (x_0+\sigma_1)e^{a(t-t_0)} + \int_{t_0}^{t} e^{a(t-\tau)} b[u(\tau)+\sigma_2]d\tau. \tag{3.55}$$

For a nonlinear control system, which, in a more general situation, can be nonautonomous, namely,

$$\dot{x}(t) = f(x(t),u(t),t), \qquad x(t_0) = x_0, \tag{3.56}$$

if we can verify that under the same control $u(t)$ the system state $x(t)$ of the nonlinear system (3.56) satisfies also the two inequalities (3.55), then we know that all the results obtained above are applicable to this nonlinear control system. Intuitively, there are many nonlinear control systems which can produce a state trajectory $x(t)$ satisfying the two inequalities (3.55) (see Figure 3.15 for the geometric meaning of the inequalities). Our goal is then to find mathematical conditions which guarantee that the state trajectory of the nonlinear system (3.56) satisfies the constraints (3.55).

To do so, suppose that the nonlinear control system (3.56) has been given, in which $f(x(\tau),u(\tau),\tau)$ is integrable with respect to τ for all x and u over $[t_0,\infty)$. Then, the system state $x(t)$ is a solution of the following integral equation:

$$x(t) = x_0 + \int_{t_0}^{t} f(x(\tau),u(\tau),\tau)d\tau, \quad t \geq t_0. \tag{3.57}$$

Note that the unknown state $x(t)$ appears on both sides of equation (3.57), and, hence, no explicit solution for it is obtained in this general formulation. However, the following result characterizes the nonlinear control systems (3.56) whose state trajectory $x(t)$ satisfies the inequality constraints (3.55).

Theorem 3.8. Suppose that for all $t_0 \leq t < \infty$ and all feasible control inputs $u(t)$, the nonlinear functional $f(x,u,t)$ is monotonically decreasing in x for fixed t and u, and, moreover, that $f(\times u(t),t)$ satisfies the following two inequalities:

$$\xi(t) \leq x_0 + \int_{t_0}^{t} f(\zeta(\tau),u(\tau),\tau)d\tau$$

and

$$x_0 + \int_{t_0}^{t} f(\xi(\tau),u(\tau),\tau)\, d\tau \leq \zeta(t)$$

for all feasible control inputs $u(t)$, where

$$\xi(t) = \sigma_1 + (x_0 - \sigma_1)e^{a(t-t_0)} + \int_{t_0}^{t} e^{a(t-\tau)} b[u(\tau) - \sigma_2]d\tau,$$

$$\zeta(t) = -\sigma_1 + (x_0 + \sigma_1)e^{a(t-t_0)} + \int_{t_0}^{t} e^{a(t-\tau)} b[u(\tau) + \sigma_2]d\tau .$$

Then, any state $x(t)$ of the nonlinear control system (3.56) satisfies

$$\xi(t) \leq x(t) \leq \zeta(t), \qquad t_0 \leq t < \infty,$$

that is, satisfies the inequality constraints (3.55).

Proof. Suppose that the claim is not true. Then there exists a $\tilde{t} \in [t_0,t^*]$ such that, say,

$$\xi(\tilde{t}) > x(\tilde{t}).$$

It then follows from the decreasing property of f and the theorem assumptions that

$$\begin{aligned}\xi(\tilde{t}) > x(\tilde{t}) &= x_0 + \int_{t_0}^{\tilde{t}} f(x(\tau),u(\tau),\tau)d\tau \\ &\geq x_0 + \int_{t_0}^{\tilde{t}} f(\zeta(\tau),u(\tau),\tau)d\tau \\ &\geq \xi(\tilde{t}),\end{aligned}$$

which is a contradiction. Similarly, if there exists a $\hat{t} \in [t_0,t^*]$ such that

$$x(\hat{t}) > \zeta(\hat{t}),$$

then we have another contradiction. Hence, the assertion of the theorem must be correct.

PROBLEMS

P3.1 Verify Part (v) of Example 3.1; namely, provide detailed derivation for the resulting three membership functions shown in Figure 3.3.

P3.2 Given a non-singular matrix $A = \begin{bmatrix} 2 & 0 & 1 \\ 1 & 1 & 0 \\ 0 & 0 & 1 \end{bmatrix}$.

(1) Compute its inverse A^{-1}.
(2) Compute its pseudo-inverse $A_1^+ = [A^T A]^{-1} A^T$.
(3) Decompose $A = LC$ such that L is 3 × 3 of full column-rank and C is 3 x 3 of full row-rank but $L \neq A$ and $C \neq A$. Then compute its pseudo-inverse

$$A_2^+ = C^T [C C^T]^{-1} [L^T L]^{-1} L^T.$$

(4) Compare your three results obtained above.

P3.3 Consider an unknown SISO system shown in Figure 3.16.

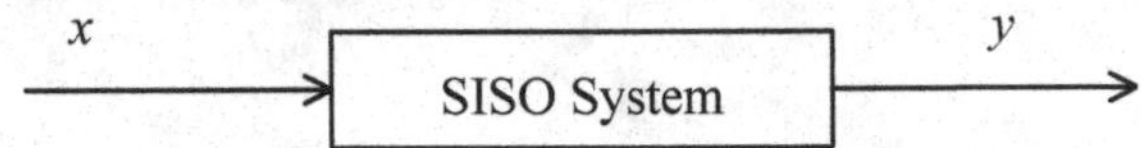

Figure 3.16 An SISO system in ***P3.3***.

Suppose that this system is described by the following two rules:

R^1: IF x is small THEN $y = a_1 x + b_1 x^2$;
R^2: IF x is large THEN $y = a_2 x + b_2 x^2$;

where the small and large membership functions are given by Figures 3.17(a) and (b), respectively.

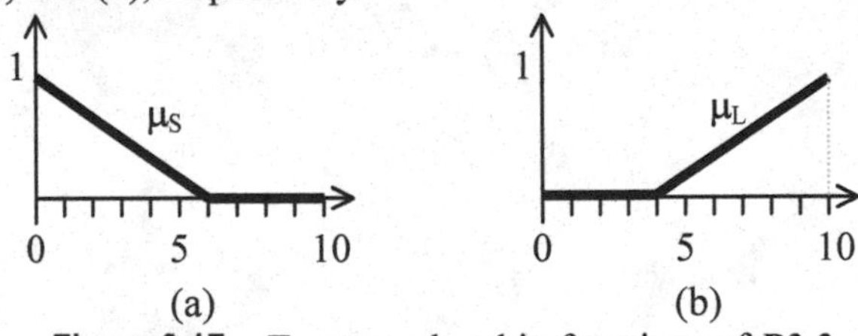

Figure 3.17 Two membership functions of ***P3.3***.

Now, a set of experimental data on the system input-output relation is available:

x	0.0	1.0	2.0	3.0	4.0	5.0	6.0	7.0	8.0	9.0	10.0
y	0.4	0.8	1.5	2.1	2.8	3.4	4.0	4.7	5.5	6.4	7.3

Use the least-squares approach to determine the unknown constant coefficients a_1, a_2, b_1, and b_2.

P3.4 Consider the discrete-time SISO linear dynamic fuzzy system described by the rules

$$R^1: \quad \text{IF } x(k-1) \text{ is small THEN } x(k+1) = -\,x(k) - \tfrac{1}{4}\,x(k-1);$$

$$R^2: \quad \text{IF } x(k-1) \text{ is large THEN } x(k+1) = +\,x(k) - \tfrac{1}{4}\,x(k-1);$$

where the small and large membership functions for inputs are the same as those given in Exercise ***P3.3*** above, and the weights for the output average are all equal to 1. Suppose that the system initial conditions are $x(-1) = x(0) = 1.0$. Let the weights in the output average be all identically equal to 1. Is this system asymptotically stable? Why? (or why not?)

P3.5 Consider a fuzzy system of the form

$$z = -x - y,$$

where the fuzzy inputs x and y have membership functions as shown in Figure 3.18 (a) and (b), respectively. Find the interval Z and membership function μ_Z for the fuzzy output z.

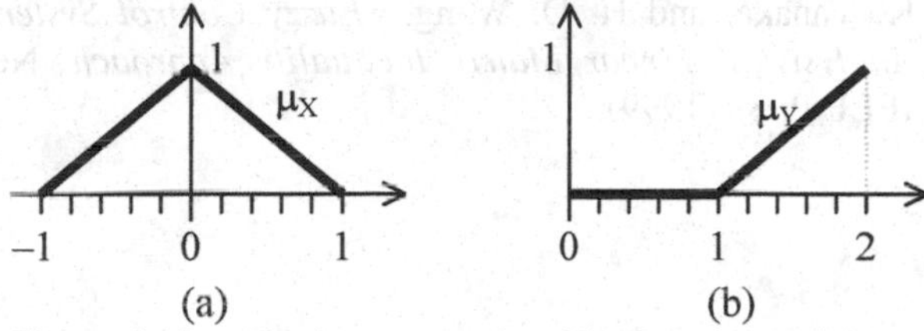

Figure 3.18 The input membership functions of ***P3.5***.

P3.6* By mimicking the discussion of Section V on the controllability of fuzzy linear state-space systems, study the *observability* issue for the same setting (see Reference [2] for definitions).

P3.7* Consider the fuzzy linear state-space system (3.43)-(3.44), along with its two boundary solutions (3.50) and (3.51) (see Figure 3.15 for their geometric meaning). Study a typical LQ optimal control problem for this setting, where LQ refers to the linear system (3.43)-(3.44) with a quadratic performance index (i.e., cost functional) to be minimized (see Reference [2] for more details).

REFERENCES

[1] S. G. Cao, N. W. Rees, and G. Feng. "Stability Analysis of Fuzzy Control Systems." *IEEE Transactions on Systems, Man, and Cybernetics* (B), Vol. 26, No. 1, pp 201-204 (1996).

[2] G. Chen, T. T. Pham, and J. J. Weiss. "Fuzzy Modeling of Control System." *IEEE Transactions on Aerospace and Electronics Systems*, Vol. 31, No. 1, pp 414-429 (1995).

[3] C. K. Chui, and G. Chen. *Linear Systems and Optimal Control*. New York, NY: Springer-Verlag (1989).

[4] T. Takagi, and M. Sugeno. "Fuzzy Identification of Systems and Its Applications to Modeling and Control." *IEEE Transactions on Systems, Man, and Cybernetics*, Vol. 15, No.1, pp 116-132 (1985).

[5] K. Tanaka, and M. Sugeno. "Stability Analysis and Design of Fuzzy Control Systems." *Fuzzy Sets and Systems*, Vol. 45, pp 135-156 (1992).

[6] K. Tanaka, and H. O. Wang. *Fuzzy Control Systems Design and Analysis: A Linear Matrix Inequality Approach*. New York, NY: IEEE Press (1999).

CHAPTER 4

Fuzzy Control Systems

Control systems theory, or what is called modern control systems theory today, can be traced back to the age of World War II, or even earlier, when the design, analysis, and synthesis of servomechanisms were essential in the manufacturing of electromechanical systems. The development of control systems theory has since gone through an evolutionary process, starting from some basic, simplistic, frequency-domain analysis for single-input single-output (SISO) linear control systems, and generalized to a mathematically sophisticated modern theory of multi-input multi-output (MIMO) linear or nonlinear systems described by differential and/or difference equations.

It is believed that the advances of space technology in the 1950s completely changed the spirit and orientation of the classical control systems theory: the challenges posed by the high accuracy and extreme complexity of the space systems, such as space vehicles and structures, stimulated and promoted the existing control theory very strongly, developing it to such a high mathematical level that can use many new concepts like state-space and optimal controls. The theory is still rapidly growing today; it employs many advanced mathematics such as differential geometry, operation theory, and functional analysis, and connects to many theoretical and applied sciences like artificial intelligence, computer science, and various types of engineering. This modern control systems theory, referred to as *conventional* or *classical* control systems theory, has been extensively developed. The theory is now relatively complete for linear control systems, and has taken the lead in modern technology and industrial applications where control and automation are fundamental. The theory has its solid foundation built on contemporary mathematical sciences and electrical engineering, as was just mentioned. As a result, it can provide rigorous analysis and often perfect solutions when a system is defined in precise mathematical terms. In addition to these advances, adaptive and robust as well as nonlinear systems control theories have also seen very rapid development in the last two decades, which have significantly extended the potential power and applicable range of the linear control systems theory in practice.

Conventional mathematics and control theory exclude vagueness and contradictory conditions. As a consequence, conventional control systems theory does not attempt to study any formulation, analysis, and control of what has been called *fuzzy systems*, which may be vague, incomplete, linguistically described, or even inconsistent. Fuzzy set theory and fuzzy logic, studied in some detail in Chapters 1 and 2, play a central role in the investigation of controlling such systems. The main contribution of *fuzzy control theory*, a new alternative and branch of control systems theory that uses fuzzy logic, is its ability to handle many practical problems that cannot be adequately managed by conventional control techniques. At the same time, the results of fuzzy control theory are consistent with the existing classical ones when the system under control reduces from fuzzy to nonfuzzy. In other words, many well-known classical results can be extended in some natural way to the fuzzy setting. In the last three chapters, we have seen many such examples: the interval arithmetic is consistent with the classical arithmetic when an interval becomes a point; the fuzzy logic is consistent with the classical logic when the multi-valued inference becomes two-valued; and the fuzzy Lyapunov stability and fuzzy controllability (and observability) become the classical ones when the fuzzy control systems become nonfuzzy.

Basically, the aim of fuzzy control systems theory is to extend the existing successful conventional control systems techniques and methods as much as possible, and to develop many new and special-purposed ones, for a much larger class of complex, complicated, and ill-modeled systems – fuzzy systems. This theory is developed for solving real-world problems. The fuzzy modeling techniques, fuzzy logic inference and decision-making, and fuzzy control methods to be studied in the following chapters, should all work for real-world problems – if they are developed correctly and appropriately. The real-world problems exist in the first place. Fuzzy logic, fuzzy set theory, fuzzy modeling, fuzzy control methods, etc. are all man-made and subjectively introduced to the scene. If this fuzzy interpretation is correct and if the fuzzy theory works, then one should be able to solve the real-world problems after the fuzzy operations have been completed in the fuzzy environment and then the entire process is finally returned to the original real-world setting. This is what is called the "fuzzification – fuzzy operation – defuzzification" routine in the fuzzy control systems theory. We will study this routine in detail for various control systems design and applications in this chapter.

Although fuzzy control systems theory is used for handling vague systems, this theory itself is not vague. On the contrary, it is rigorous: as has been seen from Chapters 1 and 2, fuzzy logic is a logic, and fuzzy mathematics is mathematics. It will be seen in this chapter that fuzzy logic control is an effective logical control technique using some simple fuzzy logic, in a way comparable to the well-known programmable logic control (PLC) using classical two-valued logic. To motivate, the PLC theory is first reviewed in the next section as an introduction to the general fuzzy logic control (FLC) theory and methodology.

I. CLASSICAL PROGRAMMABLE LOGIC CONTROL

A programmable logic controller (PLC) is a simple microprocessor-based, specialized computer that carries out logical control functions of many types, with the general structure shown in Figure 4.1.

A PLC usually operates on ordinary house current but may control circuits of large amperage and voltage of 440V and higher. A PLC typically has a

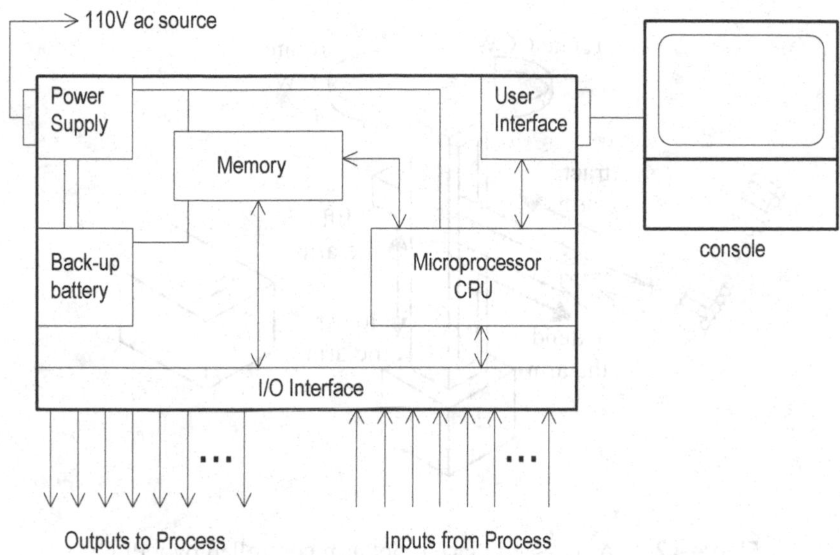

Figure 4.1 A programmable logic controller (PLC).

detachable programmer module, used to construct a new program or modify an existing one, which can be taken easily from one PLC to another. The program instructions are typed into the memory by the user via a keyboard. The central processing unit (CPU) is the heart of the PLC, which has three parts: the processor, memory, and power supply. In the input-output (I/O) modules, the input module has terminals into which the user enters outside process electrical signals and the output module has another set of terminals that send action signals to the processor. A remote electronic system for connecting the I/O modules to long-distance locations can be added, so that the actual operating process under the PLC control can be some miles away from the CPU and its I/O modules. Optional units include racks, chassis, printer, program recorder/player, etc.

To see how a PLC works, let us consider an example of a simple pick-and-place robot arm working in an automatic cycle under the control of a PLC. This simple robot arms is shown in Figure 4.2, which has three joints: one rotational type and two sliding types, as indicated by the arrows; and has a gripper that can close (to grasp an object) and open (to release the object).

To control this robot arm, according to the structure shown in Figure 4.2, we need eight actions created by the PLC, called *outputs* of the PLC, as follows:

A0: rotate the base counterclockwise (CCW),
A1: rotate the base clockwise (CW),
A2: lift the arm,

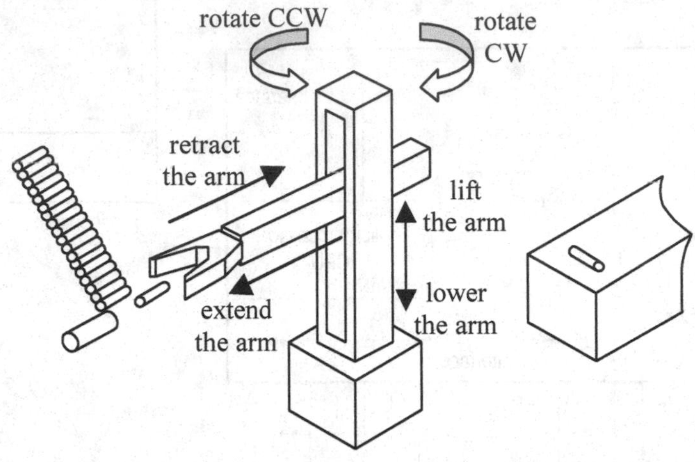

Figure 4.2 A pick-and-place robot arm controlled by a PLC.

A3: lower the arm,
A4: extend the arm,
A5: retract the arm,
A6: close the gripper,
A7: open the gripper.

Suppose that in each cycle of the control process, the entire job performance is completed in 12 steps sequentially within 40 seconds, as shown in Figure 4.3.

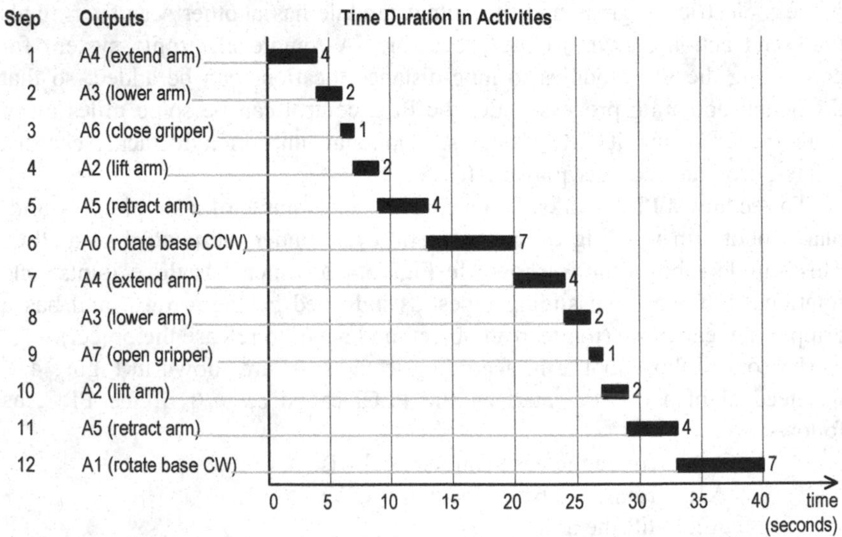

Figure 4.3 Time duration of activities of the robot arm.

Table 4.1 PLC Drum-Timer Array for the Robot Arm

Step	Counts (count/sec)	Outputs							
		A0	A1	A2	A3	A4	A5	A6	A7
1	4	0	0	0	0	1	0	0	0
2	2	0	0	0	1	0	0	0	0
3	1	0	0	0	0	0	0	1	0
4	2	0	0	1	0	0	0	0	0
5	4	0	0	0	0	0	1	0	0
6	7	1	0	0	0	0	0	0	0
7	4	0	0	0	0	1	0	0	0
8	2	0	0	0	1	0	0	0	0
9	1	0	0	0	0	0	0	0	1
10	2	0	0	1	0	0	0	0	0
11	4	0	0	0	0	0	1	0	0
12	7	0	1	0	0	0	0	0	0

In Figure 4.3, it is shown that the PLC is needed to first operate the "extend arm" action for 4 seconds (from the beginning), and then stop this action. To this end, the PLC operates the "lower arm" action for 2 seconds and then stops, and so on. For simplicity, we assume that the transitions between any two consecutive steps are continuous and smooth, so that no time is needed for any of these transitions.

To set up the PLC timer, called the *drum-timer*, so as to accomplish these sequential activities of the robot arm, we need to select the time frequency for the PLC. Since the smallest step time in Figure 4.3 is the 1-second gripper operation, we select a count frequency of 1 count per second. The entire PLC drum-timer array is shown in Table 4.1, where 1 means ON and 0 means OFF, as usual.

The above-described sequential control process is by nature an open-loop control procedure. Now, if we use sensors to provide *feedback* information at each step of the arm's motion, then we can also perform a closed-loop logic control.

For this purpose, mechanical limit switches (or proximity switches) can be used to sense the point of completion of each arm motion and, hence, provide *inputs* to the PLC. The PLC can then be programmed not to begin the next step until the previous step has been completed, regardless of how much time this might take.

We will use, by industrial convention, the number symbol 00 to represent the input from the limit switch that signals completion of the previous output A0 ("rotating the base counterclockwise"), the number symbol 01 to represent the input from the limit switch that signals completion of the previous output A1 ("rotating the base clockwise"), and, finally, 07 for the input that signals

Table 4.2 Robot Arm Control Scheme

Arm Motion	PLC Outputs	PLC Inputs	Necessary Prior Conditions
extend arm	A4	04	01, 07
lower arm	A3	03	01, 04, 07
close gripper	A6	06	01, 03
lift arm	A2	02	01, 06
retract arm	A5	05	01, 02, 06
rotate base CCW	A0	00	05, 06
extend arm	A4	04	00, 06
lower arm	A3	03	00, 04, 06
open gripper	A7	07	00, 03
lift arm	A2	02	00, 07
retract arm	A5	05	00, 02, 07
rotate base CW	A1	01	05, 07

completion of A7. The entire scheme, together with necessary prior inputs for each arm motion, is shown in Table 4.2.

Note that in this diagram, usually two previous steps are sufficient as prerequisite, but one more is needed in case two are not enough for distinction of different actions.

It is also important to remark that if we do not consider the necessary prior conditions, then when the PLC applies the logic control to the robot arm, it may make mistakes. For example, the second step in Table 4.2 is "lower arm," for which the completion is represented by 03. If 03 were to be used as the logic input to trigger the next step in the sequential operations, the "close gripper" action, the robot arm would always close its gripper after it executes the action "lower arm." However, the eighth step of the program is also "lower arm," but it is followed by "open gripper" instead. Hence, it is necessary to provide other prior conditions to the eighth step to ensure that it would not repeat the second step here.

Figure 4.4 shows the ladder logic diagram corresponding to Table 4.2, in which the standard symbols listed in Table 4.3 are used, and the input contact P0 is used to represent a power-ON switch. Here, the input contacts include switches, relays, photoelectric sensors, limit switches, and other ON-OFF signals from the logical system; the output loads include motors, valves, alarms, bells, lights, actuators, or other electrical loads to be driven by the logical system. We should not confuse the input contact symbol with the familiar circuit symbol for capacitors. Also, we should note that output loads can become input contacts in the following steps, depending on the actual system structure.

In a routine yet somewhat tedious procedure, one can verify that the PLC ladder logic diagram shown in Figure 4.4 performs the automatic control

Table 4.3 Ladder Diagram Symbols

Symbol	Meaning	Expression
─┤ ├─	Input contacts	
─( )─	Output loads	
A ─┤/├─	Logical NOT	$\overline{A}$
A ─┤ ├─┤ ├─ B	Logical AND	$A \wedge B$
A ─┤ ├─ parallel with B ─┤ ├─	Logical OR	$A \vee B$

process summarized in Table 4.2. Under the indicated necessary prior conditions for each step during the entire process, this PLC works for the designed "pick-and-place" motion-control of the robot arm described by Figure 4.3. The design procedure is time-consuming. However, once a design has been completed, the PLC can be automatically programmed over and over again, to control the robot arm to repeat the same "pick-and-place" motion for the production line. When the job assignment is changed, the PLC has the flexibility to be reprogrammed to perform a new logic control process.

II. FUZZY LOGIC CONTROL (I): A GENERAL MODEL-FREE APPROACH

The programmable logic controllers (PLCs) discussed in the last section, which have been widely used in industries, can only perform classical two-valued logic in programming some relatively simple but very precise automatic control processes. To carry out fuzzy logic-based control programming, a new type of programmable fuzzy logic controllers is needed.

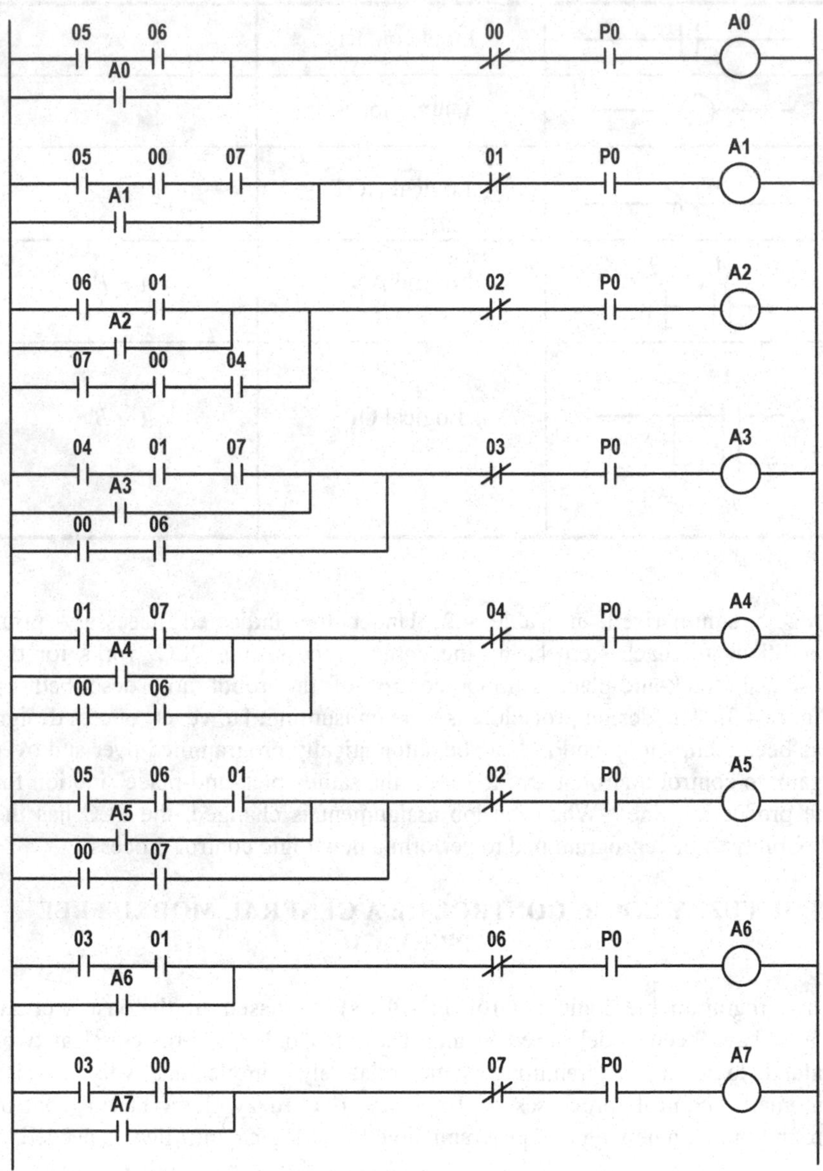

Figure 4.4 PLC ladder logic diagram for the robot arm.

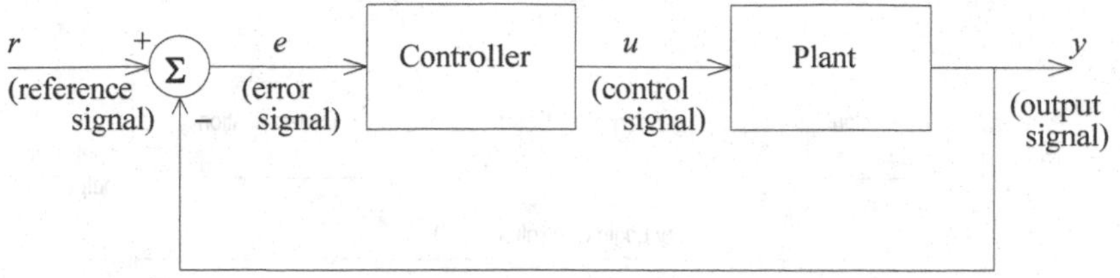

Figure 4.5 A typical closed-loop set-point tracking system.

The majority of fuzzy logic control systems are knowledge-based systems in that either their fuzzy models or their fuzzy logic controllers are described by fuzzy IF-THEN rules, which have to be established based on experts' knowledge about the systems, controllers, performance, etc. Moreover, the introduction of input-output intervals and membership functions is more or less subjective, depending on the designer's experience and the available information. However, we emphasize once again that after the determination of the fuzzy sets, all mathematics to follow are rigorous. Also, the purpose of designing and applying fuzzy logic control systems is, above all, to tackle those vague, ill-described, and complex plants and processes that can hardly be handled by classical systems theory, classical control techniques, and classical two-valued logic. This is the first type of fuzzy logic control system: the fuzzy logic controller directly performs the control actions and thus completely replaces a conventional control algorithm. Yet, there is another type of fuzzy logic control system: the fuzzy logic controller is involved in a conventional control system and thus becomes part of the mixed control algorithm, so far as to enhance or improve the performance of the overall control system. We will study these two types of fuzzy logic control systems, often alternatively, in the rest of this book.

In this section, we first discuss a general approach of fuzzy logic control for a conventional (crisp) system (plant or process) of the feedback (closed-loop) type.

A. A Closed-Loop Set-Point Tracking System

To facilitate our presentation and discussion, we consider the typical continuous-time, closed-loop, set-point tracking system shown in Figure 4.5.

In this figure, we assume that the plant is a conventional (crisp) one, which is given but its mathematical model may not be known, and that all the signals (r, e, and y) are crisp. The closed-loop set-point tracking control problem is to design the controller such that the output signal of the controlled plant, y, can track the given reference signal r (need not be a constant):

$$e(t) := r(t) - y(t) \to 0 \qquad (t \to \infty).$$

Instead of designing a conventional controller, in the following we study how to design a fuzzy logic controller for the same purpose. Recall that in

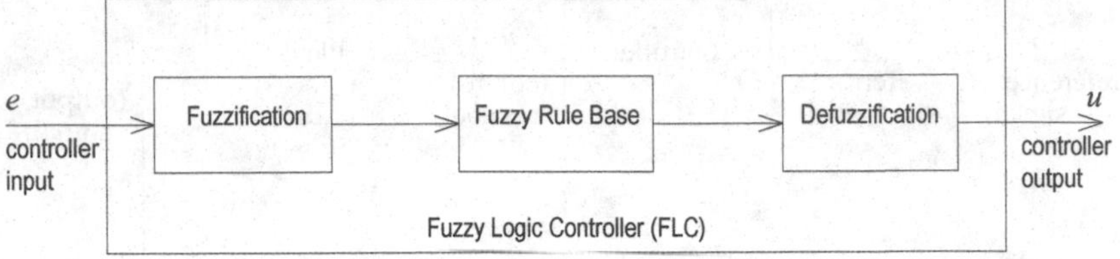

Figure 4.6 General structure of a fuzzy logic controller.

designing a conventional controller, a precise mathematical model (formulation) of the plant is usually necessary. A typical example is the conventional proportional-integral-derivative (PID) controller design, where the first- or second-order linear plant transfer function has to be first given. We will review and discuss these conventional PID controllers in Chapter 5. Here, to design a fuzzy logic controller, not necessarily of PID-type, for set-point tracking, we suppose that the mathematical formulation of the plant is completely unknown.

If the mathematical formula of the plant is unknown, for instance we don't even know if it is linear or nonlinear (and, if it is linear, we don't know what order it has; if it is nonlinear, we don't know what kind of nonlinearity it has), how can we design a controller to perform the required set-point tracking? One may think of designing a conventional controller and try to answer this question at this point, to appreciate the ease of fuzzy logic controller design to be studied below.

Before we discuss how fuzzy logic can help in completing such a design for a "black box" system, we need to clarify some basic concepts and to introduce some new terminology.

First, the general structure of a *fuzzy logic controller* (FLC), or *fuzzy controller* (FC) for short, consists of three basic portions: the *fuzzification* unit at the input terminal, the *inference engine* built on the fuzzy *logic control rule base* in the core, and the *defuzzification* unit at the output terminal, as shown in Figure 4.6.

The fuzzification module transforms the physical values of the current process signal, the error signal in Figure 4.5 which is input to the fuzzy logic controller, into a normalized fuzzy subset consisting of a subset (interval) for the range of the input values and an associate membership function describing the degrees of the confidence of the input belonging to this range. The purpose of this fuzzification step is to make the input physical signal compatible with the fuzzy control rule base in the core of the controller. Here, between the physical input signal and the fuzzy subset within the fuzzification unit, a pre-processing unit mapping the physical signal to some pointwise and crisp real values (that the fuzzy subset can accept) may be needed, depending on the nature of the underlying process. Generally speaking, a universal fuzzy logic controller for the closed-loop set-point tracking system shown in Figure 4.5 is unlikely possible. Hence, the fuzzy subset, both the subset and the membership function, has to be selected by the designer according to the particular application at hand. In other words, depending on the nature and characteristics of the given plant and reference signal, the FLC has to be designed to fit to the need, so as to make the closed-loop fuzzy control system work for that particular application. This situation is just like the design of a conventional controller for a specifically given system, where there is no universal controller in practical design.

The role of the inference engine in the FLC is key to make the controller work – and work effectively. The job of the "engine" is to create the control actions, in fuzzy terms, according to the information provided by the fuzzification module and the set-point tracking requirement (and perhaps other control performance requirements as well). A typical fuzzy logic IF-THEN rule base performing the inference is in the general form that was studied in detail in Sections II-VI of Chapter 2. More specifically, the rule base is a set of IF-THEN rules of the form

R^1: IF controller input e_1 is E_{11} AND ... AND
controller input e_n is E_{1n}
THEN controller output u_1 is U_1.

⋮

R^m: IF controller input e_1 is E_{m1} AND ... AND
controller input e_n is E_{mn}
THEN controller output u_m is U_m.

Here, as discussed in Sections II-VI of Chapter 2, the fuzzy subsets E_{11}, ..., E_{m1} share the same subset E_1 and the same membership function μ_{E_1} defined on E_1, and fuzzy subsets E_{1n}, ..., E_{mn} share the same subset E_n and the same membership function μ_{E_n} defined on E_n. In general, m rules produce m controller outputs, u_1, ..., u_m, belonging to m fuzzy subsets, U_1, ..., U_m, in which, of course, some of them may overlap. The establishment of this rule base depends heavily on the designer's work experience, knowledge about the physical plant, analysis and design skills, etc., and is, hence, more or less subjective. Thus, a good design can make the controller work; a better design can make it work more effectively. This situation is just like conventional design: any specific design is not unique in general. Yet, there are some general criteria and some routine steps for the designer to follow in a real design, which will be discussed in more detail later. Here, basically, what have to be determined are the choices of the controller's input and output variables and the IF-THEN rules.

The defuzzification module is the connection between the control rule base and the physical plant to be controlled, which plays the role of a transformer mapping the controller outputs (generated by the control rule base in fuzzy terms) back to the crisp values that the plant can accept. Hence, in a sense the defuzzification module is the inverse of the fuzzification module. The controller outputs u_1, ..., u_m, generated by the rule base above, are fuzzy signals belonging to the fuzzy subsets U_1, ..., U_m, respectively. The job of the defuzzification module is to convert these fuzzy controller outputs to a pointwise and crisp real signal, u, and then send it to the physical plant as a control action for tracking. Between the defuzzification step and the physical plant, a post-processing unit mapping the pointwise signal u to a physical signal (that the plant can accept) may be needed, depending again on the nature of the underlying process. What has to be determined in this stage is essentially a defuzzification formula. There are several commonly used, logically meaningful, and practically effective defuzzification formulas available, which are by nature weighted average formulas in various forms.

This tree-step design routine, the "fuzzification – rule base establishment – defuzzification" procedure, is further discussed in detail next.

B. Design Principle of Fuzzy Logic Controllers

Now, we are in a position to discuss some design principles of a fuzzy logic controller, using the typical structure shown in Figure 4.6, for the set-point tracking system depicted in Figure 4.5.

We only consider a continuous-time single-input/single-output (SISO) system in this study for simplicity of notation. Discrete-time and multi-input/multi-output (MIMO) cases can be similarly discussed.

Suppose that a scalar-valued reference signal r, the set-point, is given, which needs not be a constant in general. For ease of explanation, we fix it to be constant in the following discussion. A plant is also assumed to be given, whose mathematical formulation is assumed to be unavailable. The aim is to design a fuzzy logic controller, to be put into the controller block of Figure 4.5, to derive the plant output $y(t)$ to track the constant set-point r as $t\to\infty$, or, in other words, to force the error signal

$$e(t) := r - y(t) \to 0 \qquad (t\to\infty).$$

The following is a general approach for the design of the fuzzy logic controller.

(1) The Fuzzification Module. The fuzzification module performs the following functions:

(1.1) It transforms the physical values (position, voltage, degree, etc.) of the process signal, the error signal shown in Figure 4.6 which is an input to the fuzzy logic controller, into a normalized fuzzy subset consisting of a subset (interval) for the range of the input values and a normalized membership function describing the degree of confidence of the input belonging to this range.

For example, suppose that the physical error signal, the input to the fuzzy logic controller shown in Figure 4.6, is the degree of temperature. Suppose also that the distribution of this temperature error signal is within the range

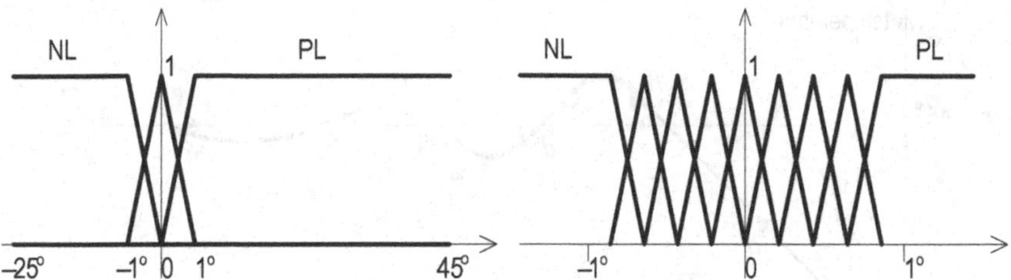

Figure 4.7 Selection examples of membership functions for error signals.

[–25°,45°] in an application in which the fuzzy logic controller needs to be designed to reduce this error to 0°±0.01°. In this case, the scale of 1° is too large to use in the measurement of the control effect within ±0.01°. Hence, we may first try to rescale the range to be [–2500,4500] in the unit of 0.01°. But this is not convenient in the calculation of fuzzy membership functions either. To have a compromise, a better choice can be a combination of two different scales (see Figure 4.7):

Use [–25°,45°] when |error| > 1;
Use [–100*,100*] when |error| ≤ 1, with 1° = 100*.

(1.2) It selects reasonable and good, ideally optimal, membership functions under certain convenient criteria meaningful to the application.

In the above-described temperature control example, one may use the membership functions shown in Figure 4.7 to describe "the error is positive large (PL)," or "the error is negative large (NL)," etc., in which one figure is in the 1° scale and the other is in the 1* = 0.01° scale. A designer may have other choices, of course, but this example should provide some idea about how the selection can be done in general.

In the fuzzification module, the input is a crisp physical signal (e.g., temperature) of the real process and the output is a fuzzy subset consisting of intervals and membership functions. This output will be the input to the next module, the fuzzy logic IF-THEN rule base for the control, which requires fuzzy-subset inputs in order to be compatible with the fuzzy logic rules.

(2) The Fuzzy Logic Rule Base. Designing a good fuzzy logic rule base is key to obtaining a satisfactory controller for a particular application. Classical analysis and control strategies should be incorporated in the establishment of a rule base. A general procedure in designing a fuzzy logic rule base includes the following:

(2.1) Determining the process states and control variables.

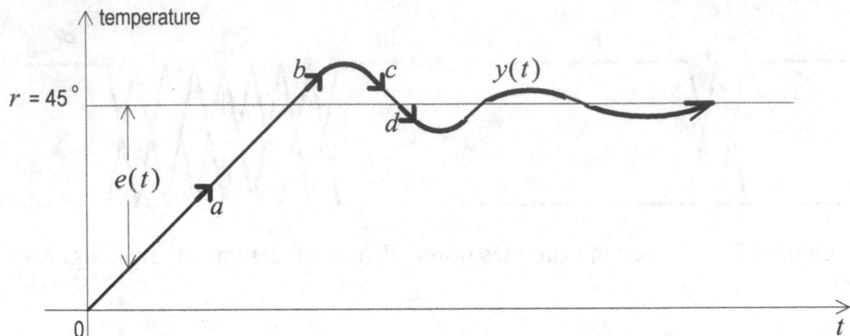

Figure 4.8 Temperature set-point tracking example.

In the set-point tracking example discussed above in Figures 4.5 and 4.6, let us suppose that the physical process is a temperature control. Thus, the set-point r is a target temperature to be reached, say $r = 45°$. In this application, the process state is the overall controlled system output, $y(t)$, which is also temperature. Finally, the error signal

$$e(t) = r - y(t), \tag{4.1}$$

as shown in Figure 4.8, is used to create the control variable u (see Figure 4.5) through the controller.

(2.2) Determining input variables to the controller.

As mentioned above, the tracking error signal $e(t)$ is an input variable to the controller (see Figures 4.5 and 4.6). Oftentimes, we need more auxiliary input variables in order to establish a complete and effective rule base. In this temperature control example, it can be easily seen that only the error signal $e(t)$ is not enough to write an IF-THEN control rule. Indeed, let us say $e > 0$ at a moment. Then we know

$$e = r - y > 0 \qquad \text{or} \qquad r > y,$$

namely, at that moment the system output y is below the set-point. This indicates that the output y is either at position a or position d in Figure 4.8. However, this information is not sufficient for determining a control strategy that can bring the trajectory of y to approach the set-point thereafter: if the output y is at position a then the controller should take action to keep the trajectory going up; if y is at position d then the controller should turn the trajectory to the opposite moving direction (from pointing down to pointing up). Namely, in the first case the controller should maintain the previous action but in the last case it should switch its previous action to the opposite. Therefore, one more input variable that can distinguish these two situations is necessary.

Recall from Calculus that if a curve is moving up its derivative is positive and if it is moving down its derivative is negative. Hence, the change of the error signal, denoted $\dot{e}$ or Δe, can help distinguish the two situations at points

a and d, as well as that at points b and c, shown in Figure 4.8. Hence, we introduce the change of error as the second input variable for this temperature set-point tracking example. More input variables can be introduced such as the second derivative $\ddot{e}$ or the sum (integral) Σe of the errors, to make the controller work more effectively. However, this will also complicate the design, and so there is a trade-off for the design engineer. In order to simplify the design and for simplicity of discussion, in this example we only use two input variables, e and $\dot{e}$, for the controller.

(2.3) Establishing a fuzzy logic IF-THEN rule base.

The input variables determined in the previous step will be used in the design of the rule base for control. As mentioned above, successful classical analysis and control strategies should also be incorporated in the design. To obtain some ideas and insights about the rule base design, we consider again the above temperature control example.

In Figure 4.8, it is clear that essentially we have four situations to consider: when the temperature output $y(t)$ is at the situations represented by the points a, b, c, and d. We thus need at least four rules for this set-point tracking application in order to make it work and work effectively. In principle, also intuitively, we can set up the following four control rules, where u is the output (control action) of the controller:

R^1: IF $e > 0$ AND $\dot{e} < 0$ THEN $u(t+) = u(t)$;
R^2: IF $e < 0$ AND $\dot{e} < 0$ THEN $u(t+) = -u(t)$;
R^3: IF $e < 0$ AND $\dot{e} > 0$ THEN $u(t+) = u(t)$;
R^4: IF $e > 0$ AND $\dot{e} > 0$ THEN $u(t+) = -u(t)$.

Otherwise (e.g., $e = 0$ or $\dot{e} = 0$), $u(t+) = u(t)$, until the next step.

In Figure 4.8, it is very important to observe that

$$\dot{e}(t) = \dot{r} - \dot{y}(t) = -\dot{y}(t). \tag{4.2}$$

Hence, the rules R^1, R^2, R^3, and R^4 correspond to the situations indicated by points a, b, c, and d, respectively, in Figure 4.8. In these rules, the notation $u(t+) = u(t)$ means that, conceptually, the controller retains its previous action unchanged, while $u(t+) = -u(t)$ means that the controller turns its previous action to the opposite (e.g., from a positive action to a negative action, or vice versa). Quantitative implementation of these actions will be further discussed below.

To this end, it can be easily realized that by following the above four rules, the controller is able to drive the temperature output $y(t)$ to track the set-point r, at least in principle. For example, if $e > 0$ and $\dot{e} < 0$ then $r > y$ and $\dot{y} > 0$, which means that the curve y is at position a of Figure 4.8. In this case, rule R^1 implies that the controller should maintain its current action (to keep pushing the system in the same way). The other three rules can be similarly analyzed.

However, some technical issues remain. Let us again look at the situation at point a in Figure 4.8, which has the control rule R^1. One problem with this rule is that if the previous control action u were small, this rule would let the

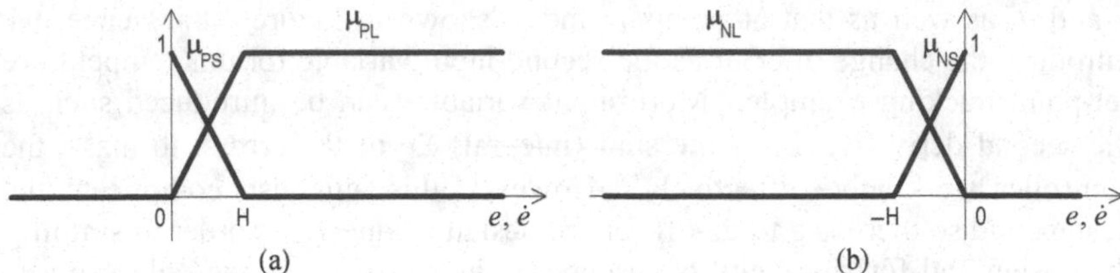

Figure 4.9 Four membership functions for both e and $\dot{e}$.

controller keep driving the output temperature up toward the set-point, but very slowly; so it may need a long time to complete the task. Another problem with this rule, on the other side of the issue, is that if the previous control action u were large, this rule would lead the controller to drive the output temperature overshooting the set-point. As a result of such rules and actions, the output temperature oscillates up and down around the set-point, which may not eventually settle at the set-point, or take a very long time to do so. Hence, these four rules have to be further improved to be practical.

Recall that we have fuzzified the error signal e in the fuzzification module. If we also fuzzify the change of error, $\dot{e}$, in the fuzzification module, then we have two fuzzified input variables, e and $\dot{e}$, for the controller, and they have corresponding membership functions to describe their properties of "positive large," "negative small," etc. These properties can be used to improve the above rules. Again, let us consider the situation at point a in Figure 4.8 and its corresponding control rule R^1. It is clear that if the error $e > 0$ is small then the controller can take a smaller action, and if the error $e > 0$ is large then the controller can take a larger action. In doing so, the output trajectory $y(t)$ will be directed to the set-point correctly and more efficiently. Therefore, we should incorporate the membership values of the two input variables e and $\dot{e}$ in the four control rules.

To further simplify the notation and discussion below, we only use the simple membership functions shown in Figure 4.9, rather than those shown in Figure 4.7, for both e and $\dot{e}$. In a real application, of course, a designer can (and likely should) employ more membership functions.

Using these membership functions as weights for the control $u(t)$, we can accomplish the following task: if y is far away from r then the control action is large, but if y is close to r then the control action is small. The improved control rule base is obtained as follows:

R^1: IF e = PL AND $\dot{e} < 0$ THEN $u(t^+) = \mu_{PL}(e) \cdot u(t)$;
R^2: IF e = PS AND $\dot{e} < 0$ THEN $u(t^+) = (1-\mu_{PS}(e)) \cdot u(t)$;
R^3: IF e = NL AND $\dot{e} < 0$ THEN $u(t^+) = -\mu_{NL}(e) \cdot u(t)$;
R^4: IF e = NS AND $\dot{e} < 0$ THEN $u(t^+) = -(1-\mu_{NS}(e)) \cdot u(t)$;
R^5: IF e = NL AND $\dot{e} > 0$ THEN $u(t^+) = \mu_{NL}(e) \cdot u(t)$;
R^6: IF e = NS AND $\dot{e} > 0$ THEN $u(t^+) = (1-\mu_{NS}(e)) \cdot u(t)$;
R^7: IF e = PL AND $\dot{e} > 0$ THEN $u(t^+) = -\mu_{PL}(e) \cdot u(t)$;

R^8: IF e = PS AND $\dot{e} > 0$ THEN $u(t+1) = -(1-\mu_{PS}(e)) \cdot u(t)$.
Here and below, "= PL" means "is PL," etc.

These rules can be easily verified by taking into account the temperature output curve $y(t)$ shown in Figure 4.8, the relation $\dot{y}(t) = -\dot{e}(t)$, and the four membership functions for e (and the same four membership functions for $\dot{e}$) shown in Figure 4.9. Note that "$\cdot$" in the formulas of $u(t+1)$ above are algebraic multiplications.

To implement these rules on a digital computer, we actually use their discrete-time version of the form

$$u(t) = u(kT) \quad \text{and} \quad u(t+) = u((k+1)T), \tag{4.3}$$

where T is the sampling time and $u(kT)$ is the value of the new change of the control action: $\Delta u(t)$ at $t = kT$, $k = 0, 1, 2, \cdots$. Thus, the actual executive rules on a computer program become

R^1: IF $e(kT)$ = PL AND $\dot{e}(kT) < 0$
THEN $u((k+1)T) = \mu_{PL}(e(kT)) \cdot u(kT)$;
R^2: IF $e(kT)$ = PS AND $\dot{e}(kT) < 0$
THEN $u((k+1)T) = (1-\mu_{PS}(e(kT))) \cdot u(kT)$;
R^3: IF $e(kT)$ = NL AND $\dot{e}(kT) < 0$
THEN $u((k+1)T) = -\mu_{NL}(e(kT)) \cdot u(kT)$;
R^4: IF $e(kT)$ = NS AND $\dot{e}(kT) < 0$
THEN $u((k+1)T) = -(1-\mu_{NS}(e(kT))) \cdot u(kT)$;
R^5: IF $e(kT)$ = NL AND $\dot{e}(kT) > 0$
THEN $u((k+1)T) = \mu_{NL}(e(kT)) \cdot u(kT)$;
R^6: IF $e(kT)$ = NS AND $\dot{e}(kT) > 0$
THEN $u((k+1)T) = (1-\mu_{NS}(e(kT))) \cdot u(kT)$;
R^7: IF $e(kT)$ = PL AND $\dot{e}(kT) > 0$
THEN $u((k+1)T) = -\mu_{PL}(e(kT)) \cdot u(kT)$;
R^8: IF $e(kT)$ = PS AND $\dot{e}(kT) > 0$
THEN $u((k+1)T) = -(1-\mu_{PS}(e(kT))) \cdot u(kT)$,

for all $k = 0, 1, 2, \cdots$, where $\dot{e}(kT) \approx \frac{1}{T}[e(kT) - e((k-1)T)]$, with the initial conditions $y(0) = 0$, $e(-T) = e(0) = r - y(0)$, $\dot{e}(0) = \frac{1}{T}[e(0) - e(-T)] = 0$.

We finally remark that another commonly used alternative for determining the weight of the control u(t) is to give it a value of PL, PS, NL, or NS. More specifically, we put the above rule base in a tabular form as shown in Table 4.4.

Table 4.4 is sometimes called a "look-up table," which can be made more accurate by dividing both e and $\dot{e}$ into more subcases. Table 4.5 gives an example of a more subtle rule-base table for fuzzy logic controller whose control action is proportional to both e and $\dot{e}$, namely,

$$u((k+1)T) = a\, e(kT) + b\, \dot{e}(kT)$$

Table 4.4 A Rule Base in the Tabular Form for Δu

e \ $\dot{e}$	<0	>0
PL	PL	NL
PS	PS	NS
NS	NS	PS
NL	NL	PL

Table 4.5 A Rule Base Table for $u = a\,e + b\,\dot{e}$

e \ $\dot{e}$	NL	NM	NS	ZO	PS	PM	PL
NL	NL	NL	NL	NL	NM	ZO	PS
NM	NL	NL	NL	NM	ZO	PS	PM
NS	NL	NL	NM	ZO	PS	PM	PL
ZO	NL	NM	ZO	PS	PM	PL	PL
PS	NM	ZO	PS	PM	PL	PL	PL
PM	ZO	PS	PM	PL	PL	PL	PL
PL	PS	PM	PL	PL	PL	PL	PL

for some constants $a > 0$ and $b > 0$, where NM and ZO mean "negative medium" and "zero," respectively, and the other abbreviations are similarly understood.

Finally, we need to select membership functions for the different control outputs, given either by the formulas shown in the rule base R^1- R^8 above, or by the linguistic terms PL, PS, NS, and NL shown in Table 4.4. Figure 4.10 is a simple, yet typical choice for the membership functions of u, where P, N, and ZO indicate positive, negative, and zero, respectively, and H is a real number.

(2.4) Establishing a fuzzy logic inference engine.

In order to complete the fuzzy logic inference embedded in the control rule base, the general fuzzy IF-THEN rule given in Definition 2.1 of Chapter 2 has to be applied to each rule in the rule base.

Take the first rule as an example:

R^1: IF $e(kT)$ = PL AND $\dot{e}(kT) < 0$

THEN $u((k+1)T) = \mu_{PL}(e(kT)) \cdot u(kT)$.

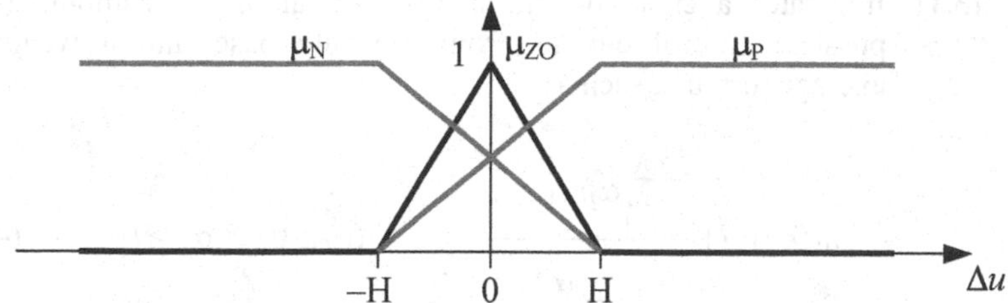

Figure 4.10 Typical membership functions for Δu.

In this rule, $e(kT)$ = PL has a membership function $\mu_{PL}(e(kT))$ shown in Figure 4.9 (a), $\dot{e}(kT) < 0$ is nonfuzzy and so has membership values $\mu_N(\dot{e}(kT)) = 1$, and $u((k+1)T)$ has three membership functions as shown in Figure 4.10. Thus, the fuzzy logic inference

$$\mu_{PL}(e(kT)) \wedge \mu_N(\dot{e}(kT)) \Rightarrow \mu(u((k+1)T))$$

yields the logical inference membership function (see Sections II, VI, Chapter 2 for more detail):

$$\mu(u((k+1)T)) = \min \{ \mu_{PL}(e(kT)) , \mu_N(\dot{e}(kT)) \} = \mu_{PL}(e(kT))$$

without using the membership functions shown in Figure 4.10.

Another approach is to use the output membership functions shown in Figure 4.10, along with the weighted average formula (3.12), Chapter 3, as shown in formula (4.5) below. This approach is quite common in practice.

Once this logical inference has a fuzzy logic membership function, it is completed as a logical formula. A fuzzy logic control rule base with all complete logical inference membership functions is called a *fuzzy logic inference engine*: it activates the rules with fuzzy logic.

We remark, however, that the logical inference membership functions are usually not used in the design of a digital fuzzy logic controller in many control engineering applications today, where only the control rule base discussed in the last part, part **(2.3)**, is essential. We will follow this common practice to use the rule base rather than the inference engine in the rest of the book. Remember, this is a design of a controller, for which the design engineer has a choice of what approach to take. It is just like a classical controller design for which the designer usually has different options.

(3) The Defuzzification Module. The defuzzification module is in a sense the reverse of the fuzzification module: it converts all the fuzzy terms created by the rule base of the controller to crisp terms (numerical values) and then sends them to the physical system (plant, process), so as to execute the control of the system.

The defuzzification module performs the following functions:

(3.1) It creates a crisp, overall control signal, u, by combining all possible control outputs from the rule base into a weighted average formula, such as

$$u((k+1)T) = \frac{\sum_{i=1}^{N} \alpha_i u_i(kT)}{\sum_{i=1}^{N} \alpha_i}, \qquad (\alpha_i \geq 0, \ \sum_{i=1}^{N} \alpha_i > 0). \tag{4.4}$$

In the control rule base R^1- R^8, established above, we obtained eight different control outputs, u, which we now denote by $u_i(kT)$, $i = 1, ..., 8$.

Suppose that we have a fuzzy logic control rule base

R^i: IF $e(kT)$ is E_i AND $\dot{e}(kT)$ is F_i THEN $u_i((k+1)T)$ is U_i,

where $i = 1, ..., N$; E_i, F_i, and U_i are fuzzy subsets consisting of some bounded intervals with their associated membership functions $\mu_{E_i}(\times)$, $\mu_{F_i}(\times)$, and $\mu_{U_i}(\times)$, respectively. Note that this control rule base may also be given in a tabular form like the one shown in Table 4.5.

There are several commonly used defuzzification formulas:

(i) The "center-of-gravity" formula:

$$u((k+1)T) = \frac{\sum_{i=1}^{N} \mu_{U_i}(u_i(kT)) \times u_i(kT)}{\sum_{i=1}^{N} \mu_{U_i}(u_i(kT))}, \tag{4.5}$$

which is by nature the same as the formula (3.12) introduced in Chapter 3. The continuous-time version of this formula is

$$u(t) = \frac{\int_U \mu_U(\tilde{u}) \times \tilde{u}\, d\tilde{u}}{\int_U \mu_U(\tilde{u})\, d\tilde{u}}, \tag{4.6}$$

where U is the value range (interval) of the control $\tilde{u}$ in the rule base (the continuous-time version):

R: IF $e(t)$ is E AND $\dot{e}(t)$ is F THEN $u(t+)$ is U.

In applications, namely, in a practical engineering design employing digital computers, the continuous-time formulation is seldom used. However, the continuous-time formula (4.6) best explains the name of the formula: if μ_U represents the density function over a rigid body of volume U and $\tilde{u}$ is a coordinate variable, say the horizontal x-coordinate (in a three-dimensional x-y-z Cartesian space), then formula (4.6) gives the x-coordinate of the center of mass (gravity) of the body (at time t), as is well known in Calculus and Mechanics.

(ii) The "center-of-sums" formula:

$$u((k+1)T) = \frac{\sum_{i=1}^{N} u_i(kT) \times \sum_{j=1}^{N} \mu_{U_j}(u_i(kT))}{\sum_{i=1}^{N}\sum_{j=1}^{N} \mu_{U_j}(u_i(kT))}. \tag{4.7}$$

This formula can be computed faster than the "center-of-gravity" formula (4.5). Its continuous-time version is

$$u(t) = \frac{\int_U \tilde{u} \times \sum_{j=1}^{N} \mu_{U_j}(\tilde{u}) d\tilde{u}}{\int_U \sum_{j=1}^{N} \mu_{U_j}(\tilde{u}) d\tilde{u}}. \tag{4.8}$$

(iii) The "mean-of-maxima" formula:

For each rule R^i, $i = 1, \ldots, N$, we first find the particular control output value $u_i(kT)$ from its value range (interval) U_i at which its membership value reaches the maximum: $\mu_{U_i}(u_i(kT))$ = maximum (= 1 in the normalized case). If there are several such values $u_i(kT)$ at which their membership values reach the maxima, then we count the multiplicities. If there is a continuum (subinterval) of such values $u_i(kT)$ at which their membership values reach the maxima, then we pick the smallest and the largest values and count them just twice. Let us say the total number of such values of $u_i(kT)$, at which their membership values reach the maxima, is M. Then we take their mean (average) in the following form:

$$u(kT) = \frac{1}{M} \sum_{i=1}^{M} \tilde{u}_i(kT). \tag{4.9}$$

Its continuous-time version is:

$$u(t) = \frac{1}{2}\{ \inf\{ u \in U \mid \mu_U(u) = \text{maximum} \} + \sup\{ u \in U \mid \mu_U(u) = \text{maximum} \} \}. \tag{4.10}$$

There are other defuzzification formulas suggested in the literature, each of which has its own rationale, either from logical argument or from geometrical reasoning, or from control principles. We will however only discuss the "center-of-gravity" formula (4.5) throughout the book.

(3.2) Just like the first step of the fuzzification module, this step of the defuzzification module transforms the afterall control output, u, obtained in the previous step, to the corresponding physical values (position, voltage, degree, etc.) that the system (plant, process) can accept. This converts the fuzzy logic controller's numerical output to a physical means that can actually drive the given plant (process) to produce the expected outputs.

The overall fuzzy logic controller that combines the "fuzzification - rule base - defuzzification" modules has been shown in Figure 4.6, which is the controller block in the closed-loop set-point control system of Figure 4.5.

Finally, it is very important to recall that the design of the fuzzy logic controller that we discussed in this section does not require any information about the system (plant, process): it doesn't matter if the system is linear or nonlinear (nor the order of linearity, structure of nonlinearity, etc.) as long as its output, y, can be measured by a sensor and used for the control. In other words, the design of the fuzzy logic controller is independent of the mathematical model of the system under control. Comparing it with the conventional controllers design, its advantages are obvious. The price that we have to pay for this success is the complexity in computation in the design: the "fuzzification - rule base - defuzzification" steps. But this is what it is supposed to be: under a worse condition (the system model is unknown, or only vaguely given), if we want to design a controller to achieve the same (or better) set-point tracking performance then we have to do more analysis and calculation in the design. This is generally true in a sensible comparison of a fuzzy logic controller with a similar conventional one: it is likely impossible to obtain a better controller with less effort for a system under even worse conditions. However, the rapid development of modern high-speed computers has made such computational tasks affordable.

C. Examples of Model-Free Fuzzy Controller Design

Two examples are given in this section to illustrate the model-free fuzzy control approach and to demonstrate its effectiveness in controlling a complex mechanical system without assuming its mathematical model.

Example 4.1. We first consider a truck-parking control problem, as shown in Figure 4.11. The objective is to design a fuzzy logic controller, without assuming a mathematical model of the truck, to park the truck anywhere on the x-axis. Suppose that the truck can move forward at a constant speed of $v = 0.5$ *m/s* and it is assumed that the truck is equipped with sensors that can measure location (x,y) and orientation (angle) θ at all times. The fuzzy logic controller is to provide an input, u, to rotate the steering wheels and, consequently, to maneuver the truck.

In the simulation, the input variables are the truck angle θ and the vertical position coordinate, y, while the output variable is the steering angle (signal), u. The variable ranges are pre-assigned as

$$-100 \le y \le 100, \qquad -180^\circ \le \theta \le 180^\circ, \qquad -30^\circ \le u \le 30^\circ.$$

Here, clockwise rotations are considered positive in θ, and, therefore, counterclockwise are negative. The linguistic terms used in this design are given as follows:

Angle θ	y-position	Steering angle u
AB: Above	AO: Above much	NB: Negative big
AC: Above center	AR: Above	NM: Negative medium
CE: Center	AH: Above horizontal	NS: Negative small
BC: Below center	HZ: Horizontal	ZE: Zero
BE: Below	BH: Below horizontal	PS: Positive small
	BR: Below	PM: Positive medium
	BO: Below much	PB: Positive big

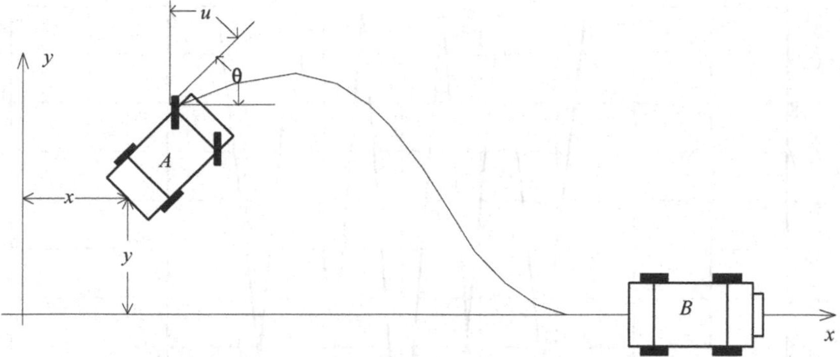

Figure 4.11 A truck-driving control example.

The first step is to choose membership functions. They are as shown in Figures 4.12 – 4.14. These choices of their shapes, ranges, and overlapping are somewhat arbitrary. We only note that narrow membership functions are used to permit fine control near the designated parking spot, while wide membership functions are used to perform fast controls when the truck is far away from the parking place.

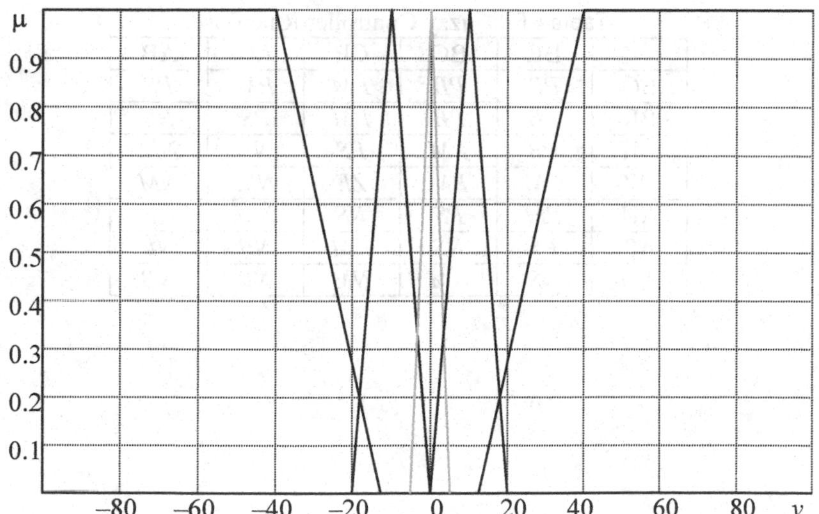

Figure 4.12 Fuzzy membership functions for the y-position.

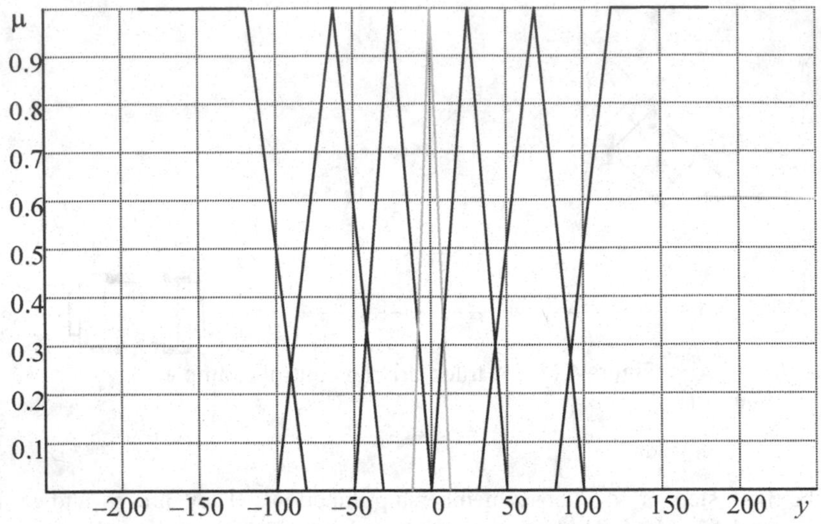

Figure 4.13 Fuzzy membership functions for the angle θ.

The rule base used in simulation is summarized in Table 4.6. Each rule has the form IF y is Y AND θ is Θ THEN u is U, as usual. A glance of these rules reveals the symmetry, an intrinsic property of this controller, which is reasonable for this parking application since the truck can move in any direction. This is not always the case, however, as can be seen from the next example where symmetry of control rules is not reasonable.

Table 4.6 Fuzzy Controller Rule Base

	BE	BC	CE	AC	AB
BO	*PB*	*PB*	*PM*	*PM*	*PS*
BR	*PB*	*PB*	*PM*	*PS*	*NS*
BH	*PB*	*PM*	*PS*	*NS*	*NM*
HZ	*PM*	*PM*	*ZE*	*NM*	*NM*
AH	*PM*	*PS*	*NS*	*NM*	*NB*
AR	*PS*	*NS*	*NM*	*NB*	*NB*
AO	*NS*	*NM*	*NM*	*NB*	*NB*

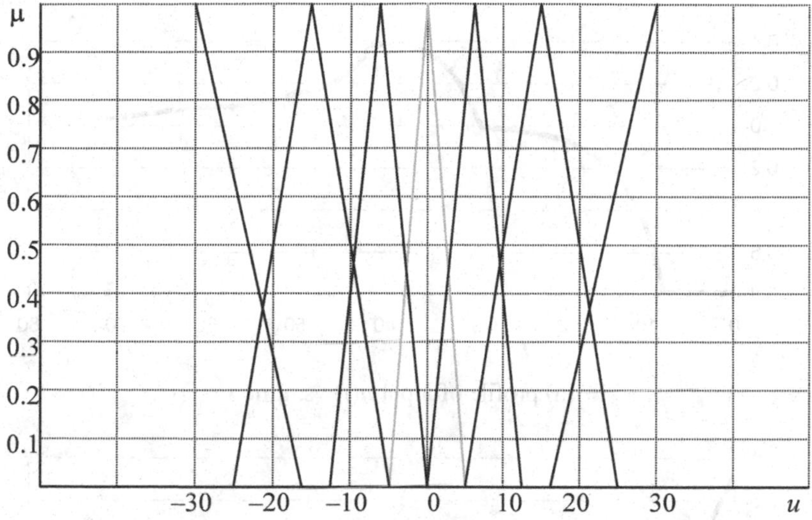

Figure 4.14 Fuzzy membership functions for the control signal u.

Finally, we need to obtain the output action under the given input conditions. The following standard weighted average defuzzification formula was used in simulation:

$$u = \frac{\sum_{i=1}^{4} \mu(u_i) u_i}{\sum_{i=1}^{4} \mu(u_i)}.$$

A total of 12 computer simulations were performed. Their initial conditions are summarized in Table 4.7, and the corresponding parking performances are shown in Figure 4.15.

The following is a simplified mathematical model for the motion of the truck. This model was used to create data for simulation but was not used in the design of the controller. This model directly follows from the geometry of the truck (see Figure 4.11):

$$\begin{aligned} \theta(k+1) &= \theta(k) + v\,T \tan(\, u(k)\,) \,/\, L, \\ x(k+1) &= x(k) + v\,T \cos(\, \theta(k)\,), \\ y(k+1) &= y(k) + v\,T \sin(\, \theta(k)\,), \end{aligned} \tag{4.11}$$

where

Table 4.7 Initial Conditions of the 12 Simulated Cases

Case	1	2	3	4	5	6	7	8	9	10	11	12
θ	0	0	0	90	90	90	180	180	180	–90	–90	–90
y	30	20	–20	30	10	–20	30	–20	–10	–10	–20	20

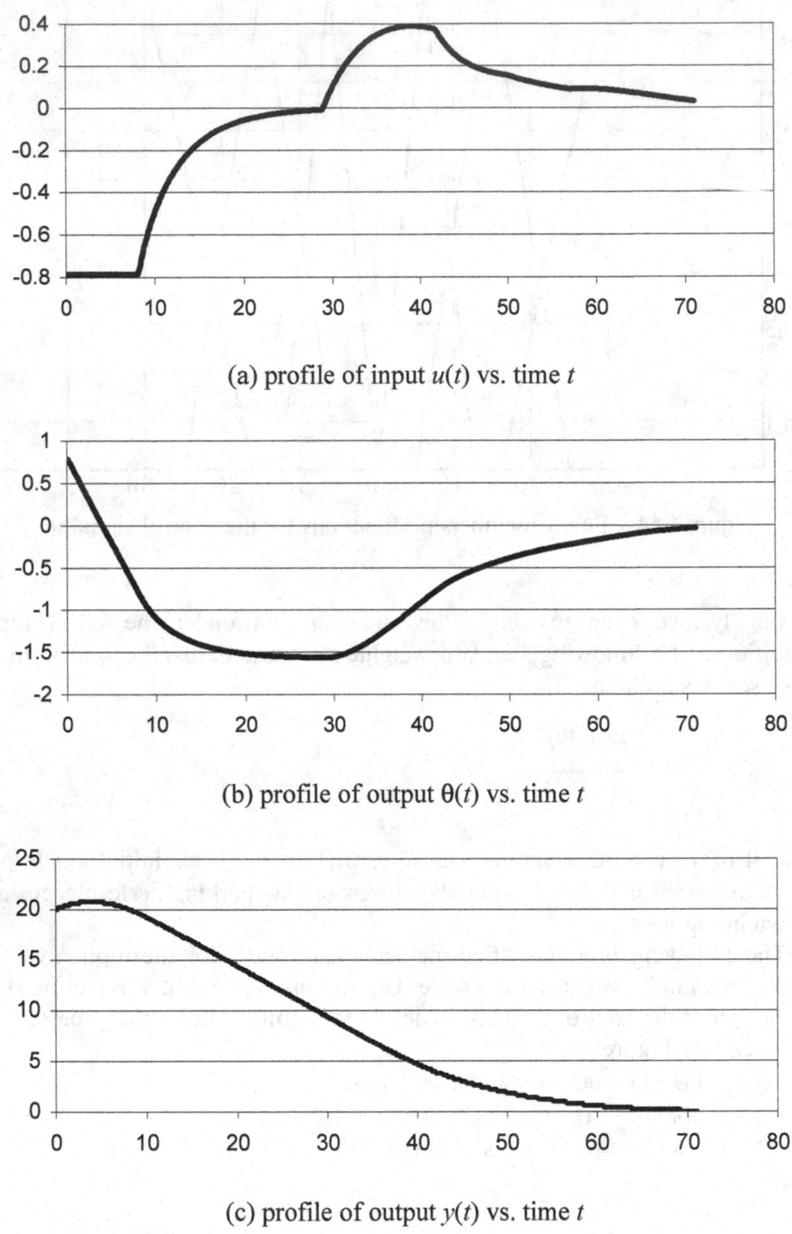

(a) profile of input $u(t)$ vs. time t

(b) profile of output $\theta(t)$ vs. time t

(c) profile of output $y(t)$ vs. time t

Figure 4.15 Model-free parking control simulation results.

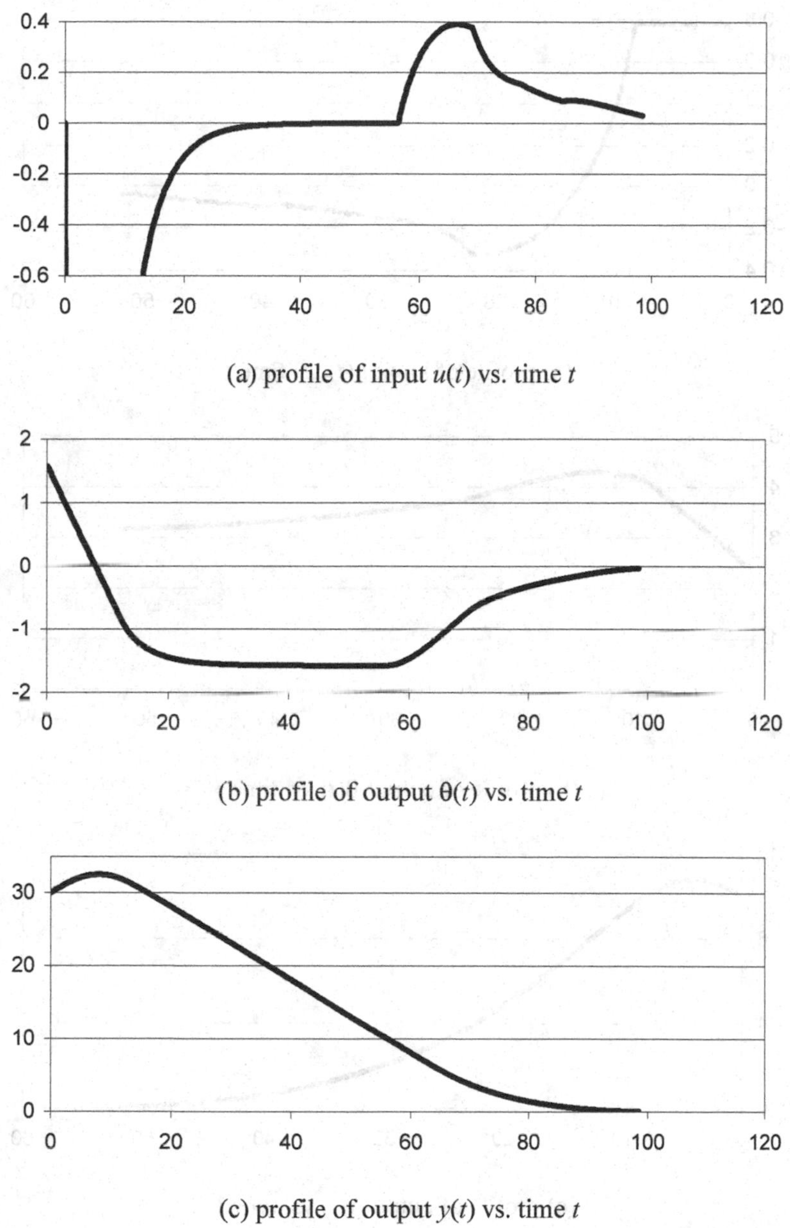

(a) profile of input $u(t)$ vs. time t

(b) profile of output $\theta(t)$ vs. time t

(c) profile of output $y(t)$ vs. time t

Figure 4.15 Model-free parking control simulation results (continued).

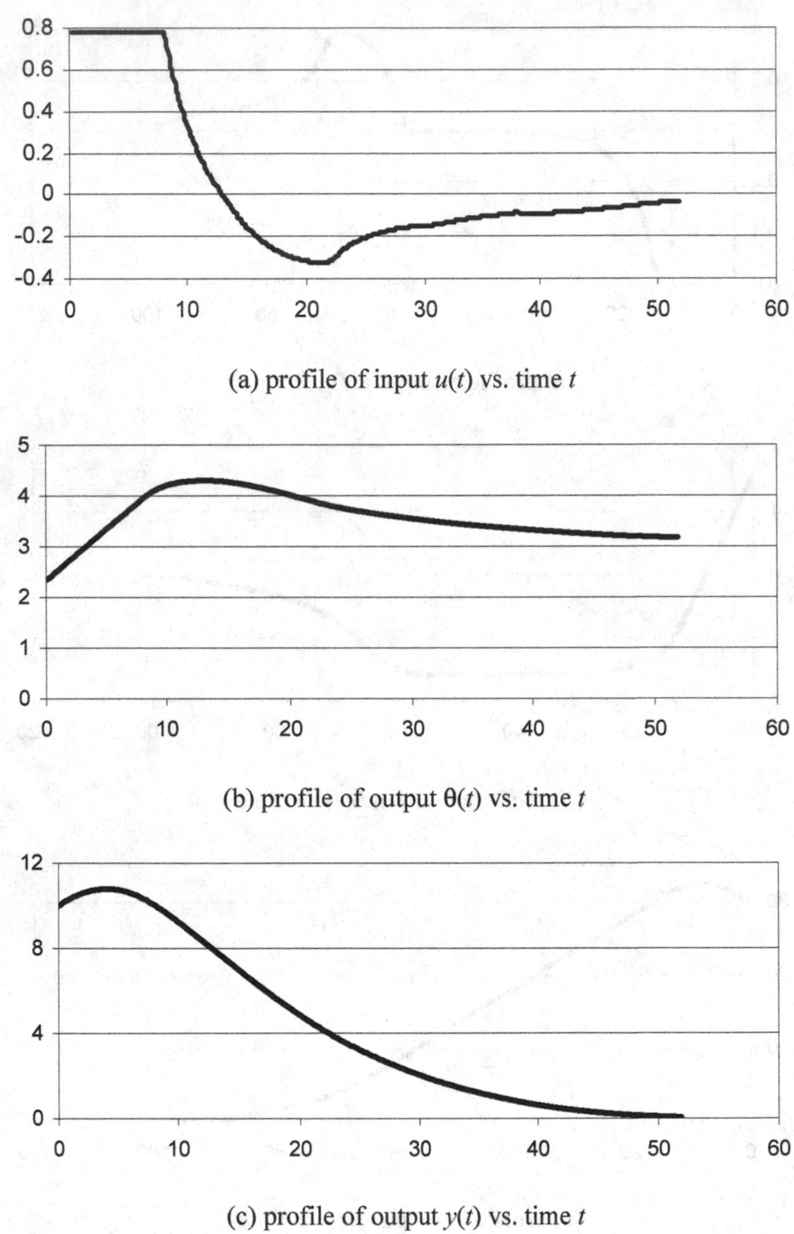

(a) profile of input $u(t)$ vs. time t

(b) profile of output $\theta(t)$ vs. time t

(c) profile of output $y(t)$ vs. time t

Figure 4.15 Model-free parking control simulation results (continued).

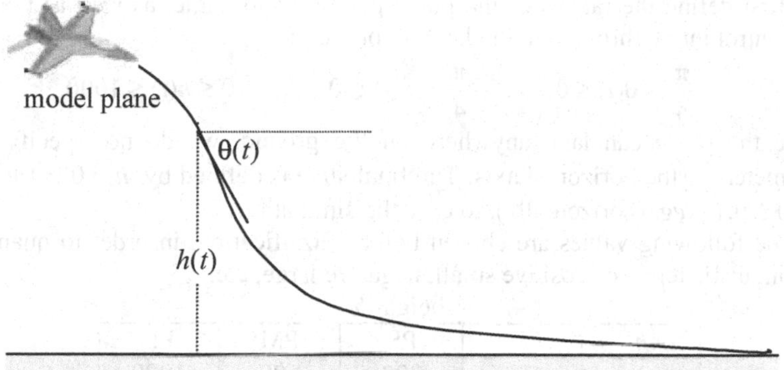

Figure 4.16 A model airplane landing control problem.

$\theta(k)$	—	angle of the truck at time k,
$x(k)$	—	horizontal position of the truck rear-end at time k,
$y(k)$	—	vertical position of the truck rear-end at time k,
$u(k)$	—	steering angle as control input to the truck at time k,
L	—	length of the truck (L=2.5m),
T	—	sampling time of the discrete model (T=0.1s),
v	—	constant speed of the truck (v=0.5m/s).

Example 4.2. In this example, we show a landing control problem of a model plane from a model-free approach. This airplane landing problem is visualized by Figure 4.16, which differs from the truck-parking control problem of Example 4.1 in that the airplane is not allowed to "overshoot" the set-point (i.e., cannot hit the ground) in order to avoid crashing. Therefore, the control rule base will not have symmetry in this example.

The objective is to design a fuzzy logic controller to land the plane safely, anywhere on the ground (the x-axis). Assume that no mathematical model for the plane is available for the design, but the plane is equipped with sensors that can measure the height $h(t)$ and the angle θ of motion of the plane. Suppose that the model plane is moving forward at a constant speed of v=5 m/s, and the controller is used to steer the angle of the plane. Let the initial conditions be $h(0)$=1000m and θ (0)=0 deg. (horizontally). The fuzzy logic controller is used to provide an input, u, which controls the angle of the plane, so as to guide it to land on the ground safely.

The parameters involved in the control system are as follows:

θ(t)	—	angle of the plane at time t,
$h(t)$	—	vertical position of the plane at time t,
$u(t)$	—	control input to the plane at time t,
T	—	sampling time for discretization (1s),
v	—	constant speed of the plane (5m/s).

We first define the ranges of the plane position and angle, as well as that of the control input, throughout the landing process:

$$-\frac{\pi}{4} \leq \theta(t) \leq 0, \qquad -\frac{\pi}{4} \leq u(t) \leq 0, \qquad 0 \leq h(t) \leq 1000.$$

Since the plane can land anywhere on the ground, we do not specify the parameter for the horizontal axis. The final state is defined by $h = 0 \pm 1m$ and $\theta = 0 \pm 0.1$ *deg.* (horizontally), to ease the simulation.

The following values are chosen before fuzzification, in order to quantify the linguistic terms of positive small, negative large, etc.

height *h*

h	Z	PS	PM	PL
value	0	300	600	1000

Angle θ

θ	NL	NM	NS	Z
value	$-\pi/4$	$-\pi/6$	$-\pi/8$	0

The membership functions for the height *h* and the angle θ are chosen as shown in Figures 4.17 and 4.18, respectively.

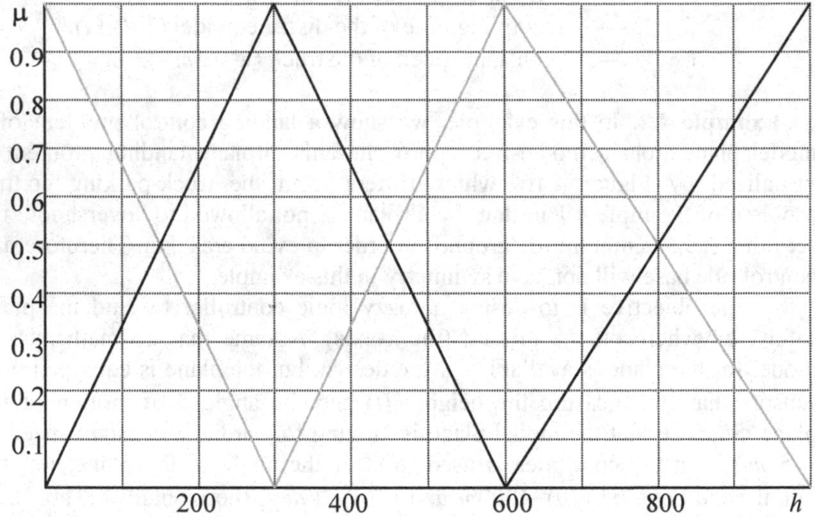

Figure 4.17 Fuzzy membership function for the height (h).

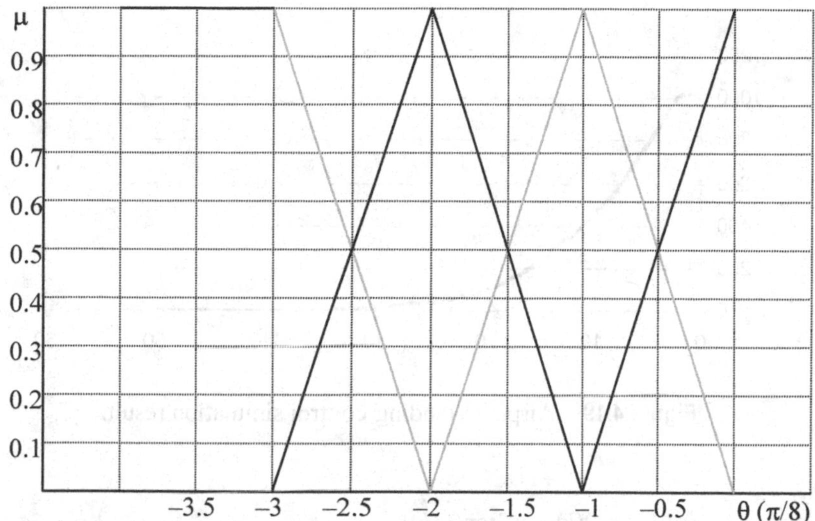

Figure 4.18 Fuzzy membership function for the angle (θ).

The fuzzy rule base used in the simulation is shown in Table 4.8, where V means "very." It is also in the form of IF h is H AND θ is Θ THEN u is U.

Table 4.8 Fuzzy Control Rule Base

θ / h	NL	NM	NS	Z
Z	PMS	PS	PVS	Z
PS	PS	PVS	Z	NVS
PM	PVS	Z	NVS	NS
PL	Z	NVS	NS	NMS

The basic idea in constructing this simple control rule base is as follows: First, if the plane is far away from the ground, we increase the degree of the plane angle (downward). Second, as the plane approaches the ground, the desired plane angle will gradually turn from downward to horizontal, so that it can land smoothly and safely.

The defuzzfication formula used is, again, the weighted average of all control actions, where the weights are the corresponding membership values:

$$u = \frac{\sum_{i=1}^{n} \mu(u_i) u_i}{\sum_{i=1}^{n} \mu(u_i)}.$$

For the purpose of computer simulation, the following simplified mathematical model of the plane was used to create data for control (but it was not used in the controller design):

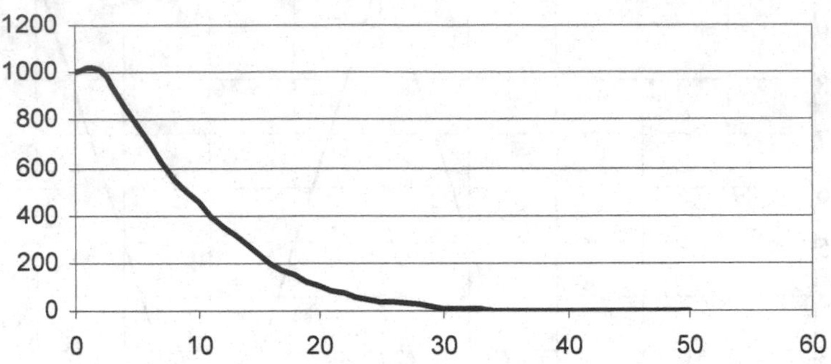

Figure 4.19 Airplane-landing control simulation result.

$$\theta(k+1) = \theta(k) + vT\tan(u(k))$$
$$y(k+1) = y(k) + vT\sin(u(k)). \quad (4.12)$$

A simulation result is shown in Figure 4.19, which demonstrates the safe landing process of the plane under the designed fuzzy logic controller.

III. FUZZY LOGIC CONTROL (II): A GENERAL MODEL-BASED APPROACH

In the last section, we have seen that fuzzy logic controllers can be designed and work even without any knowledge about the structure of the system (plant, process) for set-point tracking problems, if the system output can be measured and used on-line. The main idea was to use the error signal (the difference between the reference signal and the system output) to drive the fuzzy logic controller, so as to create a new control action each time. These control actions should be able to alter the system outputs in such a way that the error signal is reduced until satisfactory set-point tracking performance is achieved.

If the mathematical model, or a fairly good approximation of it, is available, one may expect to be able to design a better fuzzy logic controller with performance specifications and guaranteed stability. In this section, we discuss another general approach of fuzzy logic control that uses an approximate mathematical model of the given system (plant, process).

In the last section, we used the set-point tracking control problem as a prototype for illustrating the basic ideas and methods. To facilitate our discussion in this section, we use the truck-driving example discussed in Example 4.1 above as a platform.

Consider the simplified truck-driving control problem described in Example 4.1. Suppose that the truck is in the initial position A and we want to drive it to the target position B as shown in Figure 4.11. In order not to get the actuator-dynamics involved in this study of basic design principles, we again assume that the truck is driven by the actuator in a constant speed v while the control action is applied only to the steering angle of the front wheelers of the truck. Thus, a very simple mathematical model for the motion of the truck can be obtained from the geometry as shown in (4.11), which was used for simulation purposes before but is employed for controller design here. The model to be used is

$$\begin{aligned} \theta(k+1) &= \theta(k) + v\,T\tan(u(k))\,/\,L, \\ x(k+1) &= x(k) + v\,T\cos(\theta(k)), \\ y(k+1) &= y(k) + v\,T\sin(\theta(k)), \end{aligned} \tag{4.13}$$

with notation as defined in (4.11).

We note that this mathematical model, as it stands, is very simple and brief for describing the motion of the truck. Nevertheless, our design of a fuzzy logic controller $u(k)$ that controls the steering angle will be based on this approximate model.

We first discuss the general approach to this control problem. To begin, we convert this mathematical model to a fuzzy model described by the following IF-THEN rules (see Section II.B of Chapter 3):

$$\begin{aligned} R_S^i:\ & \text{IF } \theta(k) \text{ is } \Theta_i \text{ AND } x(k) \text{ is } X_i \text{ AND } y(k) \text{ is } Y_i \\ & \text{THEN } \boldsymbol{x}(k+1) = \boldsymbol{f}_i(\boldsymbol{x}(k)) + \boldsymbol{b}_i(u(k)), \quad i=1,\ldots,N, \end{aligned} \tag{4.14}$$

where

$$\boldsymbol{x}(k) = [\ \theta(k)\ \ x(k)\ \ y(k)\ ]^{T},$$

$$\boldsymbol{f}_i(\boldsymbol{x}(k)) = \begin{bmatrix} \theta(k) \\ x(k)+vT\cos(\theta(k)) \\ y(k)+vT\sin(\theta(k)) \end{bmatrix},$$

$$\boldsymbol{b}_i(u(k)) = \begin{bmatrix} vT\tan(u(k))\,/\,L \\ 0 \\ 0 \end{bmatrix}.$$

We then calculate the new state vector by the following standard weighted average formula:

$$\boldsymbol{x}(k+1) = \frac{\sum_{i=1}^{N} w_i(k)\boldsymbol{x}_i(k+1)}{\sum_{i=1}^{N} w_i(k)}, \tag{4.15}$$

where

$$w_i(k) = \min\{\ \mu_{\theta,i}(\theta(k)),\ \mu_{x,i}(x(k)),\ \mu_{y,i}(y(k))\ \},$$

$$w_i(k) \geq 0, \qquad \sum_{i=1}^{N} w_i(k) > 0,$$

in which $\mu_{\theta,i}(\cdot)$, $\mu_{x,i}(\cdot)$, $\mu_{y,i}(\cdot)$ are membership functions defined on the fuzzy subsets Θ, X, Y, and $\mu_{\theta,i}$, $\mu_{x,i}$, $\mu_{y,i}$ are the corresponding membership values of $\theta(k)$, $x(k)$, $y(k)$ in the IF-THEN rule R^i, $i = 1,\ldots, N$, respectively. Example 4.1 discussed above shows a concrete case of this description.

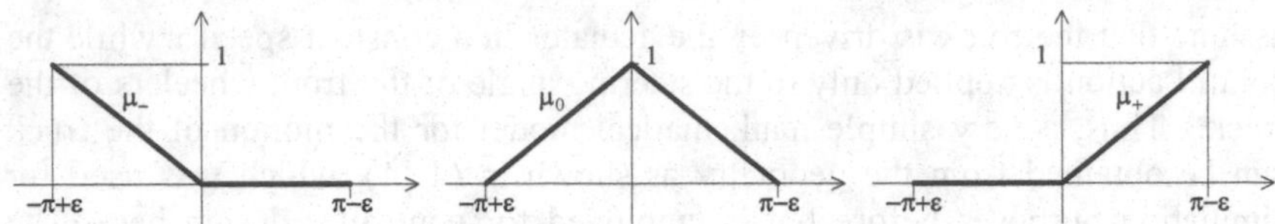

Figure 4.20 Membership functions for truck model linearization.

We now design a state-feedback fuzzy logic controller for guiding the steering angle of the front wheelers of the truck. This controller is governed by the following fuzzy IF-THEN rules:

$$R_C^i: \quad \text{IF } \theta(k) \text{ is } \Theta_i \text{ AND } x(k) \text{ is } X_i \text{ AND } y(k) \text{ is } Y_i$$
$$\text{THEN } u_i(k) = \boldsymbol{K}_i\, \boldsymbol{x}(k), \qquad i=1,\ldots,N, \tag{4.16}$$

where $\{\boldsymbol{K}_i\}$ are constant matrices (control gains) to be determined in the design, with the averaged control action given by

$$u(k) = \frac{\sum_{i=1}^{N} w_i(k)\boldsymbol{K}_i \boldsymbol{x}_i(k+1)}{\sum_{i=1}^{N} w_i(k)}, \tag{4.17}$$

where, for simplicity, we use the same weights as those obtained in the system model above. Of course, it is possible to use different weights for the controller to make it more effective in a real design.

Since this is a design, we may take different approaches. Here, to take a simple approach, we continue our design by further simplifying the models in two aspects:

(i) We only study the simple case of horizontal parking of the truck, to position B of Figure 4.11, where the final horizontal position, $x(k)$, is not specified. Hence, we can simply delete the second equation of (4.13) without altering the control effect on the process.

(ii) We linearize the rest of the mathematical model as follows:

$$\begin{aligned} \theta(k+1) &= \theta(k) + \frac{vT}{L}u(k), \\ y(k+1) &= y(k) + vT\theta(k), \end{aligned} \tag{4.18}$$

where we note that

$$-\theta(k) \le \sin(\theta(k)) \le \theta(k), \qquad -\pi \le \theta(k) \le \pi,$$

and

$$\begin{aligned} \theta(k) &\to 0 \qquad \Rightarrow \qquad \sin(\theta(k)) \to 0, \\ \theta(k) &\to \pm\pi \qquad \Rightarrow \qquad \sin(\theta(k)) \to 0. \end{aligned}$$

To avoid the singular points $\pm\pi$ at which the linearization is invalid, we only consider the range

$$-\pi + \varepsilon \le \theta(k) \le \pi - \varepsilon$$

for a small $\varepsilon > 0$.

Under these two simplications, we obtain the following two linearized models for the truck-driving control problem:

S_1: (when $\theta(k) \to 0$)

$$\begin{bmatrix} \theta(k+1) \\ y(k+1) \end{bmatrix} = \begin{bmatrix} 1 & 0 \\ vT & 1 \end{bmatrix} \begin{bmatrix} \theta(k) \\ y(k) \end{bmatrix} + \begin{bmatrix} vT/L \\ 0 \end{bmatrix} u_1(k);$$

S_2: (when $\theta(k) \to \pm\pi\mp\varepsilon$)

$$\begin{bmatrix} \theta(k+1) \\ y(k+1) \end{bmatrix} = \begin{bmatrix} 1 & 0 \\ \varepsilon vT & 1 \end{bmatrix} \begin{bmatrix} \theta(k) \\ y(k) \end{bmatrix} + \begin{bmatrix} vT/L \\ 0 \end{bmatrix} u_2(k).$$

Note that in these two linearized models, the data set $\{y(k)\}$ is actually not used (they are zeroed out by the coefficient matrix) for the changes of the angle $\theta(k)$. This is reasonable since the truck is moving to the horizontal direction, $\theta(k) \to 0$, independently of the vertical position of the truck.

Converting these mathematical models to fuzzy ones, we have

$\widetilde{R}_S^1$: IF $\theta(k)$ is approximately 0
THEN $\boldsymbol{x}_1(k+1) = \boldsymbol{A}_1\,\boldsymbol{x}(k) + \boldsymbol{b}_1\,u(k)$;

$\widetilde{R}_S^2$: IF $\theta(k)$ is approximately $\pi-\varepsilon$ OR $\theta(k)$ is approximately $-\pi+\varepsilon$
THEN $\boldsymbol{x}_2(k+1) = \boldsymbol{A}_2\,\boldsymbol{x}(k) + \boldsymbol{b}_2\,u(k)$,

where $\boldsymbol{x}(k) = [\,\theta(k)\ \ y(k)\,]^T$, and

$$\boldsymbol{A}_1 = \begin{bmatrix} 1 & 0 \\ vT & 1 \end{bmatrix}, \quad \boldsymbol{A}_2 = \begin{bmatrix} 1 & 0 \\ \varepsilon vT & 1 \end{bmatrix}, \quad \boldsymbol{b}_1 = \boldsymbol{b}_2 = \begin{bmatrix} vT/L \\ 0 \end{bmatrix},$$

with the membership functions for linearization defined by μ_-, μ_0, and μ_+ in Figure 4.20. Note that $\widetilde{R}_S^2$ is equivalent to the following two standard IF-THEN rules:

$\widetilde{R}_S^{21}$: IF $\theta(k)$ is approximately $\pi-\varepsilon$
THEN $\boldsymbol{x}_{21}(k+1) = \boldsymbol{A}_2\,\boldsymbol{x}(k) + \boldsymbol{b}_2\,u_{21}(k)$;

$\widetilde{R}_S^{22}$: IF $\theta(k)$ is approximately $-\pi+\varepsilon$
THEN $\boldsymbol{x}_{22}(k+1) = \boldsymbol{A}_2\,\boldsymbol{x}(k) + \boldsymbol{b}_2\,u_{22}(k)$.

In the above rules, $\widetilde{R}_S^1$, $\widetilde{R}_S^{21}$, and $\widetilde{R}_S^{22}$ have the associated membership functions μ_-, μ_0, and μ_+, respectively. Hence, the averaged state vector is given by

$$\boldsymbol{x}(k+1) = \frac{w_1(k)\boldsymbol{x}_1(k+1) + w_2(k)\boldsymbol{x}_{21}(k+1) + w_3(k)\boldsymbol{x}_{22}(k+1)}{w_1(k)+w_2(k)+w_3(k)}, \tag{4.19}$$

where $w_1(k) = \mu_0(\theta(k))$, $w_2(k) = \mu_+(\theta(k))$, and $w_3(k) = \mu_-(\theta(k))$.

The fuzzy logic controller for this system is described by the following two fuzzy IF-THEN rules:

$\widetilde{R}_C^1$: IF $\theta(k)$ is approximately 0
THEN $u(k) = \boldsymbol{K}_1\,\boldsymbol{x}(k) = [\,K_{11}\ \ K_{12}\,]\,\boldsymbol{x}(k)$;

$\widetilde{R}_C^2$: IF $\theta(k)$ is approximately $\pi-\varepsilon$ OR $\theta(k)$ is approximately $-\pi+\varepsilon$
THEN $u(k) = \boldsymbol{K}_2\,\boldsymbol{x}(k) = [\,K_{21}\ \ K_{22}\,]\,\boldsymbol{x}(k)$,

where the constant gains, K_{11}, K_{12}, K_{21}, and K_{22}, are to be determined in the design for the control task: driving the truck from position A to the horizontal position B as shown in Figure 4.11.

Similar to the system rules, the second control rule is equivalent to the following two standard IF-THEN rules:

$\widetilde{R}_C^{21}$: IF $\theta(k)$ is approximately $\pi-\varepsilon$
THEN $u_{21}(k) = \boldsymbol{K}_2\, \boldsymbol{x}(k) = [\, K_{21} \;\; K_{22}\,]\, \boldsymbol{x}(k)$;

$\widetilde{R}_C^{22}$: IF $\theta(k)$ is approximately $-\pi+\varepsilon$
THEN $u_{22}(k) = \boldsymbol{K}_2\, \boldsymbol{x}(k) = [\, K_{21} \;\; K_{22}\,]\, \boldsymbol{x}(k)$,

and the averaged control is given by

$$u(k) \;=\; \frac{w_1(k)u_1(k)+w_2(k)u_{21}(k)+w_3(k)u_{22}(k)}{w_1(k)+w_2(k)+w_3(k)}, \tag{4.20}$$

where, as mentioned above, we use the same weights $w_1(k)$, $w_2(k)$, and $w_3(k)$ as those used in the system rules for simplicity of discussion.

Observe that our control task in this example is to drive the truck to position B where $\theta(k^*) = 0$, $x(k^*)$ is not specified ("don't care"), and $y(k^*) = 0$, for all large values of k^*. This is guaranteed if we can design the fuzzy controller $u(k)$ such that the equilibrium point $\boldsymbol{x}^* = \boldsymbol{0}$ of the linear controlled system (described by $\widetilde{R}_S^i$ and $\widetilde{R}_C^i$, $i = 1, 2$) becomes asymptotically stable. If so, then

$$\theta(k) \to 0, \qquad y(k) \to 0 \qquad \text{as } k \to \infty.$$

Therefore, we next complete the design by determining the constant control gains, $\boldsymbol{K}_1 = [K_{11} \;\; K_{12}]$ and $\boldsymbol{K}_2 = [K_{21} \;\; K_{22}]$, such that the feedback controlled system is asymptotically stable about its equilibrium point $\boldsymbol{x}^* = [\theta^* \;\; y^*]^{\mathrm{T}} = \boldsymbol{0}$.

By substituting the control rules $\widetilde{R}_C^1$, $\widetilde{R}_C^{21}$, $\widetilde{R}_C^{22}$ into $\widetilde{R}_S^1$, $\widetilde{R}_S^2$, and then using the averaged state vector (4.19), in which they have the same weights, we obtain

$$\boldsymbol{x}(k+1) = \frac{w_1(k)\boldsymbol{H}_1 + w_2(k)\boldsymbol{H}_2 + w_3(k)\boldsymbol{H}_3}{w_1(k)+w_2(k)+w_3(k)}\, \boldsymbol{x}(k), \tag{4.21}$$

where

$$\boldsymbol{H}_1 \;=\; \boldsymbol{A}_1 + \boldsymbol{b}_1\, \boldsymbol{K}_1 = \begin{bmatrix} 1+vTK_{11}/L & vTK_{12}/L \\ vT & 1 \end{bmatrix},$$

$$\boldsymbol{H}_2 \;=\; \frac{1}{2}\,[(\boldsymbol{A}_1 + \boldsymbol{b}_1\, \boldsymbol{K}_1) + (\boldsymbol{A}_2 + \boldsymbol{b}_2\, \boldsymbol{K}_2)]$$

$$\;=\; \begin{bmatrix} 1+vT(K_{11}+K_{21})/2L & vT(K_{12}+K_{22})/2L \\ (1+\varepsilon)vT/2 & 1 \end{bmatrix},$$

$$\boldsymbol{H}_3 \;= \boldsymbol{A}_2 + \boldsymbol{b}_2\, \boldsymbol{K}_2 = \begin{bmatrix} 1+vTK_{21}/L & vTK_{22}/L \\ \varepsilon vT & 1 \end{bmatrix}.$$

It then follows from Theorem 3.3 of Chapter 3 that if we can choose K_{11}, K_{12}, K_{21}, and K_{22} such that there is a common positive definite matrix $\boldsymbol{P}$ satisfying

$$\boldsymbol{H}_i^{\mathrm{T}}\, \boldsymbol{P}\boldsymbol{H}_i - \boldsymbol{P} < 0, \qquad i = 1, 2, 3,$$

Table 4.9 Different Initial Positions of the Truck

	θ(0) [deg.]	y(0) [m]
1	0	30
2	0	20
3	0	10
4	0	–10
5	0	–20
6	0	–30
7	90	30
8	90	20
9	90	10
10	90	–10
11	90	–20
12	90	–30
13	180	30
14	180	20
15	180	10
16	180	–10
17	180	–20
18	180	–30
19	–90	30
20	–90	20
21	–90	10
22	–90	–10
23	–90	–20
24	–90	–30

then the feedback controlled system (4.21) will be asymptotically stable about its equilibrium point $\boldsymbol{x}^* = \boldsymbol{0}$.

As a numerical example, let

$$L = 2.8 \text{ m}, \qquad T = 1.0 \text{ sec.},$$
$$v = 1.0 \text{ m/sec.}, \qquad \varepsilon = 0.01\pi,$$
$$K_{11} = -0.4212, \qquad K_{12} = -0.02944,$$
$$K_{21} = -0.0991, \qquad K_{22} = -0.00967.$$

It can be verified that

$$\boldsymbol{P} = \begin{bmatrix} 989.0 & 75.25 \\ 75.25 & 26.29 \end{bmatrix} > 0,$$

which yields

$$\boldsymbol{H}_1^{\mathrm{T}} \boldsymbol{P} \boldsymbol{H}_1 - \boldsymbol{P} = \begin{bmatrix} -120.4 & 6.008 \\ 6.008 & -1.468 \end{bmatrix} < 0,$$

$$\boldsymbol{H}_2^{\mathrm{T}} \boldsymbol{P} \boldsymbol{H}_2 - \boldsymbol{P} = \begin{bmatrix} -68.31 & -5.873 \\ -5.873 & -0.5079 \end{bmatrix} < 0,$$

$$H_3^{\mathrm{T}} P H_3 - P = \begin{bmatrix} -100.0 & -0.000114 \\ -0.000114 & -1.0 \end{bmatrix} < 0.$$

The computer simulation results of this fuzzy control, for the linearized truck-driving system, with the target angle θ=0, and with 24 different initial positions (the position A in Figure 4.11) listed in Table 4.9, are shown in Figures 4.21 (a)-(d).

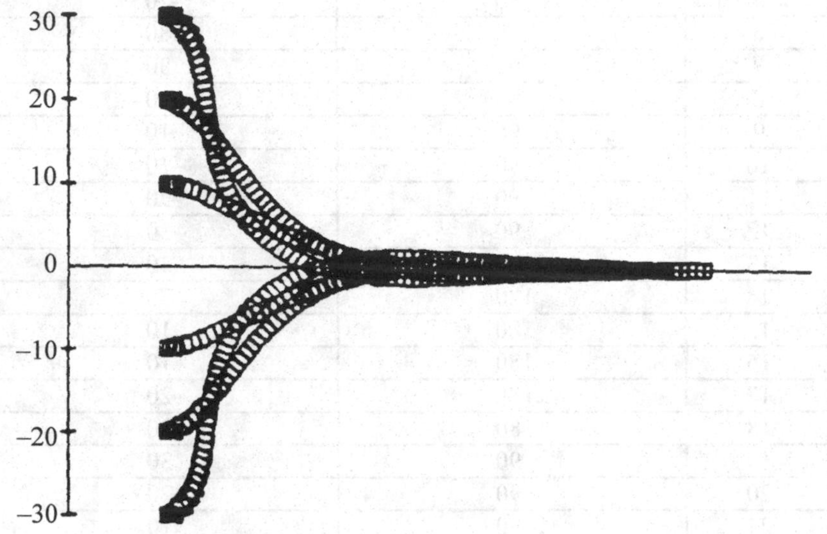

Figure 4.21 (a) Control simulation results for Cases 1 - 6.

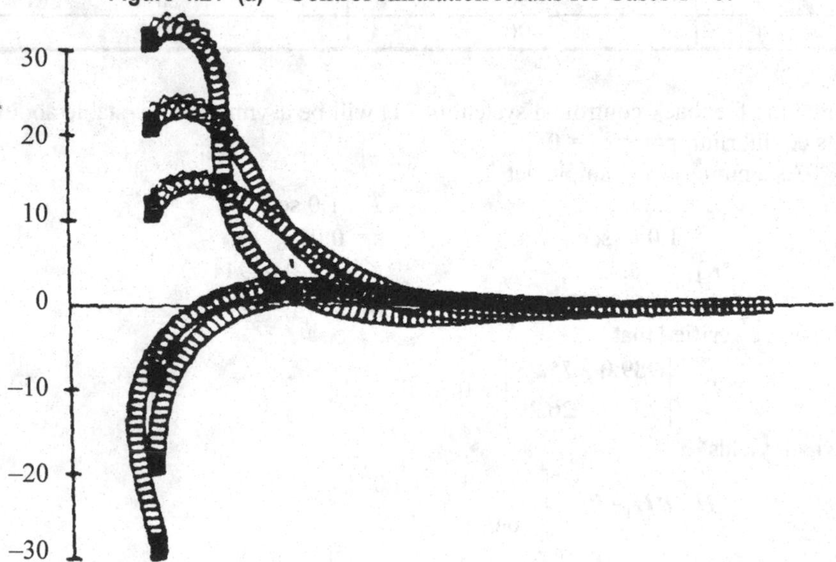

Figure 4.21 (b) Control simulation results for Cases 7 - 12.

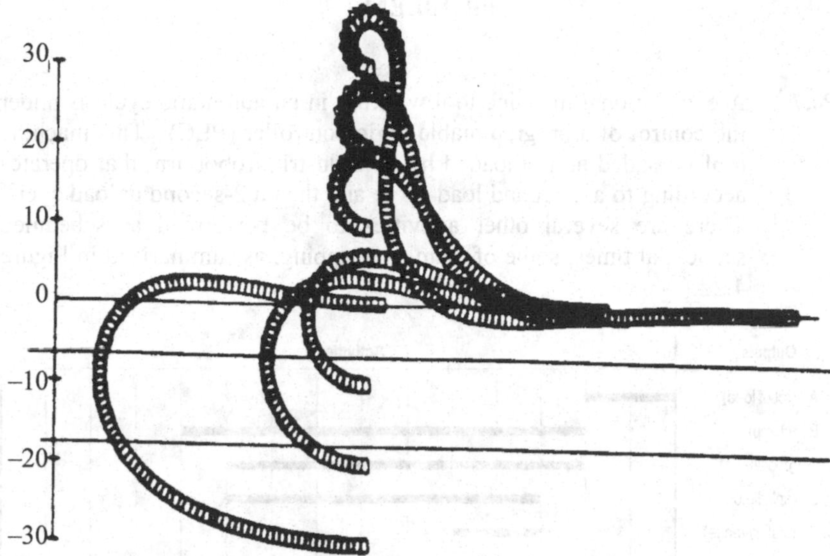

Figure 4.21 (c) Control simulation results for Cases 13 - 18.

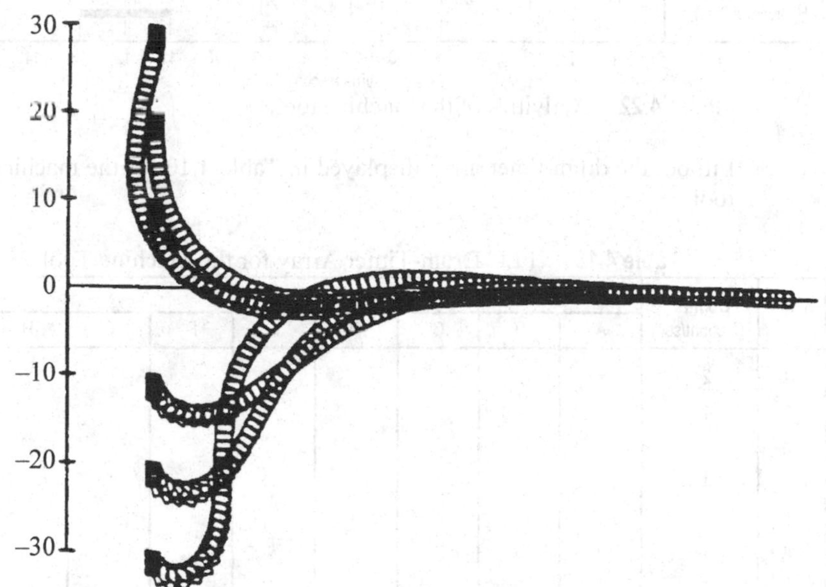

Figure 4.21 (d) Control simulation results for Cases 19 - 24.

PROBLEMS

P4.1 A conventional machine tool working in an automatic cycle is under the control of a programmable logic controller (PLC). This machine tool is loaded and unloaded by an industrial robot arm that operates according to a 2-second load cycle and then a 2-second unload cycle. There are several other activities to be performed at scheduled sequential times, some of them overlapping, as summarized in Figure 4.22.

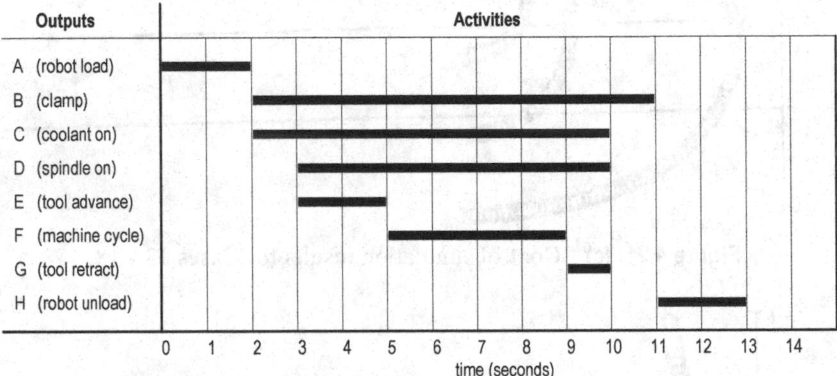

Figure 4.22 Activities of the machine tool.

Fill out the drum-timer array displayed in Table 4.10 for the machine tool.

Table 4.10 PLC Drum-Timer Array for the Machine Tool

Steps	Counts (1 count/sec)	Outputs							
		A	B	C	D	E	F	G	H
1	2								
2	1								
3	2								
4	4								
5	1								
6	1								
7	2								

P4.2 Consider the ladder logic diagram shown in Figure 4.23. Show what the following logical operations in this diagram perform (in Boolean algebraic notation):

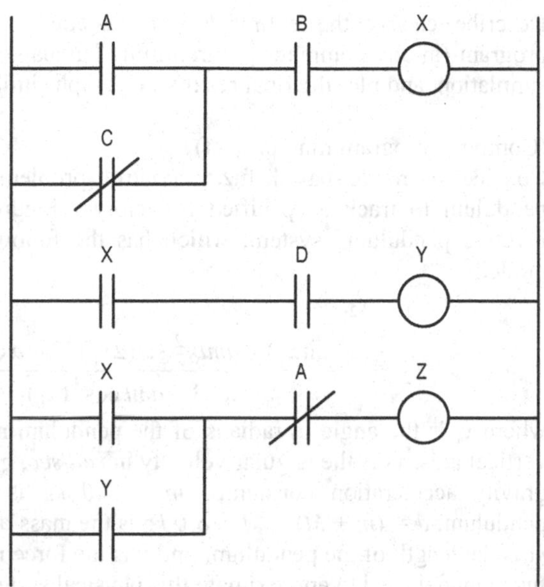

Figure 4.23 A ladder logic diagram.

P4-3 (Computer programming project)
Consider a unitless set-point tracking system shown in Figure 4.5, where the set-point $r = 2.0$; the plant is given by the nonlinear model

$$y((k+1)T) = \frac{T}{2}[y(kT)]^2 + u(kT), \tag{4.22}$$

where $T = 0.1$ is the sampling time, and $u(kT)$ is the control input to the plant created by the fuzzy logic controller shown in Figure 4.6. Follow the general design principle discussed in Section IV.B to design a fuzzy logic controller for this set-point tracking task (see Figure 4.8).

In the design, use the four membership functions shown in Figure 4.9 for both the error signal $e(kT)$ and the change of error $\dot{e}(kT)$, and the membership functions shown in Figure 4.10 for the control $u(kT)$ and in the fuzzy logic control rule base. The common real constant $H > 0$ in the membership functions in Figures 4.9 and 4.10 can be used as a tuning parameter for improving the tracking performance when programming the control. Use the following initial conditions:

$$y(0) = 0, \qquad e(0) = r - y(0) = 2.0,$$

and

$$\dot{e}(0) = \frac{1}{T}[e(0) - e(-T)] = \frac{1}{0.1}[2.0 - 2.0] = 0.$$

Describe clearly the entire design procedure, write a computer program (in any computer programming language) for the described simulation, and plot the final results in a graph similar to Figure 4.8.

P4-4 (Computer programming project)
Consider a model-based fuzzy control problem for an inverted pendulum to track a specified trajectory. Figure 4.24 shows the inverted pendulum system, which has the following mathematical model:

$$\dot{x}_1 = x_2,$$

$$\dot{x}_2 = \frac{g\sin(x_1) - amlx_2^2\sin(2x_1)/2 - a\cos(x_1)u}{4l/3 - aml\cos^2(x_1)}, \quad (4.23)$$

where x_1 is the angle in radians of the pendulum measured from the vertical axis, x_2 is the angular velocity in *rad/sec*, g =9.8 *m/sec*^{2} is the gravity acceleration constant, m = *2.0 kg* is the mass of the pendulum, $a = (m + M)^{-1}$, M = *8.0 kg* is the mass of the cart, $2l$ = 1.0 *m* is the length of the pendulum, and u is the force input to the cart. A fuzzy model used to approximate this physical system is as follows:

R^1: IF x_1 is about 0, THEN $\dot{x} = A_1x + B_1u$
R^2: IF x_1 is about $\pm\pi/2$, THEN $\dot{x} = A_1x + B_1u$,

where

$$A_1 = \begin{bmatrix} 0 & 1 \\ \dfrac{g}{4l/3 - aml} & 0 \end{bmatrix}, \qquad B_1 = \begin{bmatrix} 0 \\ -\dfrac{a}{4l/3 - aml} \end{bmatrix},$$

$$A_2 = \begin{bmatrix} 0 & 1 \\ \dfrac{2g}{\pi(4l/3 - aml\beta^2)} & 0 \end{bmatrix}, \qquad B_2 = \begin{bmatrix} 0 \\ -\dfrac{a\beta}{4l/3 - aml\beta^2} \end{bmatrix}.$$

The membership functions for Rules 1 and 2 are shown in Figure 4.25. Design a model-based fuzzy logic controller of the form $u_i = K_ix_i$ to stabilize this inverted pendulum system such that $x_1 \to 0$ and $x_2 \to 0$.
[Hint: First, discretize the model, and then follow the design procedure discussed in Section III. One possible solution is

$$K_1 = [-120.7 \quad -22.7] \qquad K_2 = [-255.6 \quad -746.0].$$

But this solution is by no means unique.]

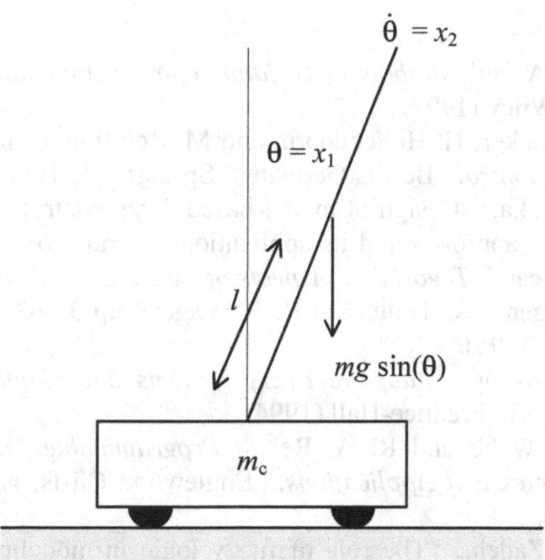

Figure 4.24 The inverted pendulum system.

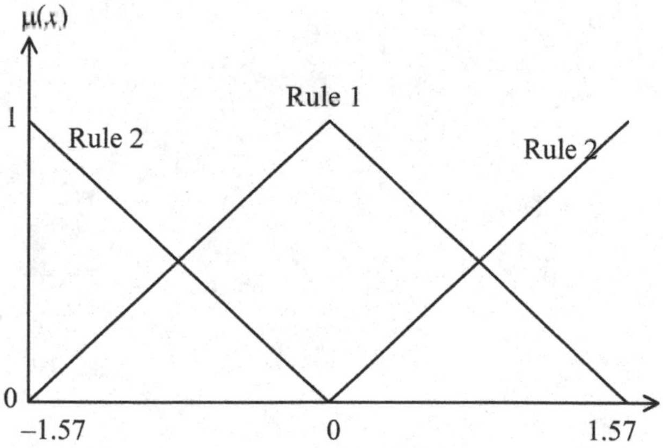

Figure 4.25 The membership functions for the inverted pendulum.

REFERENCES

[1] C. R. Asfahl. *Robots and Manufacturing Automation.* New York, NY: Wiley (1992).

[2] D. Drianker, H. Hellendoorn, and M. Reinfrank. *An Introduction to Fuzzy Control.* Berlin, Germany: Springer-Verlag (1993).

[3] K. Tanaka. "Design of model-based fuzzy control using Lyapunov's stability approach and its applications to trajectory stabilization of a model car." *Theoretical Aspects of Fuzzy Control* (ed. H. T. Nguyen, M. Sugeno, R. Tong, and R. R. Yager), pp 31-63, New York, NY: Wiley (1995).

[4] L. X. Wang. *Adaptive Fuzzy Systems and Control.* Englewood Cliffs, NJ: Prentice-Hall (1994).

[5] J. W. Webb and R. A. Reis. *Programmable Logic Controllers: Principles and Applications.* Englewood Cliffs, NJ: Prentice-Hall (1995).

[6] L. A. Zadeh. "The role of fuzzy logic in modeling, identification, and control." *Modeling, Identification, and Control*, Vol. 15, No. 3, pp 191-203 (1994).

CHAPTER 5

Fuzzy PID Controllers

Conventional (classical) proportional-integral-derivative (PID) controllers are perhaps the most well-known and most widely used controllers in modern industries: statistics has shown that more than 90% controllers used in industries todays are PID or PID-type of controllers.

Generally speaking, PID controllers have the merits of being simple, reliable, and effective: they consume lower cost but are very easy to operate. Besides, for lower-order linear systems (plants, processes), PID controllers have remarkable set-point tracking performance and guaranteed stability. Therefore, PID controllers are very popular in real-world applications.

In this chapter, we first give a brief review on the conventional PID controllers, and then introduce the design and stability analysis of a new type of fuzzy PID controller. We will also compare these two classes of PID controllers and discuss their advantages as well as limitations. It will be seen that the fuzzy PID controllers are generally superior to the conventional ones, particularly for higher-order, time-delayed, and nonlinear systems, and for those systems that have only vague mathematical models which the conventional PID controllers are difficult, if not impossible, to handle. For lower-order linear systems, it will be seen that both conventional and fuzzy PID controllers work equally well, so in these simple cases conventional PID controllers are recommended duc to their simple structures. The price to pay for the success of the fuzzy PID controllers is that their design methods are slightly more advanced and their resulting formulas are somewhat more complicated (e.g., containing variable control gains in contrast to the conventional PID controllers where the control gains are constant). It will be seen that although the fuzzy PID controllers are designed by fuzzy mathematics, their final form as controllers are conventional controllers. As such, they can be used to directly replace the conventional ones in applications. Moreover, their variable control gains are self-tuned formulas that have adaptive capability to handle time-delay effects, nonlinearities, and uncertainties of the given system. This chapter will make these points clear.

I. CONVENTIONAL PID CONTROLLERS

In this section, we provide a brief review on several basic types of conventional PID controllers: their configuration, design methods, and stability analysis, which will be needed in introducing the fuzzy PID controller later, where the fuzzy PID controllers are natural extensions of the conventional ones: they have the same structure but are defined based on fuzzy mathematics and fuzzy control strategies.

Individually, the conventional proportional (P), integral (I), and derivative (D) controllers for controlling a given system (plant, process) have the

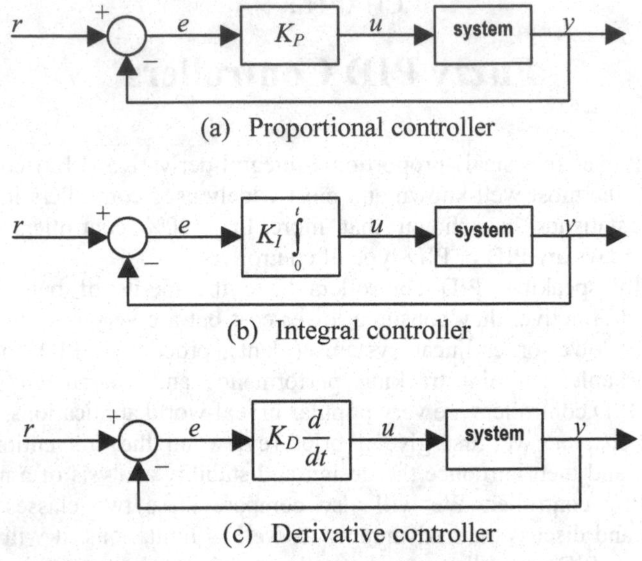

Figure 5.1. Proportional-integral-derivative controllers.

structures shown in Figure 5.1(a), (b), and (c), respectively, where $r = r(t)$ is the reference input (set-point), $y = y(t)$ is the controlled system's output, $e = e(t) := r(t) - y(t)$ is the set-point tracking error, and $u = u(t)$ is the control action (output of the controller) which is used as the input to the system.

In the time domain, they have the following solutions:

(i) The P-controller $u(t) = K_P\, e(t);$

(ii) The I-controller $u(t) = K_I \int_0^t e(\tau)\, d\tau;$

(iii) The D-controller $u(t) = K_D \frac{d}{dt} e(t).$

Here, the three control gains, K_P, K_I, and K_D, are constants to be determined in the design for the set-point tracking performance and stability consideration. To increase their control capabilities, they are usually used in combinations, which will be further discussed below.

In the frequency domain, we have the following corresponding relations for the individual P, I, and D controllers:

(i) The P-controller $U(s) = K_P\, E(s);$

(ii) The I-controller $U(s) = \frac{K_I}{s} E(s);$

(iii) The D-controller $U(s) = K_D\, s\, E(s).$

Here and throughout the chapter, we use capital letters for the Laplace transform $L\{x\}$ of a continuous-time signal, or the z-transform $Z\{x\}$ of a discrete-time signal. Thus,

$$U(s) = L\{\, u(t) \,\} \qquad \text{and} \qquad E(s) = L\{\, e(t) \,\},$$

with zero initial conditions.

The following are some basic, typical combinations of P, I, and D controllers (see Figure 5.2 for their configurations):

In the time domain:

(iv) The PI-controller $$u(t) = K_P\, e(t) + K_I \int_0^t e(\tau)\, d\tau;$$

(v) The PD-controller $$u(t) = K_P\, e(t) + K_D \frac{d}{dt} e(t);$$

(vi) The PID-controller $$u(t) = K_P\, e(t) + K_I \int_0^t e(\tau)\, d\tau + K_D \frac{d}{dt} e(t);$$

(vii) The PI+D-controller $$u(t) = K_P\, e(t) + K_I \int_0^t e(\tau)\, d\tau - K_D \frac{d}{dt} y(t).$$

Correspondingly, in the frequency domain:

(iv) The PI-controller $$U(s) = K_P\, E(s) + \frac{K_I}{s} E(s);$$

(v) The PD-controller $$U(s) = K_P\, E(s) + K_D\, s\, E(s);$$

(vi) The PID-controller $$U(s) = K_P\, E(s) + \frac{K_I}{s} E(s) + K_D\, s\, E(s);$$

(vii) The PI+D-controller $$U(s) = K_P\, E(s) + \frac{K_I}{s} E(s) - K_D\, s\, Y(s).$$

In the above,

$$Y(s) = L\{\, y(t) \,\} = L\{\, r(t) - e(t) \,\} = R(s) - E(s).$$

We note that the PID controller shown in Figure 5.2(c) is not a good combination of the three controllers in practice, since if the error signal $e(t)$ has discontinuities then the D-controller will produce very bad (even unbounded) responses. This combination is good only for illustration since it has a neat formulation, and hence is called a "textbook PID controller." A practical combination of the three controllers is the PI+D controller shown in Figure 5.2(d), where the system output signal y(t) is usually smoother than the error signal (through the I-controller and the system). More importantly, this structure of closed-loop control systems has been validated to be efficient by many practical examples and case studies.

To show how a PID-type of controller works and how to design such a controller to perform set-point tracking with guaranteed stability, we study two simple examples below.

Example 5.1. Consider the PI-control system shown in the frequency domain by Figure 5.3, where the given linear system has a first-order transfer function, $1/(as + b)$, with known constants $a > 0$ and b. The reference r is a given constant (set-point). The design of the PI controller is to determine the

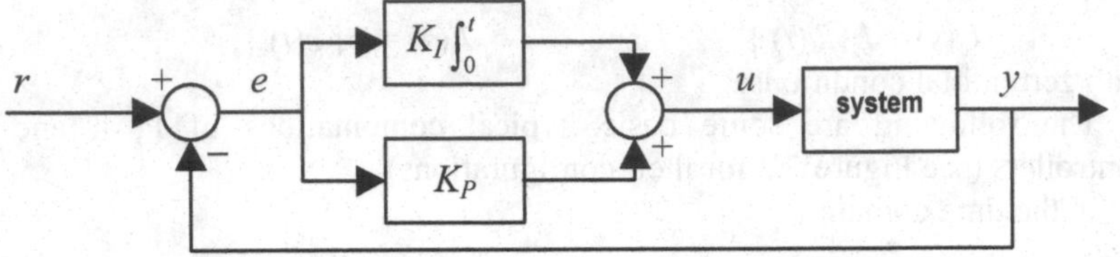

(a) PI controller

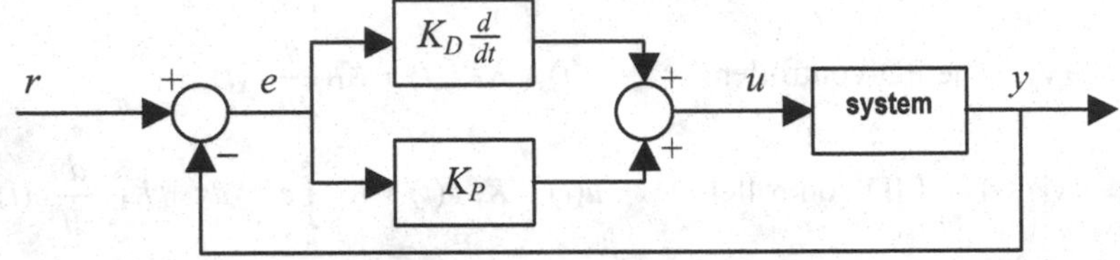

(b) PD controller

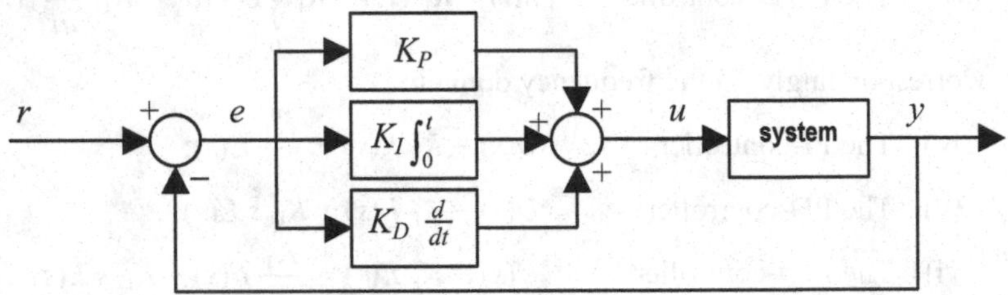

(c) PID controller

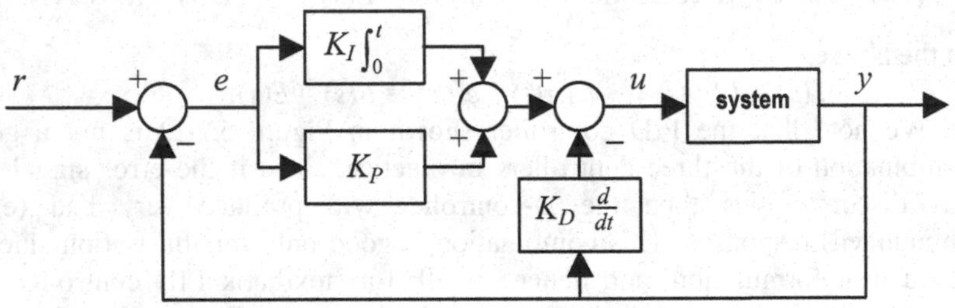

(d) PI+D controller

Figure 5.2 Some typical combinations of P, I, and D controllers.

two constant control gains, K_P and K_I, such that the system output $y(t)$ can track the reference:

$$y(t) \to r \qquad \text{as} \qquad t \to \infty,$$

while the entire feedback control system is stable even if the given transfer function $1/(as + b)$ is unstable.

We first observe from Figure 5.3 that

$$U(s) = K_P\, E(s) + \frac{K_I}{s} E(s),$$

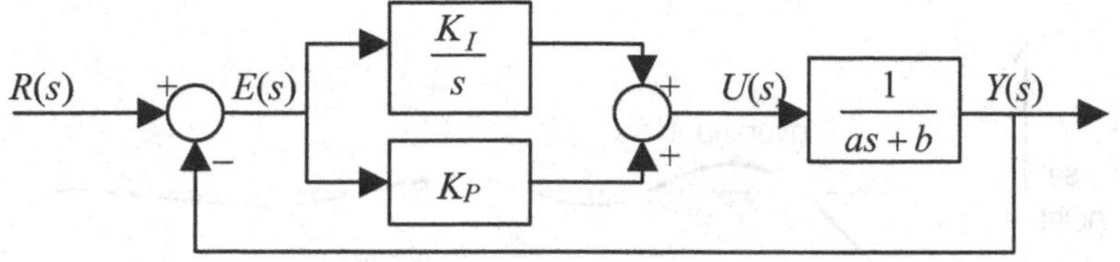

Figure 5.3 A PI control system.

$$E(s) = R(s) - Y(s),$$

$$Y(s) = \frac{1}{as+b} U(s).$$

By combining these relations together, we can obtain the following overall system input-output relation:

$$Y(s) = H(s)\,R(s) = \frac{K_P s + K_I}{as+b} R(s), \tag{5.1}$$

where $H(s)$ is the transfer function of the overall feedback control system. It is easy to see that this transfer function has two poles:

$$s_{1,2} = \frac{-(b+K_P) \pm \sqrt{(b+K_P)^2 - 4aK_I}}{2a}. \tag{5.2}$$

Thus, in our design, if we choose (note: $a > 0$)

$$K_P > -b, \tag{5.3}$$

then we can guarantee that these two poles have negative real parts, so that the overall controlled system is stable.

What we want for set-point tracking is

$$\lim_{t\to\infty} e(t) = 0. \tag{5.4}$$

It follows from the Terminal-Value Theorem of Laplace transforms that

$$\begin{aligned} \lim_{t\to\infty} e(t) &= \lim_{|s|\to 0} s\,E(s) \\ &= \lim_{|s|\to 0} s\,[\,R(s) - Y(s)\,] \\ &= \lim_{|s|\to 0} s\,[\,1 - H(s)\,]\,R(s) \\ &= \lim_{|s|\to 0} s \times \frac{as^2 + bs}{as^2 + (b+K_P)s + K_I} \times \frac{r}{s} \\ &= 0. \end{aligned}$$

It may appear that the set-point tracking task can always be done no matter how we choose the control gain K_I, provided that the other control gain, K_P, satisfies the condition (5.3), and this is true for any given constants a and b in the given system (plant). This observation is correct for this example. However, in so doing it often happens that the system output $y(t)$ tracks the set-point r with higher-frequency oscillations caused by the pure imaginary parts of the two poles given in (5.2). Hence, to eliminate such undesirable oscillations so as to obtain better tracking performance, we can select the

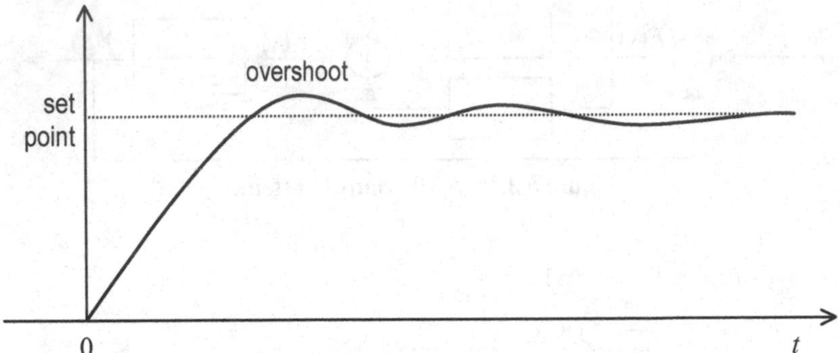

Figure 5.4 A typical tracking performance of the PI controller.

control gain K_I to zero out the pure imaginary parts of the poles s_1 and s_2. Namely, we can force

$$(b + K_P)^2 - 4\,a\,K_I = 0,$$

which yields

$$K_I = \frac{(b + K_P)^2}{4a},$$

where K_P has been determined by (5.3). Thus we complete the design of the PI controller for the given plant $1/(as + b)$, which can be originally unstable (i.e., $b < 0$). We thus obtain an overall stable feedback control system whose output can track the set-point without oscillations, at least in theory. Its output $y(t)$ generally has the shape as shown in Figure 5.4.

As seen, a PI controller so designed can completely eliminate both the steady-state tracking error and the transient oscillations, but may not be able to reduce the maximum overshoot in the output. On the contrary, a PD controller can generally improve the tracking performance by reducing the maximum overshoot of the output, but may not be able to eliminate the steady-state tracking error, as can be seen from the next example.

Example 5.2. Consider the PD-control system shown in the frequency domain by Figure 5.5, where the given linear system has a second-order transfer function, $1/(as^2 + bs + c)$, with known constants $a > 0$, b, and c. The design of the PD controller is again to determine the two constant control gains, K_P and K_D, such that the system output can track the set-point while the entire feedback control system is stable even if the given transfer function $1/(as^2 + bs + c)$ is unstable.

It follows from Figure 5.5 that

$$U(s) = K_P\,E(s) + K_D\,s\,E(s),$$

$$E(s) = R(s) - Y(s),$$

$$Y(s) = \frac{1}{as^2 + bs + c}\,U(s),$$

so that

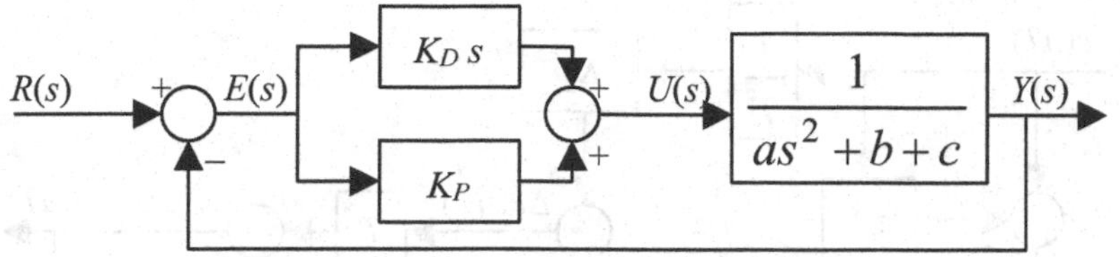

Figure 5.5 A PD control system.

$$Y(s) = H(s)\, R(s) = \frac{K_P + K_D s}{as^2 + (b + K_D)s + (c + K_P)}\, R(s).$$

The transfer function $H(s)$ of the overall feedback control system has two poles:

$$s_{1,2} = \frac{-(b+K_D) \pm \sqrt{(b+K_D)^2 - 4a(c+K_P)}}{2a}, \tag{5.5}$$

and so the selection (note $a > 0$)

$$K_D > -b \tag{5.6}$$

and

$$K_P = \frac{(b+K_D)^2}{4a} - c \tag{5.7}$$

can guarantee the controlled system be stable and have no oscillations on the output trajectory during the set-point tracking process. However, the asymptotic tracking error for this PD controlled system is

$$\begin{aligned}
\lim_{t\to\infty} e(t) &= \lim_{|s|\to 0} s\,E(s) \\
&= \lim_{|s|\to 0} s[I - H(s)]R(s) \\
&= \lim_{|s|\to 0} s \times \frac{as^2 + bs + c}{as^2 + (b+K_D)s + (c+K_P)} \times \frac{r}{s} \\
&= \frac{cr}{c+K_P},
\end{aligned}$$

which will not be zero if $c \neq 0$ and $r \neq 0$. This implies that the PD controller generally cannot eliminate the steady-state error in set-point tracking.

The PD controller, as compared to the PI controller, has its advantages: it can produce smaller maximum overshoot and is more sensitive (easier to tune) in general. A typical set-point tracking performance of the PD control is similar to that of the PI controller shown in Figure 5.4.

To implement a PI or PD controller on a computer, we need the digital version of the analog one discussed above. To digitize an analog controller, the following three discretization formulas can be used:

(i) The forward divided difference: $s = \dfrac{z-1}{T}$;

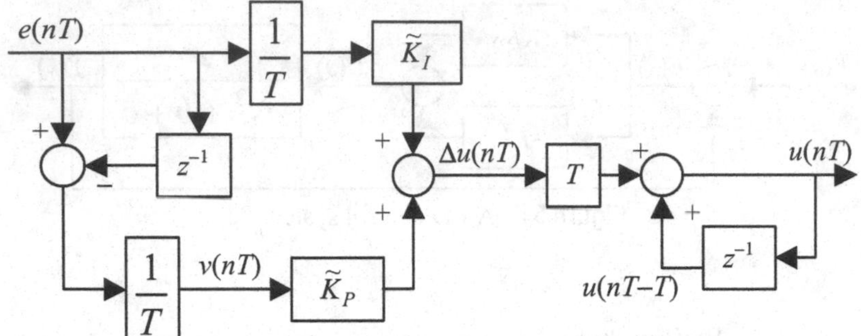

Figure 5.6 The digital PI controller.

(ii) The backward divided difference: $s = \dfrac{1-z^{-1}}{T}$;

(iii) The trapezoidal formula: $s = \dfrac{2}{T}\dfrac{z-1}{z+1}$,

where T > 0 is the sampling period. These formulas transform the analog controller and the controlled system from the continuous-time frequency domain (in the Laplace-transform s-variable) to the discrete-time frequency domain (in the *z*-transform *z*-variable). In particular, the last formula maps the entire left-half (right-half) *s*-plane into the entire inner (outer) unit circle in the *z*-plane in such a way that the mapping is one-to-one and the entire imaginary axis corresponds to the entire unit circle. Hence, this formula preserves all basic properties, particularly the stability, of the original controller and controlled system, and hence is the one used most frequently. This trapezoidal formula is also known as the *bilinear transform* in engineering and the *conformal mapping* in mathematics.

Using the bilinear transform

$$s = \frac{2}{T}\frac{z-1}{z+1} \tag{5.8}$$

in the continuous-time PI controller, we obtain (see Figure 5.3):

$$\frac{K_I}{s} \rightarrow \frac{K_I T}{2}\frac{z-1}{z+1} = \frac{K_I T}{2}\frac{2-(1-z^{-1})}{1-z^{-1}} = -\frac{K_I T}{2} + \frac{K_I T}{1-z^{-1}},$$

$$U(s) = (K_P + \frac{K_I}{s})\,E(s) \rightarrow \widetilde{U}(z) = (K_P - \frac{K_I T}{2} + \frac{K_I T}{1-z^{-1}})\,\widetilde{E}(z).$$

Let

$$\widetilde{K}_P = K_P + \frac{K_I}{s} \qquad \text{and} \qquad \widetilde{K}_I = K_I\,T. \tag{5.9}$$

Then we have

$$(1-z^{-1})\widetilde{U}(z) = \widetilde{K}_P\,(1-z^{-1})\,\widetilde{E}(z) + \widetilde{K}_I\,\widetilde{E}(z).$$

It then follows from the inverse *z*-transform that

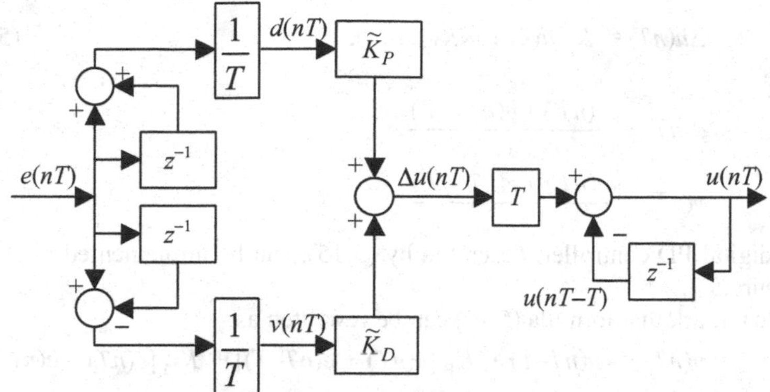

Figure 5.7 The digital PD controller.

$$u(nT) - u(nT-T) = \widetilde{K}_{\rm P}\,[\,e(nT) - e(nT-T)\,] + \widetilde{K}_{\rm I}\,e(nT),$$

so that

$$\frac{u(nT)-u(nT-T)}{T} = \widetilde{K}_{\rm P}\,\frac{e(nT)-e(nT-T)}{T} + \frac{\widetilde{K}_{\rm I}}{T}\,e(nT).$$

Let, furthermore,

$$\Delta u(nT) = \frac{u(nT)-u(nT-T)}{T} \tag{5.10}$$

be the incremental control and

$$v(nT) = \frac{e(nT)-e(nT-T)}{T}. \tag{5.11}$$

We arrive at

$$u(nT) = u(nT-T) + T\Delta u(nT) \tag{5.12}$$

$$\Delta u(nT) = \widetilde{K}_{\rm P}\,v(nT) + \frac{\widetilde{K}_{\rm I}}{T}\,e(nT), \tag{5.13}$$

which can be implemented as shown in Figure 5.6.

Similarly, using the bilinear transform (5.8) in the continuous-time PD controller (see Figure 5.5), we obtain

$$s\,K_{\rm D} \rightarrow \frac{2}{T}\,\frac{z-1}{z+1}\,K_{\rm D} = \frac{2}{T}\,\frac{1-z^{-1}}{1+z^{-1}}\,K_{\rm D},$$

$$U(s) = (K_{\rm P} + sK_{\rm D})\,E(s) \rightarrow \widetilde{U}(z) = (K_{\rm P} + \frac{2}{T}\,\frac{1-z^{-1}}{1+z^{-1}}\,K_{\rm D})\,\widetilde{E}(z).$$

Let

$$\widetilde{K}_{\rm P} = K_{\rm P} \quad \text{and} \quad \widetilde{K}_{\rm D} = \frac{2}{T}\,K_{\rm D}. \tag{5.14}$$

Then we have

$$(1+z^{-1})\,\widetilde{U}(z) = (1+z^{-1})\,\widetilde{K}_{\rm P}\,\widetilde{E}(z) + \widetilde{K}_{\rm D}\,(1+z^{-1})\,\widetilde{E}(z),$$

so that the inverse z-transform gives

$$\Delta u(nT) = \widetilde{K}_P\, d(nT) + \widetilde{K}_D\, v(nT), \tag{5.15}$$

where

$$d(nT) = \frac{e(nT) + e(nT-T)}{T},$$

$$v(nT) = \frac{e(nT) - e(nT-T)}{T}.$$

This digital PD controller, described by (5.15), can be implemented as shown in Figure 5.7.

We remark that formula (5.15) can be rewritten as

$$u(nT) = -u(nT-T) + \widetilde{K}_P\,[e(nT) + e(nT-T)] + \widetilde{K}_D\,[e(nT) - e(nT-T)],$$

so that

$$\begin{aligned} u(nT-T) \;&=\; -u(nT-2T) + \widetilde{K}_P\,[e(nT-T) + e(nT-2T)] \\ &\quad + \widetilde{K}_D\,[e(nT-T) - e(nT-2T)]. \end{aligned}$$

Thus, by substituting $u(nT-T)$ into $u(nT)$, we obtain

$$\begin{aligned} u(nT) \;&=\; u(nT-2T) + \widetilde{K}_P\,[e(nT) - e(nT-2T)] \\ &\quad + \widetilde{K}_D\,[e(nT) - 2e(nT-T) + e(nT-2T)] \\ &=\; u(nT-2T) + \widetilde{K}_P\, T\,\frac{e(nT) - e(nT-2T)}{T} \\ &\quad + \widetilde{K}_D\, T^2\,\frac{e(nT) - 2e(nT-T) + e(nT-2T)}{T^2}, \end{aligned}$$

where the two finite divided differences are the discretization of $\dot{e}(t)$ and $\ddot{e}(t)$, respectively, of the continuous-time error signal $e(t)$, under the bilinear transform. This formula shows clearly that the digital PD controller implicitly uses the discretization of both $\dot{e}(t)$ and $\ddot{e}(t)$, as is known in conventional control theory. Since this formula is equivalent to, but more complicated than, formula (5.15), we will not use it in the following.

The fuzzy PI, PD, and PI+D controllers to be introduced below employ the digital PI controller (5.12) and the digital PD controller (5.15), whose implementations are shown in Figures 5.6 and 5.7, respectively.

II. FUZZY PID CONTROLLERS

We are now in a position to introduce the fuzzy PID controllers: their design methods, performance evaluation, and stability analysis.

We first study the fuzzy PD controller design in detail, in which all the basic ideas, design principles, and step-by-step derivation and calculations are discussed. The fuzzy PI controller design will be discussed briefly next, followed by the fuzzy PI+D controller design. Having this background, many other types of fuzzy PID controllers can be designed by following similar procedures. The stability analysis of these fuzzy PID controllers will be investigated in the last section of the chapter.

Although it is possible to design a fuzzy logic type of PID controller by a simple modification of the conventional ones, via inserting some meaningful fuzzy logic IF-THEN rules into the control system (e.g., to self-tune the PID control gains with the help of a "look-up" table), these approaches in general complicate the overall design and do not come up with new fuzzy PID controllers that capture the essential characteristics and nature of the conventional PID controllers. Besides, they generally do not have analytic formulas to use for control specification and stability analysis.

The fuzzy PD, PI, and PI+D controllers to be introduced below are natural extensions of their conventional versions, which preserve the linear structures of the PID controllers, with simple and conventional analytical formulas as the final results of the design. Thus, they can directly replace the conventional PID controllers in any operating control systems (plants, processes). The main difference is that these fuzzy PID controllers are designed by employing fuzzy logic control principles and techniques, which have been studied in the last few chapters, to obtain new controllers that possess analytical formulas very similar to the conventional digital PID controllers. After the design is completed, all the fuzzy logic IF-THEN rules, membership functions, defuzzification formulas, etc. will not be needed any more in applications: what one can see is a conventional controller with a few simple formulas similar to the familiar PID controllers. Thus, in operations the controllers do not use any "look-up" table at any step, and so can be operated in real time. A control engineer who doesn't have any knowledge about fuzzy logic and/or fuzzy control systems can use them just like the conventional ones, particularly for higher-order, time-delayed, and nonlinear systems, and for those systems that have only vague mathematical models or contain significant uncertainties. The key reason, which is the price to pay, for such success is that these fuzzy PID controllers are slightly more complicated than the conventional ones, in the sense that they have variable control gains in their linear structures. These variable gains are nonlinear functions of the errors and changing rates of the error signals. The main contribution of these variable gains in improving the control performance is that they are self-tuned gains and can adapt to the rapid changes of the errors and the (changing) rates of the error signals caused by the time-delayed effects, nonlinearities, and uncertainties of the underlying system (plant, process).

A. Fuzzy PD Controller

The overall fuzzy PD set-point tracking control system is shown in Figure 5.8, where the process under control is a discrete-time system (or a discretized continuous-time system), and $r(nT)$ is the reference signal which can be a constant (set-point). The fuzzy PD controller inside the dashed box differs from the conventional digital PD controller (shown in Figure 5.7) in that there is an extra "fuzzy controller" in the path of the incremental control signal $\Delta u(nT)$. Moreover, a constant multiplication block has been changed from the sampling period T to an adjustable constant control gain K_u in order to enable the new controller one more degree of freedom in the control process (but this

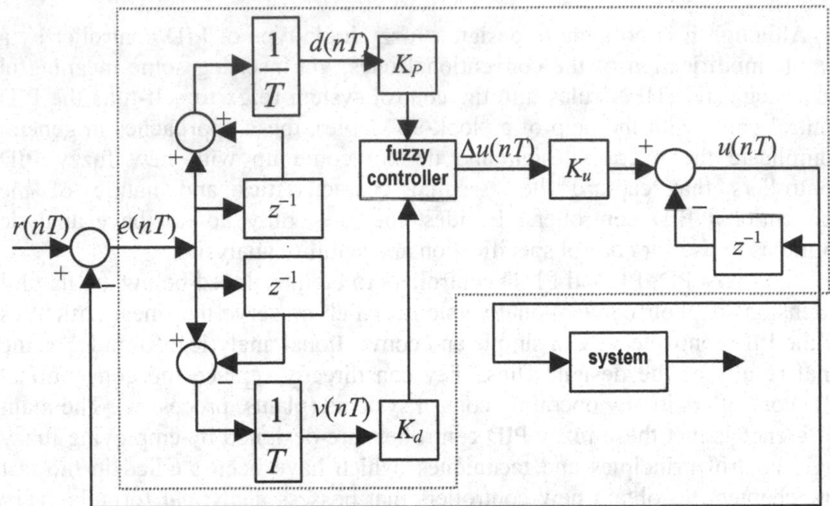

Figure 5.8 The fuzzy PD controller.

is not necessary). In this fuzzy PD controller, the "fuzzy controller" block is the key that improves the conventional digital PD controller's capabilities and performance.

To simplify the notation, we have let the adjustable control gains be

$$K_p = \widetilde{K}_{\mathrm{P}} \quad \text{and} \quad K_d = \widetilde{K}_{\mathrm{D}}$$

in Figure 5.8 (compared to Figure 5.7), and similarly let $K_i = \widetilde{K}_{\mathrm{I}}$ when the fuzzy PI controller is discussed later.

To illustrate how the "fuzzy controller" block works, we first introduce a constant parameter (an adjustable scalar) $L > 0$, and decompose the plane by L as twenty input-combination regions (IC1-IC20) as shown in Figure 5.9, where the horizontal axis is the input signal $K_pd(nT)$, and the vertical axis the input signal $K_dv(nT)$, to the "fuzzy controller." Then, according to which region the input signals $(K_pd(nT), K_dv(nT))$ belong, the "fuzzy controller" block produces the following incremental outputs:

$$\Delta u(nT) = \frac{L[K_p d(nT) - K_d v(nT)]}{2(2L - K_p \,|\, d(nT)\,|)}, \quad \text{in IC1, IC2, IC5, IC6,} \tag{5.16}$$

$$= \frac{L[K_p d(nT) - K_d v(nT)]}{2(2L - K_d \,|\, v(nT)\,|)}, \quad \text{in IC3, IC4, IC7, IC8,} \tag{5.17}$$

$$= \frac{1}{2}[\, L - K_d v(nT) \,], \quad \text{in IC9, IC10,} \tag{5.18}$$

$$= \frac{1}{2}[\, -L + K_p d(nT) \,], \quad \text{in IC11, IC12,} \tag{5.19}$$

$$= \frac{1}{2}[\, -L - K_d v(nT) \,], \quad \text{in IC13, IC14,} \tag{5.20}$$

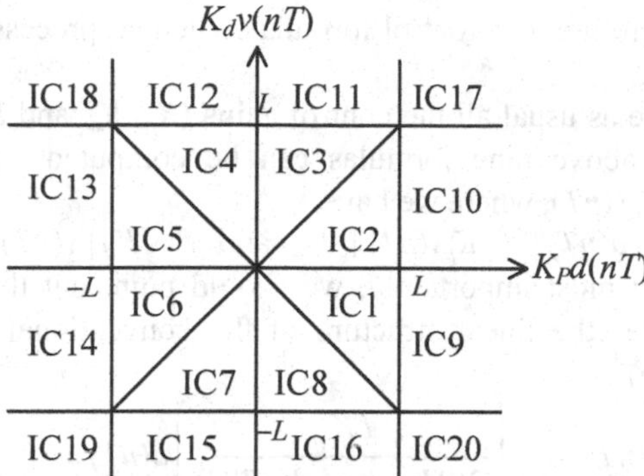

Figure 5.9 Regions of the "fuzzy controller" input-combination values.

$$= \frac{1}{2}[\,L + K_p d(nT)\,], \qquad \text{in IC15, IC16,} \qquad (5.21)$$
$$= 0, \qquad \text{in IC17, IC19,} \qquad (5.22)$$
$$= -L, \qquad \text{in IC18,} \qquad (5.23)$$
$$= L, \qquad \text{in IC20.} \qquad (5.24)$$

Here, if the input signals ($K_p d(nT)$, $K_d v(nT)$) belong to a boundary line, then either of the two neighboring regions can be used, since they will be the same (namely, all these control functions are continuous on the boundaries).

This completes the description for implementation of the fuzzy PD controller in a set-point tracking system, regardless of any knowledge about the system (plant, process) under control. The initial conditions for this control system are the following nature ones:

$$y(0) = 0, \qquad \Delta u(0) = 0, \qquad e(0) = r, \qquad v(0) = 0. \qquad (5.25)$$

Before we discuss how to derive these formulas for the "fuzzy controller" block and in what sense this design is good, a few remarks on the above formulas are in order.

First, we note that the above nine pieces of formulas are all conventional (crisp) analytical formulas, from which one does not see any fuzzy contents (membership functions, fuzzy logic IF-THEN rules, etc.). Therefore, the "fuzzy controller" block, as well as the entire fuzzy PD controller shown in Figure 5.8, are conventional controllers: the overall control system works in the conventional manner despite the name "fuzzy." As a result, this fuzzy PD controller can be used to replace the conventional PD controller anywhere, and a control engineer can operate this fuzzy PD controller in a way completely analogous to the conventional one: what he needs to do is to tune the control gains and parameters, K_p, K_d, K_u, and L, without the need of knowledge of the fuzzy mathematics, fuzzy logic, and fuzzy control theory.

Second, the above nine pieces of formulas are continuously connected as a whole (on the boundaries between different regions shown in Figure 5.9). This can be verified by direct calculation of any two adjacent formulas on a boundary. Therefore, the control formula switching process does not have any jumps.

Third, since as usual all the control gains (K_p, K_d, and K_u) are positive real numbers, the above nine formulas can be computed from the two inputs $K_pd(nT)$ and $K_dv(nT)$, where we have

$$K_p \mid d(nT) \mid = \mid K_p v(nT) \mid \qquad \text{and} \qquad K_d \mid v(nT) \mid = \mid K_d v(nT) \mid .$$

Finally, but most importantly, we should point out that the first formula (5.16) preserves the linear structure of the conventional PD controller (see formula (5.15)):

$$\Delta u(nT) = \left[\frac{LK_p}{2(2L - K_p \mid d(nT) \mid)} \right] d(nT) + \left[\frac{-LK_d}{2(2L - K_p \mid d(nT) \mid)} \right] v(nT),$$

except that the two control gains here are variable gains: they are nonlinear functions of the signal $d(nT) = \frac{1}{T}[e(nT) + e(nT-T)]$. Similarly, the second formula, (5.17), has the same linear structure but with two variable gains that are nonlinear functions of the signal $v(nT) = \frac{1}{T}[e(nT) - e(nT-T)]$. It is very important to note that it is exactly these variable control gains that improve the performance of the controller: when the error signal $e(nT)$ increases, for instance, the signal $d(nT)$ increases, so that the control gain $L/2(2L-K_p|d(nT)|)$ also increases automatically. This means that the controller takes larger actions accordingly and, hence, has certain self-tuning and adaptive capabilities in the control process. Here, it is also important to note that $K_p|d(nT)|$ will not exceed the constant L in the denominator of the control gain according to the definition of the input-combination (IC) regions shown in Figure 5.9; otherwise, the formula will switch to another one of the nine formulas.

We next discuss the design principle of the fuzzy PD controller described above, and give detailed derivations to the resulting formulas (5.16)-(5.24).

We first recall that in a standard procedure, a fuzzy controller design consists of three components: (i) fuzzification, (ii) rule base establishment, and (iii) defuzzification. Our design of the fuzzy PD controller follows this procedure.

In the fuzzification step, we employ two inputs: the error signal $e(nT)$ and the rate of change of the error signal $v(nT)$, with only one control output $u(nT)$ (to be fed to the system under control). The input to the fuzzy PD controller, namely, the "error" and the "rate" signals, have to be fuzzified before being fed into the controller. The membership functions for the two inputs (error and rate) and the output of the controller that is used in our design are shown in Figure 5.10, which are likely the simplest possible functions to use for this purpose. Both the error and the rate have two membership values: positive and negative, while the output has three (singleton functions): positive, negative,

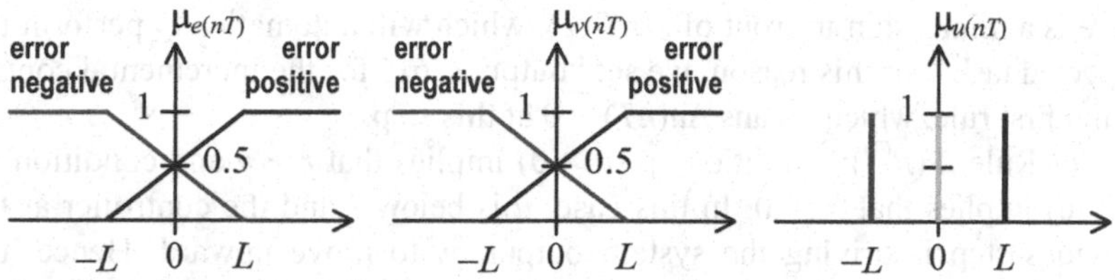

Figure 5.10 The membership functions of $e(nT)$, $v(nT)$, and $u(nT)$.

and zero. The constant $L > 0$ used in the definition of the membership functions is chosen by the designer according to the value ranges of the error, rate, and output, which is used as a tunable parameter but can also be fixed after being determined. Note that the constant L used in these three membership functions can be different in general (according to the physical meaning of the signals in the application), but we let them be the same here in order to simplify the design.

Based on these membership functions, the fuzzy control rules that we used are the following:

$R^{(1)}$ IF error = ep AND rate = vp THEN output = oz.
$R^{(2)}$ IF error = ep AND rate = vn THEN output = op.
$R^{(3)}$ IF error = en AND rate = vp THEN output = on.
$R^{(4)}$ IF error = en AND rate = vn THEN output = oz.

Here, "output" is the fuzzy control action $\Delta u(nT)$, "ep" means "error positive," "oz" means "output zero," etc. The "AND" is the logical AND defined by

$$\mu_A \text{ AND } \mu_B = \min\{\, \mu_A, \mu_B \,\}$$

for any two membership values μ_A and μ_B on the fuzzy subsets A and B, respectively.

The reason for establishing the rules in such formulation can be understood in the same way as that described in Chapter 4, Section II, which is briefly repeated here for clarity. First, it is important to observe that since the error signal is defined to be $e = r - y$, where r is the reference (set-point) and y is the system output (see Figure 5.4 for the general situation in the continuous-time setting), we have $\dot{e} = \dot{r} - \dot{y} = -\dot{y}$ in the case that the set-point r is constant.

For Rule 1 ($R^{(1)}$): condition ep (the error is positive, $e > 0$) implies that $r > y$ (the system output is below the set-point) and condition vp ($\dot{e} > 0$) implies that $\dot{y} < 0$ (the system output is decreasing). In this case, the controller at the previous step is driving the system output, y, to move downward. Hence, the controller should turn around and drive the system output to move upward. It is very important to observe, however, that in our control law (see formula (5.12) and Figure 5.8):

$$u(nT) = -u(nT-T) + K_u \Delta u(nT), \tag{5.26}$$

there is a minus sign in front of $u(nT-T)$, which will automatically perform the expected task. For this reason, we set "output = oz" for the incremental control as the first rule, which means $\Delta u(nT) = 0$ at this step.

For Rule 2 ($R^{(2)}$): condition ep ($e > 0$) implies that $r > u$ and condition vn ($\dot{e} < 0$) implies that $\dot{y} > 0$. In this case, u is below r and the controller at the previous step is driving the system output, y, to move upward. Hence, the controller needs not to take any action. But, in control law (5.26), there is a minus sign with $u(nT-T)$ which will turn the control action to the opposite. To compensate for this, we should let Δu be positive, i.e., "output = op."

For Rule 3 ($R^{(3)}$): condition en ($e < 0$) implies that $r < y$ and condition vp ($\dot{e} > 0$) implies that $\dot{y} < 0$. In this case, y is over r and the controller at the previous step is driving the system output to move downward. Therefore, the controller needs not to take any action. Similar to Rule 2, to compensate the minus sign in (5.26), we let Δu be negative, i.e., "output = on."

For Rule 4 ($R^{(4)}$): condition en ($e < 0$) implies that $r < y$ and condition vn ($\dot{e} < 0$) implies that $\dot{y} > 0$. In this case, y is over r and the controller at the previous step is driving the system output to move upward. Hence, the controller should turn its output around to drive the system output to move downward. Because of the minus sign in (5.26), similar to Rule 1, the controller needs not to take any action: "output = oz."

Here, we remark that one may try to let the controller take some action to speed up the system output in cases of Rules 1 and 4, but this will somewhat complicate the design of the controller.

In the defuzzification step, we use the same logical AND as mentioned above, and the membership functions shown in Figure 5.10 for the error $e(nT)$, the rate $v(nT)$, and the output $\Delta u(nT)$ of the "fuzzy controller" block. Because we have two (positive and negative) membership values for the error and rate, the commonly used weighted average formula is used for defuzzification, leading to

$$\Delta u(nT) = \frac{\sum(\text{membership value of input} \times \text{corresponding value of output})}{\sum(\text{membership value of input})} \tag{5.27}$$

The defuzzification procedure and its corresponding results are now analyzed and summarized in the following.

First, we observe from (5.16)-(5.24) that, instead of the error signal $e(nT)$, we will actually use the average error signal $d(nT)$. We will simply call $K_p d(nT)$ and $K_d r(nT)$ the "error signal" and the "rate signal," respectively. Note that using the average error will not alter the above reasoning for the control rules ($R^{(1)}$)-($R^{(4)}$), at least in principle.

Second, observe that the membership functions of the average error and rate signals decompose their value ranges into twenty adjacent input-combination (IC) regions, as shown in Figure 5.9. This figure is understood as follows.

We put the membership function of the error signal (given by the first picture of Figure 5.10) over the horizontal $K_p d(nT)$-axis in Figure 5.9, and put the membership function of the rate of change of the error signal (given by the second picture of Figure 5.10) over the vertical $K_d r(nT)$-axis in Figure 5.9. These two membership functions then overlap and form the third-dimensional picture (which is not shown in Figure 5.9) over the 2-D regions shown in Figure 5.9. When we look at region IC1, for example, if we look upward to the $K_p d(nT)$-axis, we see the domain $[0,L]$ and the membership function (in the third dimension) over $[0,L]$ of the error signal; if we look leftward to the $K_d r(nT)$-axis, we see the domain $[-L,0]$ and the membership function (in the third dimension) over $[-L,0]$ of the rate of change of the error signal.

Then, we consider the locations of the error $K_p d(nT)$ and the rate $K_d r(nT)$ in the region IC1 and IC2 (see Figure 5.9). Let us look at region IC1, for example, where we have ep > 0.5 > vp (see the first two pictures in Figure 5.10). Hence, the logical AND used in ($R^{(1)}$) leads to

$$\{\text{"error = ep AND rate = vp"}\} = \min\{\text{ ep, vp }\} = \text{vp},$$

so that Rule 1 ($R^{(1)}$) yields

$$R^{(1)}: \begin{cases}\text{the selected input membership value is vp;}\\ \text{the corresponding output value is oz.}\end{cases}$$

Similarly, in region IC1, Rules 2-4, ($R^{(2)}$)-($R^{(4)}$), and the logical AND used in ($R^{(2)}$)-($R^{(4)}$) together yield

$$R^{(2)}: \begin{cases}\text{the selected input membership value is vn;}\\ \text{the corresponding output value is op.}\end{cases}$$

$$R^{(3)}: \begin{cases}\text{the selected input membership value is en;}\\ \text{the corresponding output value is on.}\end{cases}$$

$$R^{(4)}: \begin{cases}\text{the selected input membership value is en;}\\ \text{the corresponding output value is oz.}\end{cases}$$

It can be verified that the above are true for the two regions IC1 and IC2. Thus, in regions IC1 and IC2, it follows from the defuzzification formula (5.27) that

$$\Delta u(nT) = \frac{\text{vp}\times\text{oz} + \text{vn}\times\text{op} + \text{en}\times\text{on} + \text{en}\times\text{oz}}{\text{vp} + \text{vn} + \text{en} + \text{en}}.$$

It is very important to note that if one follows the above procedure to work through the two cases, then it is found that both the last two cases give the same result of en (i.e., the two en in the above formula are not the misprint of en and ep!). To this end, by applying op = L, on = $-L$, oz = 0 (obtained from Figure 5.10), and the following straight line formulas from the geometry of the membership functions associated with Figure 5.9:

$$\text{ep} = \frac{K_p d(nT) + L}{2L}, \qquad \text{en} = \frac{-K_p d(nT) + L}{2L},$$

$$\text{vp} = \frac{K_d v(nT) + L}{2L}, \qquad \text{vn} = \frac{-K_d v(nT) + L}{2L},$$

we obtain

$$\Delta u(nT) = \frac{L}{2[2L - K_p d(nT)]}\,[K_p d(nT) - K_d v(nT)].$$

Here, we note that $d(nT) \geq 0$ in regions IC1 and IC2.

In the same way, one can verify that in regions IC5 and IC6,

$$\Delta u(nT) = \frac{L}{2[2L - K_p d(nT)]}\,[K_p d(nT) - K_d v(nT)],$$

where it should be noted that $d(nT) \leq 0$ in regions IC5 and IC6.

Therefore, by combining the above two formulas, we arrive at the following result for the four regions IC1, IC2, IC5, and IC6:

$$\Delta u(nT) = \frac{L}{2[2L - K_p\,|\,d(nT)\,|\,]}\,[K_p d(nT) - K_d v(nT)],$$

which is (5.16). Similarly, if $K_p d(nT)$ and $K_d v(nT)$ are located in the regions IC3, IC4, IC7, and IC8, we have

$$K_p\,|\,d(nT)\,| \leq K_d\,|\,v(nT)\,| \leq \text{L},$$

and, in this case,

$$\Delta u(nT) = \frac{L}{2[2L - K_d\,|\,d(nT)\,|\,]}\,[K_p d(nT) - K_d v(nT)],$$

which is (5.17). Finally, in the regions IC9-IC20, we have the corresponding formulas shown as in (5.18)-(5.24).

To this end, we have determined all the control rules and formulas for the fuzzy PID controller, with the control law (5.26) and the fuzzy control action $\Delta u(nT)$ calculated by (5.16)-(5.24) according to the different locations in Figure 5.9 of the error signal $K_p d(nT)$ and the rate of the change of the error signal $K_d v(nT)$. The initial conditions for the overall control system are the following natural values: for the fuzzy control action $\Delta u(0) = 0$, for the system output, $y(0) = 0$, for the original error and rate signals, $e(0) = r$ (the set-point) and $v(0) = 0$, respectively, as shown in (5.25).

Finally, we remark that in the steady-state situation, $|e(nT)| = 0$, so that $|v(nT)| = |d(nT)| = 0$ in the denominators of the coefficients of $\Delta u(nT)$. Thus, we obtain the steady-state relations between the conventional PD control gains K_d^c and K_p^c and the fuzzy PD control gain K_p and K_d as follows:

$$K_d^c = \frac{K_u K_d}{4} \qquad \text{and} \qquad K_p^c = \frac{K_u K_p}{4}. \tag{5.28}$$

Example 5.3. In order to compare the fuzzy PD controller with the conventional one, we first consider a first-order linear system with transfer function

$$H(s) = \frac{1}{s+1},$$

and the reference (set-point) r = 10.0. For this sytem, the fuzzy controller parameters are: T = 0.1, K_d = 0.5, K_p = 0.5, K_u = 1.0, and L = 361.0. The control performance is shown in Figure 5.11.

We then consider a second-order linear system with transfer function

$$H(s) = \frac{1}{s^2 + 4s + 3},$$

and the reference (set-point) r = 10.0. The system response, controlled by the fuzzy PD controller, is shown in Figure 5.12. The controller parameters are T = 0.1, K_p = 0.51, K_d = 0.02, K_u = 0.232, and L = 1000.0.

In the above two cases, both conventional and fuzzy PD controllers work equally well.

To show one more case of the second-order linear system, consider one with the transfer function

$$H(s) = \frac{1}{s(s+100)},$$

which is only marginally stable. This time, let's make it more difficult by setting the reference as a ramp signal, $r(t) = t$. The controller parameters used are T – 0.01, K_p = 25.0, K_d = –50.0, K_u = 0.5, and L = 3000.0, and the tracking tolerance is 5% of the steady-state error. For this example, it is easy to verify that the steady-state error for the conventional PD controller is given analytically by

$$e_{ss}(t) = \frac{100}{K_p^c}.$$

Therefore, we have to use a high gain of K_p^c = 2000.0 (along with K_d^c = 10.0) to obtain specified performance (of 5% steady-state tracking error). The fuzzy control result is shown in Figure 5.13, where the solid curve is the set-point while the dashed curve is the system output. This result demonstrates the advantage of the fuzzy PD controller over the conventional one (the former uses very small control gains) even for a second-order linear process.

Next, consider a lower-order linear system with time-delay, with transfer function

$$H(s) = \frac{1}{(100s+1)^2} e^{-3s}.$$

The comparison is shown in Figure 5.14. The conventional PD controller, with K_p^c = 66.0, K_d^c = 25.0, and T = 0.1, produces the solid curve. On the contrary, the fuzzy PD controller, with T = 0.1, K_p = 49.3, K_d = 5.3, K_u = 0.8, and L = 19.0, yields the dashed curve in the same figure.

Finally, the conventional and fuzzy PD controllers are compared using two nonlinear systems. The first one has the simple nonlinear model

$$\dot{y}(t) = 0.0001\,|y(t)| + u(t).$$

We used two different references: the constant set-point r = 1.0 and the ramp signal $r(t) = t$. For the constant set-point case, the fuzzy PD controller has the parameters $T = 0.1$, $K_p = 19.5$, $K_d = 0.5$, $K_u = 0.1$, and $L = 20.0$. The result is shown in Figure 5.15(a). The conventional PD controller, on the other hand, cannot handle this nonlinear system no matter how one changes its two constant gains. One control performance is shown in Figure 5.15(b), with $K_p^c = 3.0$, $K_d^c = 0.1$, and $T = 0.1$. For the ramp signal reference case, the fuzzy PD controller produces a good tracking result with very small transient oscillation, shown in Figure 5.15(c), where $K_p = 19.0$, $K_d = 0.5$, $K_u = 0.1$, $L = 40.0$, and $T = 0.1$. However, although the conventional PD controller also performs well after a long transient period, as shown in Figure 5.15(d), its transient behavior is poorer (see Figure 5.15(e)), as compared to the fuzzy controller (Figure 5.15(c)).

The second nonlinear example is shown in Figure 5.16, where the system is described by

$$\dot{y}(t) = -y(t) + 0.5y^2(t) + \mathrm{u}(t)$$

with the constant set-point $r = 2.0$. The fuzzy controller is designed with $K_p = 41.9$, $K_d = 15.4$, $K_u = 0.1$, $L = 239.0$, and $T = 0.1$, which produces the tracking response shown in Figure 5.16. However, no matter how one adjusts the two constant gains of the conventional PD controller, it does not show any reasonable tracking results.

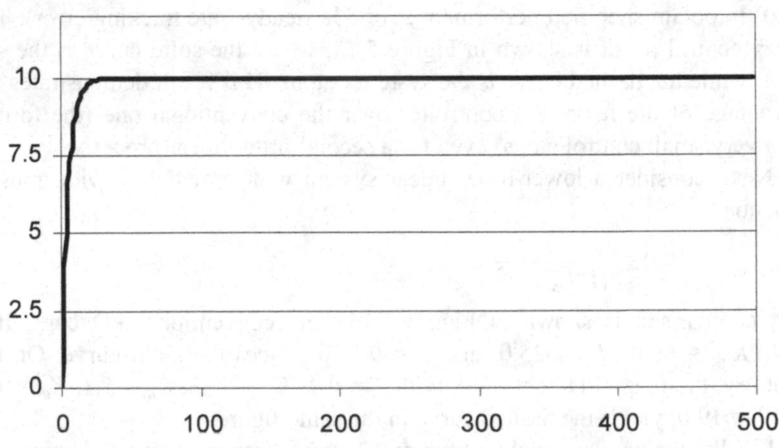

Figure 5.11 Output of a first-order linear fuzzy PD control system.

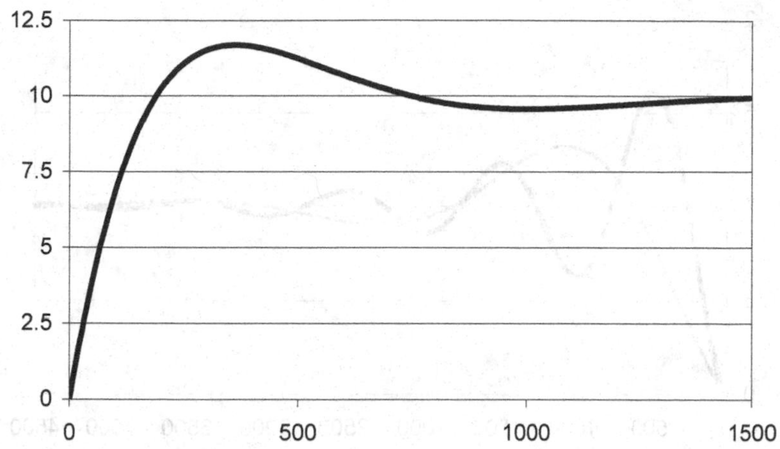

Figure 5.12 Output of a second-order linear fuzzy PD control system.

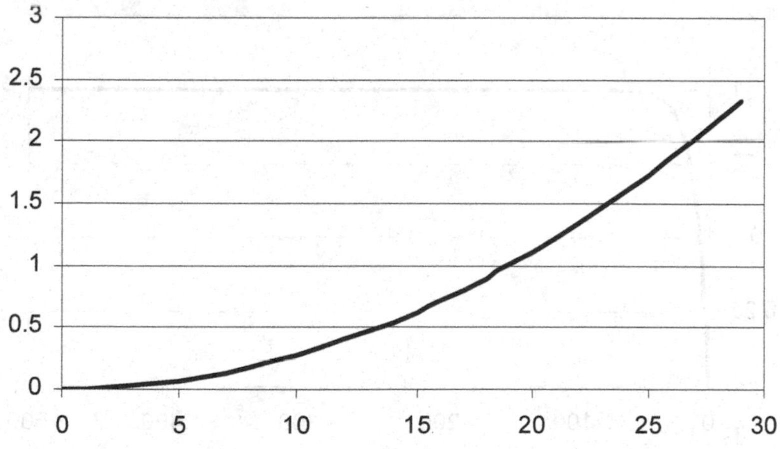

Figure 5.13 Output of a second-order linear fuzzy PD control system.

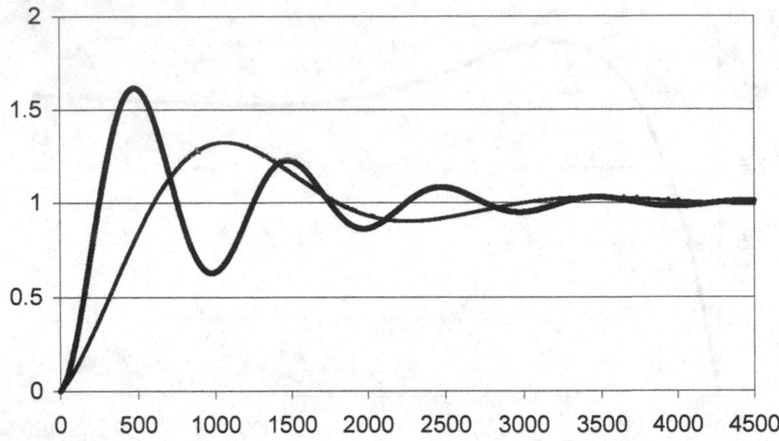

Figure 5.14 Comparison of outputs for a second-order linear fuzzy PD control system with a time delay.

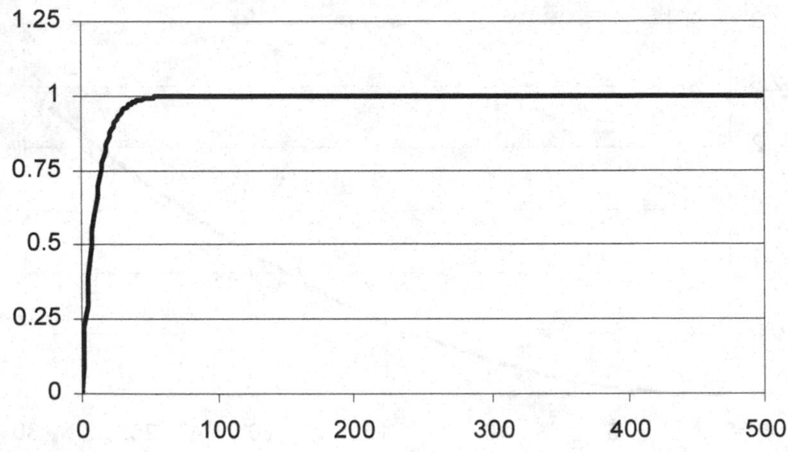

Figure 5.15(a) Output of a nonlinear fuzzy PD control system with a constant set-point.

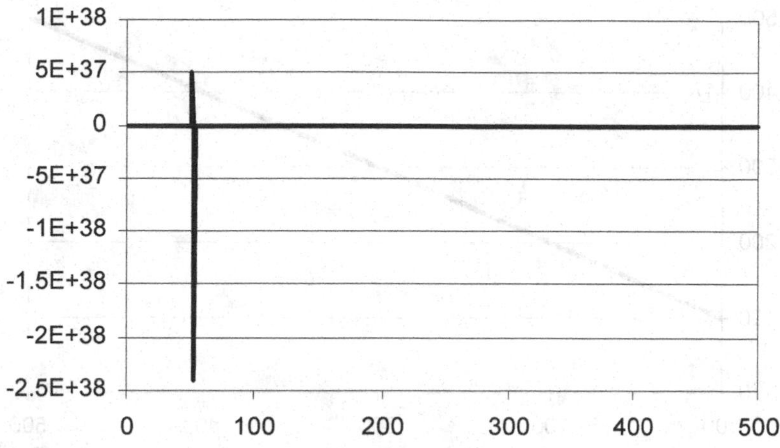

Figure 5.15(b) Output of a nonlinear conventional PD control system (with a constant set-point).

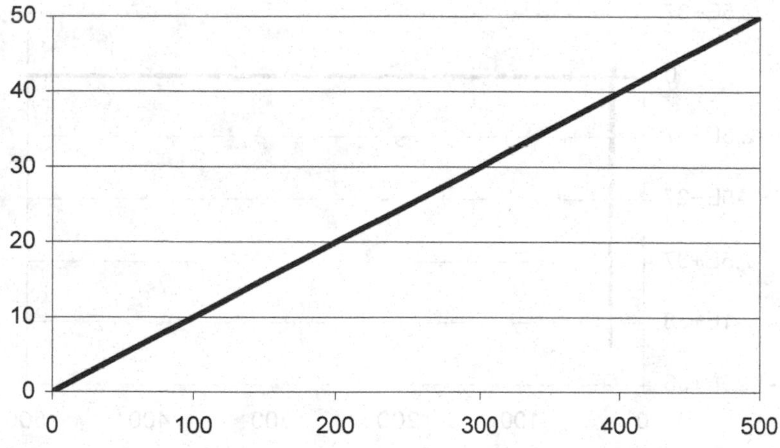

Figure 5.15(c) Output of a nonlinear fuzzy PD control system (with a ramp set-point).

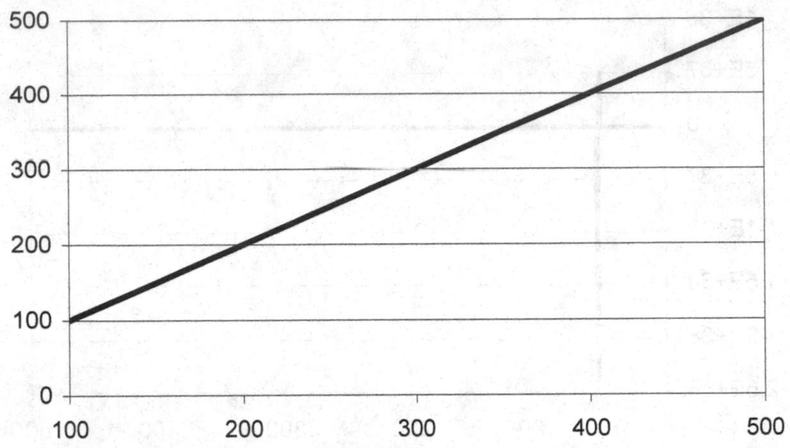

Figure 5.15(d) Steady-state output of a nonlinear conventional PD control system (with a ramp set-point).

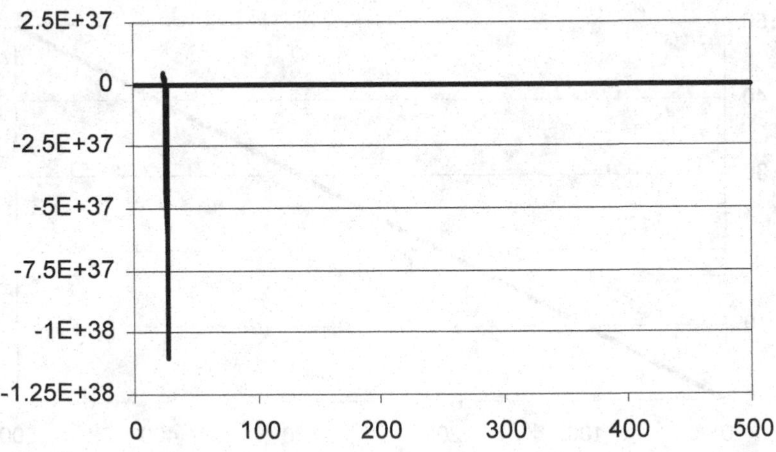

Figure 5.15(e) Transient output of a nonlinear conventional PD control system (with a ramp set-point).

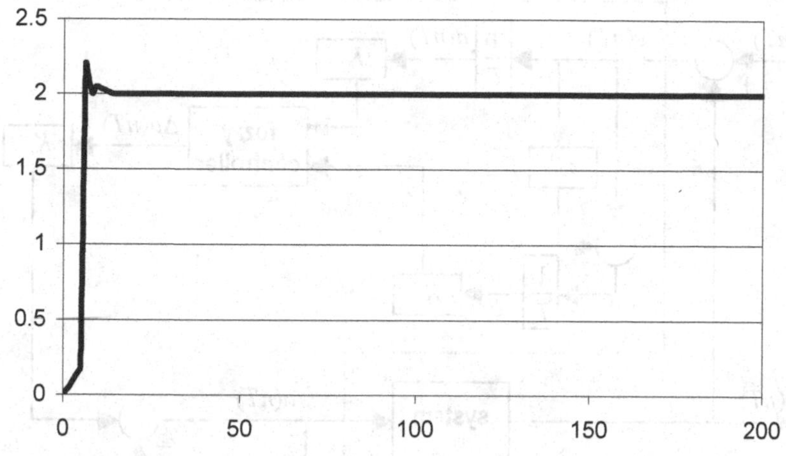

Figure 5.16 Output of a highly nonlinear PD control system.

B. Fuzzy PI Controller

Design of the fuzzy PI controller is similar to that for the fuzzy PD controller, and will be introduced briefly in this subsection.

The overall PI controller system is shown in Figure 5.17 (see Figure 5.6). In Figure 5.17, we let $K_p = \widetilde{K}_P$ and $K_i = \widetilde{K}_I$, as mentioned before. The "fuzzy controller" works in a way similar to that of the fuzzy PD controller. We first decompose the plane with a scalar $L > 0$ into twenty input-combination (IC) regions for the inputs $K_i e(nT)$ and $K_p v(nT)$, where $v(nT) = \frac{1}{T}[e(nT) - e(nT-T)]$, as shown in Figure 5.18. Here, $K_i e(nT)$ and $K_p v(nT)$ are called the error signal and the rate of change of error signal, respectively, for convenience. Then, according to the location of the inputs $(K_i e(nT), K_p v(nT))$ to the "fuzzy controller" block, the corresponding incremental control output are computed by the following formulas:

$$\Delta u(nT) = \frac{L[K_i e(nT) + K_p v(nT)]}{2(2L - K_i\,|\,e(nT)\,|)}, \quad \text{in IC1, IC2, IC5, IC6,} \tag{5.29}$$

$$= \frac{L[K_i e(nT) + K_p v(nT)]}{2(2L - K_p\,|\,e(nT)\,|)}, \quad \text{in IC3, IC4, IC7, IC8,} \tag{5.30}$$

$$= \frac{1}{2}[\,L + K_p v(nT)\,], \quad \text{in IC9, IC10,} \tag{5.31}$$

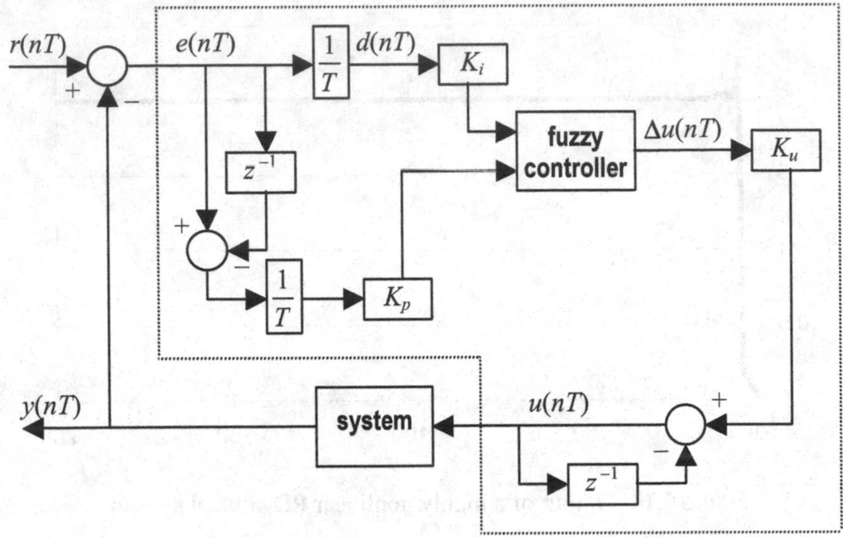

Figure 5.17 The fuzzy PI controller.

$$= \frac{1}{2}\,[\,L + K_i e(nT)\,], \qquad \text{in IC11, IC12,} \tag{5.32}$$

$$= \frac{1}{2}\,[\,-L + K_p v(nT)\,], \qquad \text{in IC13, IC14,} \tag{5.33}$$

$$= \frac{1}{2}\,[\,-L + K_i e(nT)\,], \qquad \text{in IC15, IC16,} \tag{5.34}$$

$$= 0, \qquad \text{in IC18, IC20,} \tag{5.35}$$

$$= -L, \qquad \text{in IC17,} \tag{5.36}$$

$$= L, \qquad \text{in IC19.} \tag{5.37}$$

Then, the control action output of the fuzzy PI controller, which is the control input to the system (plant, process), is given by

$$u(nT) = u(nT-T) + K_u \Delta u(nT), \tag{5.38}$$

where K_u is an adjustable constant control gain, which may be fixed to be $K_u = T$ to simplify the design (but one will then lose a degree of freedom in the tuning of the controller).

In a comparison of the fuzzy PI and PD control laws (5.38) and (5.26), one can find the main difference: there is a minus sign in front of $u(nT-T)$ in the fuzzy PD controller. This minus sign makes the rule base design quite different. Of course, their incremental controls $\Delta u(nT)$ are given by formulas (5.16)-(5.24) and (5.29)-(5.37), respectively, which are also different.

Similarly, the initial conditions for the fuzzy PI controller are the following natural ones:

$$y(0) = 0, \qquad \Delta u(0) = 0, \qquad e(0) = r, \qquad v(0) = 0. \tag{5.39}$$

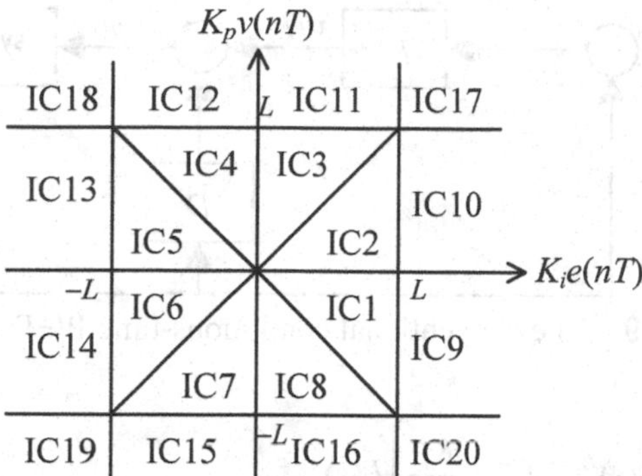

Figure 5.18 Regions of the "fuzzy controller" input-combination values.

A verification of the above fuzzy PI controller formulas can be carried out in a step-by-step procedure by mimicking the fuzzy PD controller's design given in the last subsection. This will be discussed in more detail in the next subsection within the PI+D controller.

We remark once again that the nine pieces of formulas (5.29)-(5.37) are conventional (crisp) formulas. So the "fuzzy controller" block as well as the entire fuzzy PI controller shown in Figure 5.17 are conventional controllers: the overall control system works in the conventional (crisp) manner despite the name "fuzzy." Therefore, this fuzzy PI controller can be used to replace the conventional digital PI controller anywhere. A control engineer can operate it without knowledge of fuzzy mathematics, fuzzy logic, and fuzzy control theory: what he needs is to tune the control gains and parameters: K_p, K_i, K_u, and L.

Finally, we remark that computer simulations for comparison of the fuzzy and conventional PI controllers show similar results to those demonstrated in Example 4.3. More simulations will be shown in the next subsection.

C. Fuzzy PI+D Controller

The conventional analog PI+D controller is shown in Figure 5.2(d). Similar to the fuzzy PD controller design discussed in Section II.A, we will first discretize it by applying the bilinear transform, then design the fuzzy PI and a fuzzy D controller separately, and finally combine them together as a whole in the closed-loop system.

In so doing, the design of the fuzzy PI controller is the same as that mentioned briefly above, but the design of the fuzzy D controller is rather different.

Starting with the conventional analog PI+D control system shown in Figure 5.19, the output of the conventional analog PI controller in the frequency s-domain, as can be verified easily from Figure 5.3, is given by

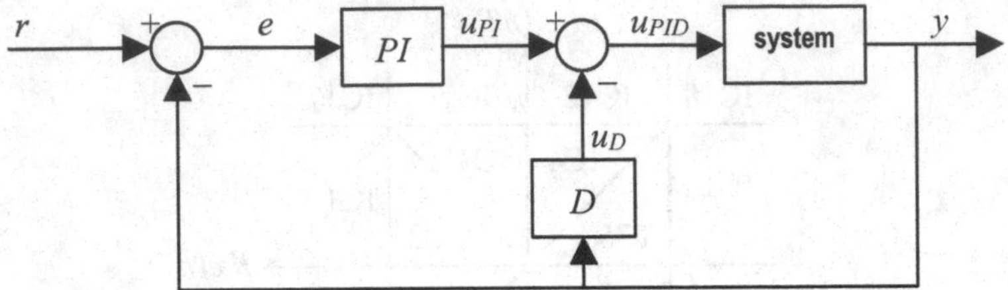

Figure 5.19 The conventional continuous-time PI+D control system.

$$U_{\rm PI}(s) = \left[K_p^c + \frac{K_i^c}{s} \right] E(s),$$

where K_p^c and K_i^c are the proportional and integral gains, respectively, and $E(s)$ is the tracking error signal, after taking the Laplace transform (with zero initial conditions). This equation can be transformed into the discrete version by applying the bilinear transformation (5.8), which results in the following form:

$$U_{\rm PI}(z) = \left[K_p^c - \frac{K_i^c T}{2} + \frac{K_i^c T}{1-z^{-1}} \right] E(z).$$

Letting

$$K_p = K_p^c - \frac{K_i^c T}{2} \quad \text{and} \quad K_i = K_i^c\, T, \tag{5.40}$$

and then taking the inverse z-transform, we obtain

$$u_{\rm PI}(nT) - u_{\rm PI}(nT-T) = K_p[\, e(nT) - e(nT-T)\,] + K_i T e(nT).$$

Dividing this equation by T gives

$$\Delta u_{\rm PI}(nT) = K_p v(nT) + K_i e(nT), \tag{5.41}$$

where

$$\Delta u_{\rm PI}(nT) = \frac{u_{\rm PI}(nT) - u_{\rm PI}(nT-T)}{T},$$

$$v(nT) = \frac{e(nT) - e(nT-T)}{T}.$$

It follows that

$$u_{\rm PI}(nT) = u_{\rm PI}(nT-T) + T\Delta u_{\rm PI}(nT).$$

In the design of the fuzzy PI controller to be discussed later, we will replace the coefficient T by a fuzzy control gain $K_{u,\rm PI}$, so that

$$u_{\rm PI}(nT) = u_{\rm PI}(nT-T) + K_{u,\rm PI}\Delta u_{\rm PI}(nT). \tag{5.42}$$

The D controller in the PI+D control system, as shown in Figure 5.19, has y as its input and $u_{\rm D}$ as its output. It is clear that

$$U_{\rm D}(s) = s\, K_d^c\, Y(s),$$

where K_d^c is the control gain and $Y(s)$ is the output signal. Under the bilinear transformation, the above equation becomes

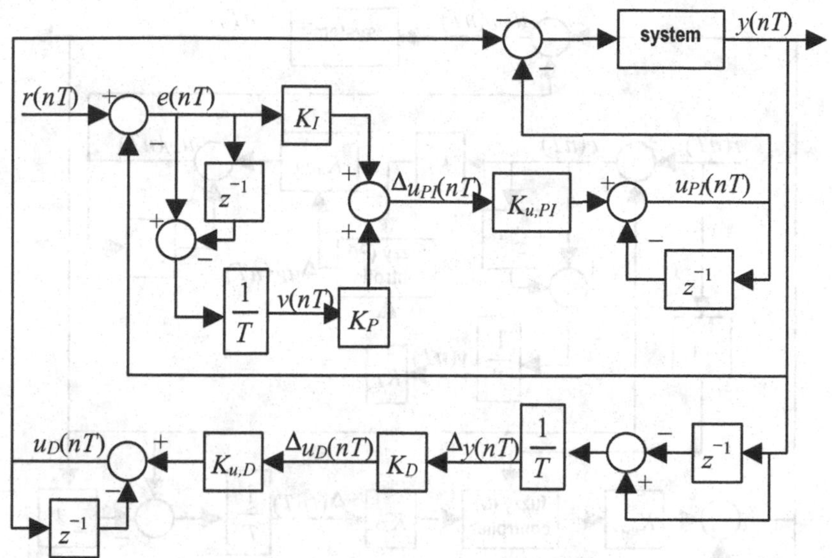

Figure 5.20 The conventional digital PI+D control system.

$$U_D(z) = \frac{2}{T}\,\frac{z-1}{z+1}\,K_d^c\,Y(z),$$

or

$$U_D(z) = K_d^c\,\frac{2}{T}\,\frac{1-z^{-1}}{1+z^{-1}}\,Y(z).$$

Consequently,

$$u_D(nT) + u_D(nT-T) = \frac{2K_d^c}{T}\,[\,y(nT) - y(nT-T)\,].$$

Then, dividing this equation by T yields

$$\Delta u_D(nT) = K_d \Delta y(nT) \tag{5.43}$$

where

$$\Delta u_D(nT) = \frac{u_D(nT) + u_D(nT-T)}{T}$$

is the incremental control output of the fuzzy D controller,

$$\Delta y(nT) = \frac{y(nT) - y(nT-T)}{T}$$

is the rate of change of the output y, and

$$K_d = \frac{2K_d^c}{T}. \tag{5.44}$$

As will be further described in the fuzzification step below, we modify (5.43) by adding the signal $K y_d(nT)$ to its right-hand side, where

$$y_d(nT) = y(nT) - r(nT) = -\,e(nT),$$

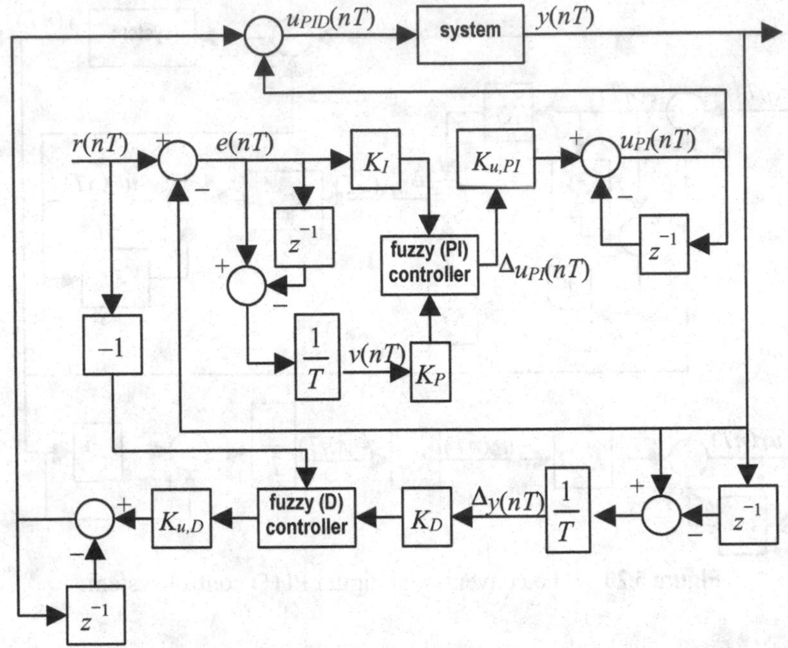

Figure 5.21 The fuzzy PI+D control system.

in order to obtain correct control rules. Thus, (5.43) becomes

$$\Delta u_{\mathrm{D}}(nT) = K_d \Delta y(nT) + K y_d(nT).$$

We always use $K = 1$ to simplify the discussion in this section, which is not necessary in a real design. Note that from (5.43)

$$\Delta u_{\mathrm{D}}(nT) = \frac{u_{\mathrm{D}}(nT) + u_{\mathrm{D}}(nT - T)}{T},$$

so that

$$u_{\mathrm{D}}(nT) = -\, u_{\mathrm{D}}(nT{-}T) + T\Delta u_{\mathrm{D}}(nT).$$

When $\Delta u_{\mathrm{D}}(nT)$ becomes a fuzzy control action later in the design, we use $K_{u,\mathrm{D}}$ as this fuzzy control gain (which will be determined later in the design). Thus, we can rewrite the above formula as

$$u_{\mathrm{D}}(nT) = -\, u_{\mathrm{D}}(nT{-}T) + K_{u,\mathrm{D}}\Delta u_{\mathrm{D}}(nT). \tag{5.45}$$

Finally, the overall fuzzy PI+D control law can be obtained by algebraically summing the fuzzy PI control law (5.42) and fuzzy D law (5.45) together. The result is

$$u_{\mathrm{PID}}(nT) = u_{\mathrm{PI}}(nT) - u_{\mathrm{D}}(nT),$$

or more precisely,

$$\begin{aligned} u_{\mathrm{PID}}(nT) \;&=\; u_{\mathrm{PI}}(nT{-}T) + K_{u,\mathrm{PI}}\Delta u_{\mathrm{PI}}(nT) + \\ & \quad\; u_{\mathrm{D}}(nT{-}T) - K_{u,\mathrm{D}}\Delta u_{\mathrm{D}}(nT). \end{aligned} \tag{5.46}$$

This equation will be referred to as the fuzzy PI+D control law below.

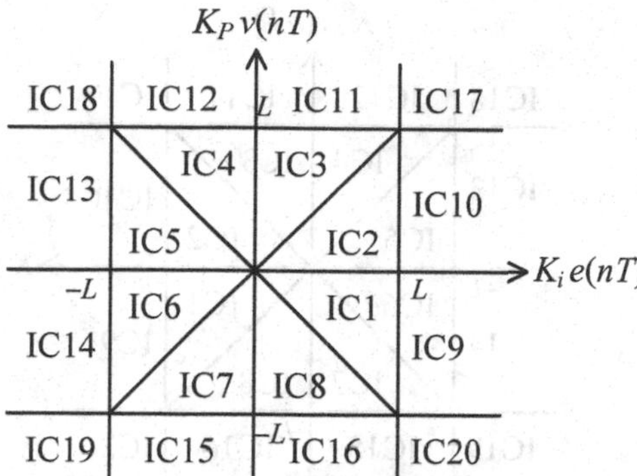

Figure 5.22 Regions of the "fuzzy (PI) controller" input-combination values.

The overall conventional PI+D control system is shown in Figure 5.20. To this end, the fuzzy PI and fuzzy D controllers will be inserted into Figure 5.20, resulting in the configuration shown in Figure 5.21.

In summary, the fuzzy PI+D control system is implemented by Figure 5.21, in which there are five constant control gains that can be tuned: K_i, K_p, $K_{u,\text{PI}}$, K_D, and $K_{u,\text{D}}$, where one may set $K_{u,\text{PI}} = K_{u,\text{D}} = T$ to simplify the design. In this implementation, the two incremental control actions $\Delta u_\text{PI}(nT)$ and $\Delta u_\text{D}(nT)$ have the analytical formulas as shown below.

Similar to Figure 5.18, we first divide the plane as twenty regions of the "fuzzy (PI) controller" input-combination (IC) for $(K_ie(nT),K_pv(nT))$, as shown in Figure 5.22.

Then, the incremental control of the fuzzy PI controller is calculated by formulas (5.29)-(5.37), namely,

$$\Delta u_\text{PI}(nT) = \frac{L[K_ie(nT)+K_pv(nT)]}{2(2L-K_i\,|\,e(nT)\,|)}, \quad \text{in IC1, IC2, IC5, IC6,} \tag{5.47}$$

$$= \frac{L[K_ie(nT)+K_pv(nT)]}{2(2L-K_p\,|\,e(nT)\,|)}, \quad \text{in IC3, IC4, IC7, IC8,} \tag{5.48}$$

$$= \frac{1}{2}[\,L+K_pv(nT)\,], \quad \text{in IC9, IC10,} \tag{5.49}$$

$$= \frac{1}{2}[\,L+K_ie(nT)\,], \quad \text{in IC11, IC12,} \tag{5.50}$$

$$= \frac{1}{2}[\,-L+K_pv(nT)\,], \quad \text{in IC13, IC14,} \tag{5.51}$$

$$= \frac{1}{2}[\,-L+K_ie(nT)\,], \quad \text{in IC15, IC16,} \tag{5.52}$$

$$= 0, \quad \text{in IC18, IC20,} \tag{5.53}$$

$$= -L, \quad \text{in IC17,} \tag{5.54}$$

$$= L, \quad \text{in IC19.} \tag{5.55}$$

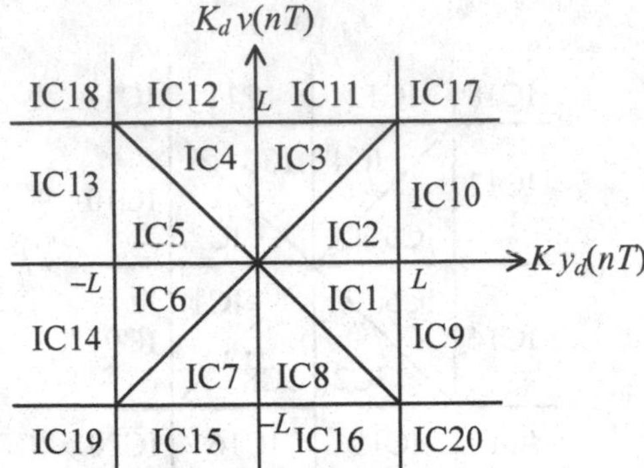

Figure 5.23 Regions of the "fuzzy (D) controller" input-combination values.

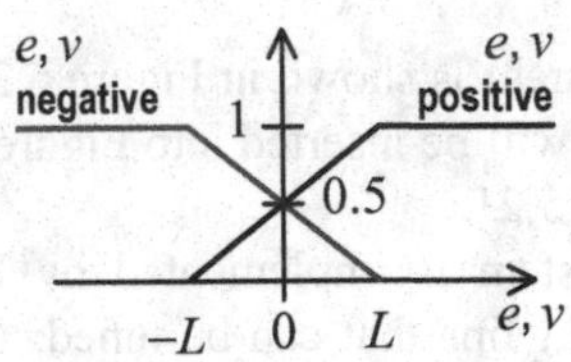

Figure 5.24 Input membership functions for the PI component.

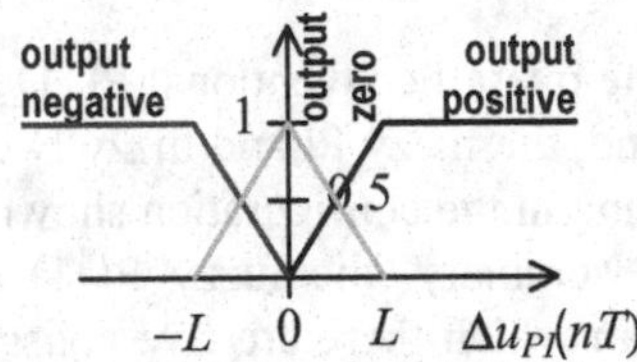

Figure 5.25 Output membership functions for the PI component.

For the fuzzy D controller, the twenty regions of the input-combination are shown in Figure 5.23. The incremental control of the fuzzy D controller is calculated by the following formulas:

$$\Delta u_D(nT) = \frac{L[Ky_d(nT) + K_d\Delta y(nT)]}{2(2L - K\,|\,y_d(nT)\,|)}, \quad \text{in IC1, IC2, IC5, IC6,} \tag{5.56}$$

$$= \frac{L[Ky_d(nT) + K_d\Delta y(nT)]}{2(2L - K_d\,|\,y_d(nT)\,|)}, \quad \text{in IC3, IC4, IC7, IC8,} \tag{5.57}$$

$$= \frac{1}{2}[\,L - K_d\Delta y(nT)\,], \quad \text{in IC9, IC10,} \tag{5.58}$$

$$= \frac{1}{2}[\,-L + Ky_d(nT)\,], \quad \text{in IC11, IC12,} \tag{5.59}$$

$$= \frac{1}{2}[\,-L - K_d\Delta y(nT)\,], \quad \text{in IC13, IC14,} \tag{5.60}$$

$$= \frac{1}{2}[\,L + Ky_d(nT)\,], \quad \text{in IC15, IC16,} \tag{5.61}$$

$$= 0, \quad \text{in IC18, IC20,} \tag{5.62}$$

$$= -L, \quad \text{in IC17,} \tag{5.63}$$

$$= L, \quad \text{in IC19.} \tag{5.64}$$

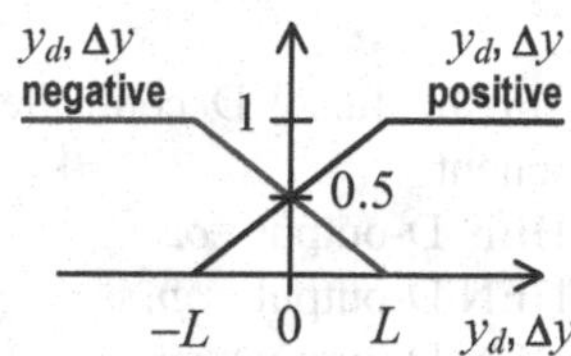

Figure 5.26 Input membership functions for the D component.

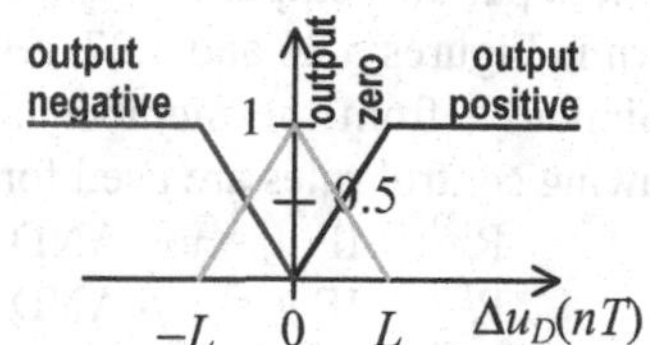

Figure 5.27 Output membership functions for the D component.

We next give detailed derivations of the above formulas of the fuzzy PI and D controllers.

We first fuzzify the PI and D components of the PI+D control system individually and then establish the desired fuzzy control rules for each of them taking into consideration the interconnected PI+D fuzzy control law given in equation (5.46). The input and output membership functions of the PI component are shown in Figures 5.24 and 5.25, respectively.

The fuzzy PI controller employs two inputs: the error signal $e(nT)$ and the rate of change of the error signal $v(nT)$. The fuzzy PI controller has a single output, $\Delta u_{\mathrm{PI}}(nT)$, as shown in Figure 5.21, where the constant $L > 0$ is a tunable parameter (which can also be fixed after being determined).

Using these membership functions, the following control rules are established for the fuzzy PI controller:

$\mathrm{R}^{(1)}$: IF $e = en$ AND $v = vn$ THEN PI-output $= on$.
$\mathrm{R}^{(2)}$: IF $e = en$ AND $v = vp$ THEN PI-output $= oz$.
$\mathrm{R}^{(3)}$: IF $e = ep$ AND $v = vn$ THEN PI-output $= oz$.
$\mathrm{R}^{(4)}$: IF $e = ep$ AND $v = vp$ THEN PI-output $= op$.

In these rules, $e := r - y$ is the error, $v = \dot{e} = 0 - \dot{y} = -\dot{y}$ is the rate of change of the error, "PI-output" is the fuzzy PI control output $\Delta u_{\mathrm{PI}}(nT)$, "*ep*" means "error positive," "*op*" means "output positive," etc. Also, AND is the logical AND operator defined as before.

If we look at the fuzzy control law for the D component, equation (5.43), the only information it contains that is relevant to the output performance is $\Delta y(nT)$. Based on this signal alone, it is impossible to come up with a useful fuzzy control law. We therefore look for another control signal that can be used in conjunction with $\Delta y(nT)$, to provide information about the output (above or below the reference signal). For this purpose, a logical and natural choice is the negative error signal

$$y_d(nT) = -e(nT). \tag{5.65}$$

Here, it is important to observe that y_d positive (resp. negative) means that the system output y is above (resp. below) the reference r. This y_d control signal is implemented as shown by the path with the −1 block in Figure 5.21 (compared with Figure 5.20).

The input and output membership functions for the fuzzy D controller are shown in Figures 5.26 and 5.27, respectively.

Similarly, from the membership functions of the fuzzy D controller, the following control rules are used for the D component:

$R^{(5)}$: IF $y_d = y_d p$ AND $\Delta y = \Delta yp$ THEN D-output =*oz*.
$R^{(6)}$: IF $y_d = y_d p$ AND $\Delta y = \Delta yn$ THEN D-output =*op*.
$R^{(7)}$: IF $y_d = y_d n$ AND $\Delta y = \Delta yp$ THEN D-output =*on*.
$R^{(8)}$: IF $y_d = y_d n$ AND $\Delta y = \Delta yn$ THEN D-output =*oz*.

In the above rules, "D-output" is the fuzzy D control output $\Delta u_D(nT)$, and the other terms are defined similar to the PI components.

These eight rules altogether yield the control actions for the fuzzy PI+D control law.

Formulation of these rules can be understood as follows. For Rule 1 ($R^{(1)}$): if we look at this rule for the PI controller, condition "*en*" (error is negative) implies that the system output, y, is above the set-point, and "*vn*" (rate of error is negative) implies $\dot{y} > 0$ (meaning that the controller at the previous step is driving the system output to move upward). Since the $\Delta u_{PI}(nT)$ component of formula (5.46) contains more control terms with gain parameters than the D controller, we set this term to be negative and set the $\Delta u_D(nT)$ component to be zero. Thus, the combined control action will drive the system output to move downward by Rules 1 and 5 of both controllers.

We note that one could have set the output of the D controller to be positive in order to drive the output to move downward faster. However, having both controllers' outputs being nonzero at the same time will complicate the design of the controller. Computer simulations have demonstrated that the simple design described above performs sufficiently well in all numerical examples tested, so more sophisticated design is not discussed here.

Similarly, for Rule 2 ($R^{(2)}$), since the output is above the set-point and is moving downward, we set the "larger" component of formula (5.46), namely, the term $\Delta u_{PI}(nT)$, to be zero, and set the "smaller" component $\Delta u_D(nT)$ to be positive. Thus, the combined controller will tend to drive the system output to move downward faster by the combined action of these two rules.

Rules 3 and 4 are similarly determined.

In the defuzzification step, for both fuzzy PI and D controllers, the same weighted average formula is employed to defuzzify the incremental control of the fuzzy control law:

$$\Delta u(nT) = \frac{\sum(\text{membership value of input} \times \text{corresponding value of output})}{\sum(\text{membership value of input})} \tag{5.66}$$

For the fuzzy PI controller, the value ranges of the two inputs, the error, and the rate of change of the error are actually decomposed into twenty adjacent input-combination (IC) regions, as shown in Figure 5.22. This figure is understood as follows: We put the membership function of the error signal (given by the curves for e in Figure 5.24) over the horizontal $K_i e(nT)$-axis on Figure 5.22, and put the membership function of the rate of change of the error signal (given by the same curves in Figure 5.24 for v) over the vertical $K_p v(nT)$-axis on Figure 5.22. These two membership functions then overlap and form the third-dimensional picture (which is not shown in Figure 5.22) over the two-dimensional regions shown in Figure 5.22. When we look at region IC1, for example, if we look upward to the $K_i e(nT)$-axis, we see the domain $[0,L]$ and the membership function (in the third dimension) over $[0,L]$ of the error signal; if we look leftward to the $K_p v(nT)$-axis, we see the domain $[-L,0]$ and the membership function (in the third dimension) over $[-L,0]$ of the rate of change of the error signal. This situation is completely analogous to the analysis of the fuzzy PD controller studied in Section II.A.

The control rules for the fuzzy PI controller ($R^{(1)}$-$R^{(4)}$), with membership functions and IC regions together, are used to evaluate appropriate fuzzy control formulas for each region.

In doing so, we consider the locations of the error $K_i e(nT)$ and the rate $K_p v(nT)$ in the regions IC1 and IC2 (see Figure 5.22). Let us look at region IC1, for example, where we have $e > 0.5 > v(nT)$ (see Figure 5.24). Hence, the logical AND used in ($R^{(1)}$) leads to

$$\{\text{"error} = e \text{ AND rate} = v\text{"}\} = \min\{\, e, v \,\} = e,$$

so that Rule 1 ($R^{(1)}$) yields

$$R^{(1)}: \begin{cases} \text{the selected input membership value is } en; \\ \text{the corresponding output value is } on. \end{cases}$$

Similarly, in region IC1, Rules 2-4, ($R^{(2)}$)-($R^{(4)}$), are obtained as

$$R^{(2)}: \begin{cases} \text{the selected input membership value is } en; \\ \text{the corresponding output value is } oz. \end{cases}$$

$$R^{(3)}: \begin{cases} \text{the selected input membership value is } vn; \\ \text{the corresponding output value is } oz. \end{cases}$$

$$R^{(4)}: \begin{cases} \text{the selected input membership value is } vp; \\ \text{the corresponding output value is } op. \end{cases}$$

It can be verified that the above are true for the two regions IC1 and IC2. Thus, in regions IC1 and IC2, it follows from the defuzzification formula (5.66) that

$$\Delta u(nT) = \frac{en \times on + en \times oz + vn \times oz + vp \times op}{en + en + vn + vp}.$$

It is very important to note that if one follows the above procedure to work through the two cases, then it is found that both the last two cases give the same result of en (i.e., the two en in the above formula are not the misprint of en and ep!). To this end, by applying $op = L$, $on = -L$, $oz = 0$ (obtained from

Figure 5.25), and the following straight line formulas from the geometry of the membership functions associated with Figure 5.22:

$$ep = \frac{K_i e(nT) + L}{2L}, \qquad en = \frac{-K_i e(nT) + L}{2L},$$

$$vp = \frac{K_p v(nT) + L}{2L}, \qquad vn = \frac{-K_p v(nT) + L}{2L},$$

we obtain

$$\Delta u_{\mathrm{PI}}(nT) = \frac{L[K_i e(nT) + K_p v(nT)]}{2[2L - K_i e(nT)]}.$$

Here, we note that $e(nT) \geq 0$ in regions IC1 and IC2. In the same way, one can verify that in regions IC5 and IC6,

$$\Delta u_{\mathrm{PI}}(nT) = \frac{L[K_i e(nT) + K_p v(nT)]}{2[2L + K_i e(nT)]},$$

where it should be noted that $e(nT) \leq 0$ in regions IC5 and IC6. Thus, by combining the above two formulas, we arrive at the following control formula for the four regions IC1, IC2, IC5, and IC6:

$$\Delta u_{\mathrm{PI}}(nT) = \frac{L[K_i e(nT) + K_p v(nT)]}{2[2L - K_i \,|\, e(nT) \,|]}.$$

Working through all regions in the same way, we obtain the PI control formulas (5.47)-(5.55) for the twenty IC regions.

Similarly, defuzzification of the fuzzy D controller follows the same procedure as described above for the PI component, except that the input signals in this case are different. The IC combinations of these two inputs are decomposed into twenty similar regions, as shown in Figure 5.23.

Similarly, by applying the value $op = L$, $on = -L$, $oz = 0$, and the following straight line formulas obtained from the geometry of Figure 5.27:

$$y_d p = \frac{K y_d(nT) + L}{2L}, \qquad y_d n = \frac{-K y_d(nT) + L}{2L},$$

$$\Delta yp = \frac{K_d \Delta y(nT) + L}{2L}, \qquad \Delta yn = \frac{-K_d \Delta y(nT) + L}{2L},$$

we obtain the D control formulas (5.56)-(5.64) for the twenty IC regions.

Note that the constant K that multiplies the signal $y_d(nT)$ is used as a parameter for generality here, and in the derivation of $\Delta u_{\mathrm{D}}(nT)$. Although it could be used as a control gain, its value is permanently set to one throughout the computer simulations shown in the next example.

Example 5.4. We first apply the fuzzy PI+D controller to a lower-order linear system, to see how well it performs for such simple cases. Recall that the conventional PI+D controller is designed for linear systems, for which it works very well. We show that the fuzzy PI+D controller is as good as, if not better than, the conventional one for such lower-order linear systems.

The first example is a first-order linear system, with transfer function

$$H(s) = \frac{1}{s+1}$$

and with controller parameters $T = 0.1$, $K = 1.0$, $K_p = 1.2$, $K_i = 0.1$, $K_{u,\mathrm{PI}} = 0.2$, $K_{u,\mathrm{D}} = 0.01$, $L = 360.0$. The set-point is $r = 5.0$. The response of the fuzzy PI+D controller for a step input is shown in Figure 5.28.

The second example is a second-order linear system with transfer function

$$H(s) = \frac{1}{s^2 + 4s + 3},$$

where the controller parameters are $T = 0.01$, $K = 1.0$, $K_p = 8.0$, $K_d = 0.01$, $K_i = 1.0$, $K_{u,\mathrm{PI}} = 0.2$, $K_{u,\mathrm{D}} = 0.01$, $L = 1000.0$, and the set-point is $r = 5.0$. The response of the fuzzy PI+D controller is shown in Figure 5.29.

Finally, we compare the performance of both the conventional and the fuzzy PI+D controllers, using two nonlinear systems. Although the fuzzy PI+D controller has the same linear structure as the conventional PI+D controller, the fuzzy PI+D gains are nonlinear with self-tuning capability and, therefore, have better performance in this simulation and in general.

The first nonlinear system has the following simple model:

$$\dot{y}(t) = 0.0001\,|y(t)| + u_{\mathrm{PID}}(t),$$

with fuzzy PI+D parameters $T = 0.1$, $K = 1.0$, $K_p = 1.5$, $K_d = 0.1$, $K_i = 2.0$, $K_{u,\mathrm{PI}} = 0.11$, $K_{u,\mathrm{D}} = 1.0$, $L = 45.0$, and the set-point is $r = 5.0$. The result is shown in Figure 5.30. On the contrary, the conventional PI+D controller is unable to track the set-point, no matter how one changes its parameters. A typical response of the conventional PI+D controller is shown in Figure 5.31, with parameters $T = 0.1$, $K_p^c = 19.5$, $K_i^c = 1.0$, for $r = 5.0$.

The second nonlinear system used in the simulation is

$$\dot{y}(t) = y(t) + \sqrt{y(t)} + u_{\mathrm{PID}}(t).$$

In this case, the fuzzy PI+D parameters are $T = 0.1$, $K = 1.0$, $K_p = 2.0$, $K_d = 1.942$, $K_i = 1.0$, $K_{u,\mathrm{PI}} = 0.1$, $K_{u,\mathrm{D}} = 0.27$, $L = 350.0$, and the set-point $r = 5.0$. The response of this fuzzy PI+D control system to a step input is shown in Figure 5.32. The conventional PI+D controller, however, cannot yield any reasonable response, no matter how one adjusts its gains. A typical response of the conventional PI+D controller (with parameters $T = 0.1$, $K_p^c = 2.0$, $K_i^c = 1.0$, and $r = 5.0$) is shown in Figure 5.33. We remark that this result is due to the fact that the conventional PI+D controller usually has difficulties in controlling higher-order and time-delayed linear systems as well as nonlinear systems, because they are designed only for lower-order linear systems, for which they can work very well as has been widely experienced.

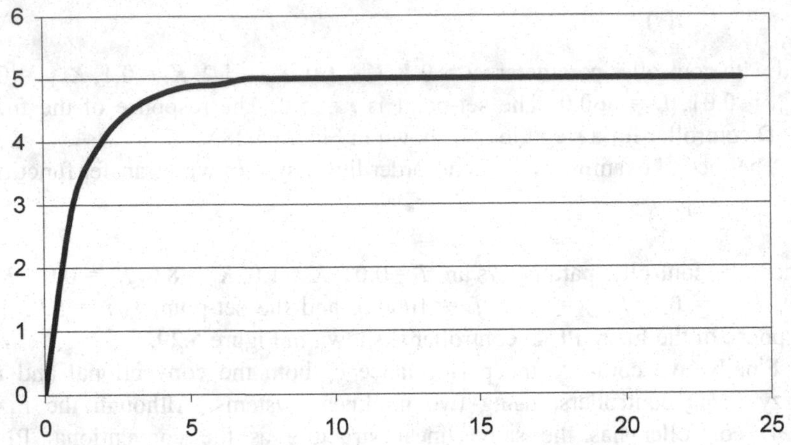

Figure 5.28 Output of a first-order linear fuzzy PI+D control system.

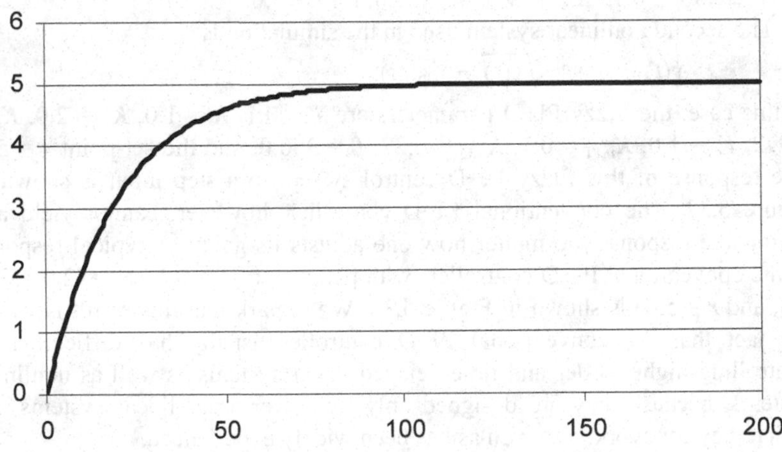

Figure 5.29 Output of a second-order linear fuzzy PI+D control system.

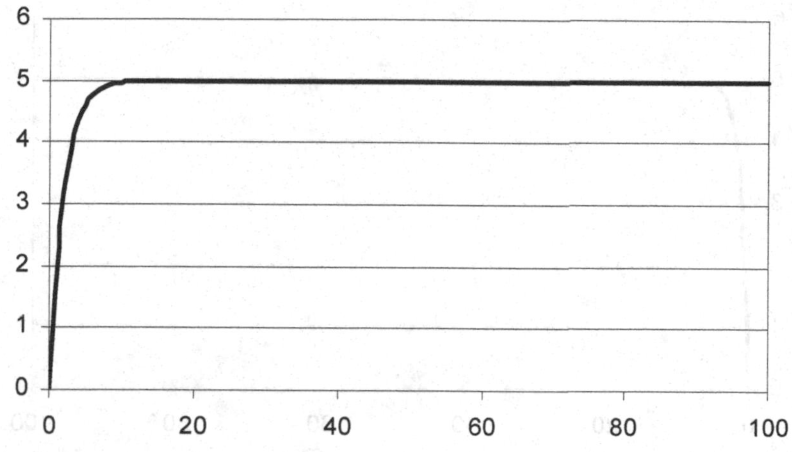

Figure 5.30 Output of a nonlinear fuzzy PI+D control system.

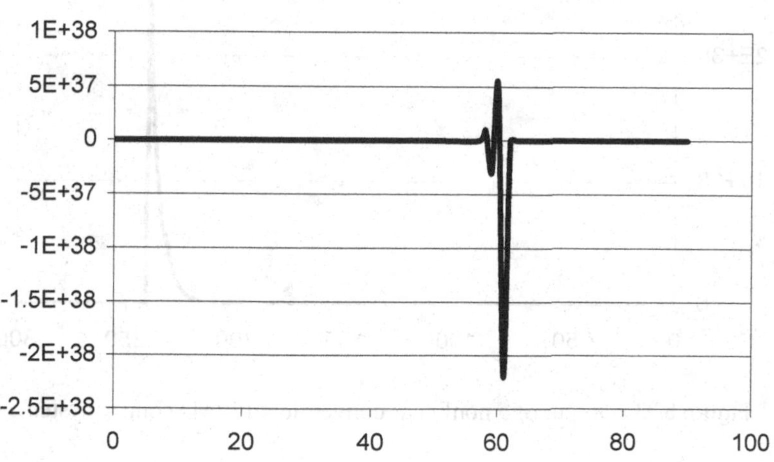

Figure 5.31 Output of a nonlinear conventional PI+D control system.

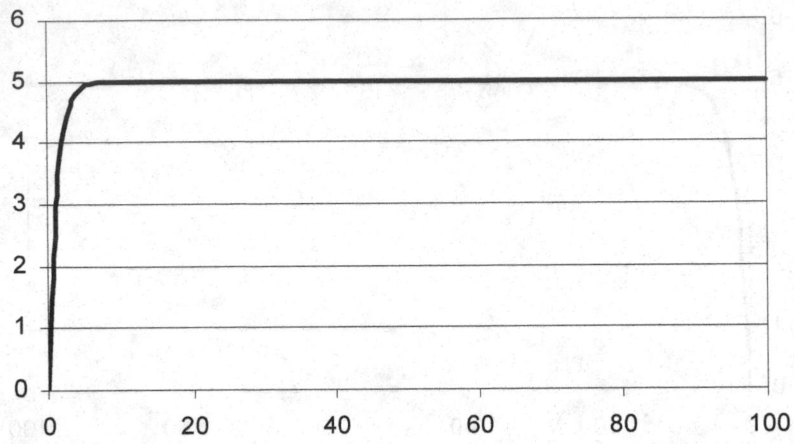

Figure 5.32 Output of a nonlinear fuzzy PI+D control system.

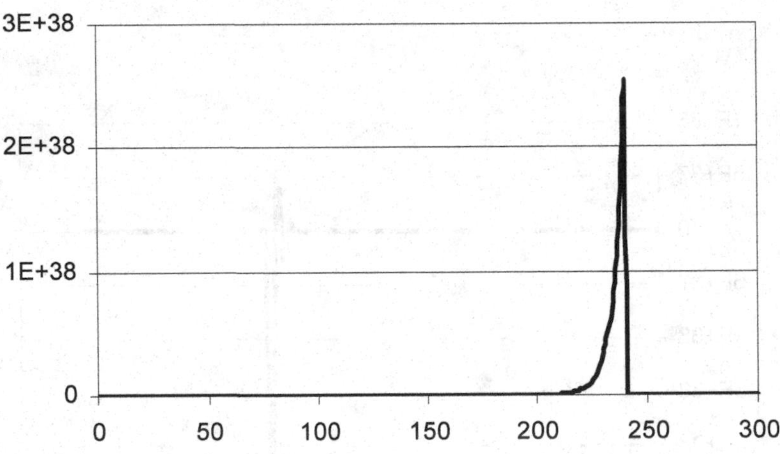

Figure 5.33 Output of a nonlinear conventional PI+D control system.

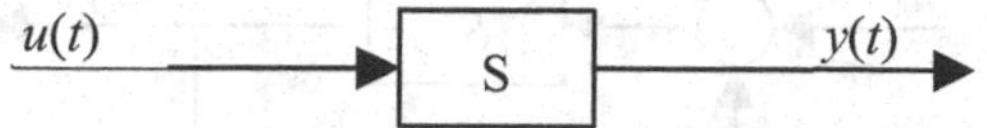

Figure 5.34 Input-output relation of a system.

III. FUZZY PID CONTROLLERS: STABILITY ANALYSIS

In this section, we analyze the stability of the fuzzy PI, PD, and PI+D control systems designed and studied in the last section.

Recall the Lyapunov asymptotic stability for fuzzy modeling discussed in Chapter 3, for both discrete-time and continuous-time fuzzy dynamic systems. In order to show another important type of stability, namely, the bounded-input bounded-output (BIBO) stability of a control system, we introduce the Small Gain Theorem in this section and discuss the BIBO stability of the fuzzy PI, PD, and PI+D control systems. We first note that differing from the Lyapunov asymptotic stability, which is usually local (in a neighborhood of an equilibrium point), the BIBO stability is global and is particularly suitable for nonlinear systems described by input-output maps. We should also note that both the Lyapunov asymptotic stability and the BIBO stability analyses provide conservative sufficient conditions, especially for nonlinear systems. From a theoretical point of view, the larger the region of stability can be found, the better the result is. However, from the design point of view, relatively conservative stability region is actually safer and more reliable in applications. BIBO stability theory turns out to be appropriate for this purpose.

A. BIBO Stability and the Small Gain Theorem

Definition 5.1. A (linear or nonlinear) control system is said to be *bounded-input bounded-output* (*BIBO*) *stable* if a bounded control input to the system always produces a bounded output through the system.

Here, the boundedness is defined in the norm (l_2, l_∞, etc.) of the function space which we consider in the design.

Let S denote a (linear or nonlinear) system. S may be considered as a *mapping* which maps a control input, $u(t)$, to the corresponding system output $y(t)$, as shown in Figure 5.34, where

$$S: u(t) \to y(t) \quad \text{or} \quad y(t) = S\{u(t)\}.$$

Recall the standard L_p-spaces of signals:

$$1 \le p < \infty: \qquad L_p = \{f(t) |\ \int_0^\infty |f(t)|^p \, dt < \infty \},$$

$$p = \infty: \qquad L_p = \{f(t) |\ \operatorname*{ess\,sup}_{0 \le t < \infty} |f(t)| < \infty \},$$

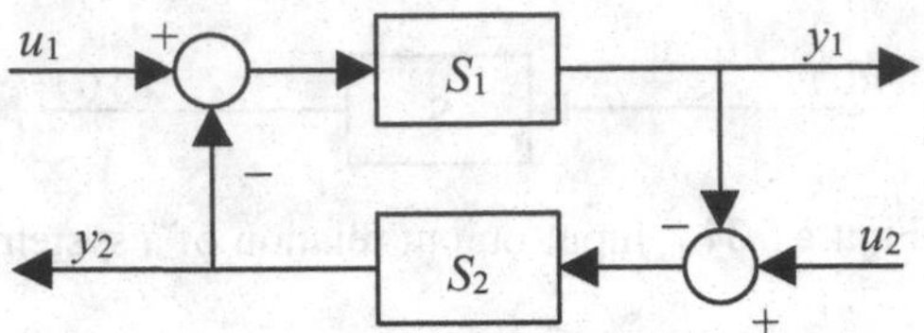

Figure 5.35 A nonlinear feedback system.

where "*ess*" means "essential," namely, the supirum holds except over a set of measure zero. For piecewise continuous signals, essential supirum and supirum are the same, so "*ess*" can be dropped from the above.

Consider a nonlinear (including linear) feedback system shown in Figure 5.35, where for simplicity it is assumed that all signals $u, e, y_1, u_2, e_2, y_2 \in R^n$. It is clear from Figure 5.35 that

$$\begin{cases} e_1 = u_1 - S_2(e_2), \\ e_2 = u_2 + S_1(e_1), \end{cases} \tag{5.67}$$

or, equivalently,

$$\begin{cases} u_1 = e_1 + S_2(e_2), \\ u_2 = e_2 - S_1(e_1). \end{cases} \tag{5.68}$$

For this system, we have the following so-called Small Gain Theorem, which gives sufficient conditions under which a "bounded input" yields a "bounded output," where the norm $\| \times \|$ is the standard Euclidean norm ("length" of a vector).

Theorem 5.1. Consider the nonlinear feedback system shown in Figure 5.35, which is described by the relationship (5.67)-(5.68). Suppose that there exist constants L_1, L_2, M_1, M_2, with $L_1L_2 < 1$, such that

$$\begin{cases} \| S_1(e_1) \| \le M_1 + L_1 \| e_1 \|, \\ \| S_2(e_2) \| \le M_2 + L_2 \| e_2 \|. \end{cases} \tag{5.69}$$

Then, we have

$$\begin{cases} \| e_1 \| \le (1 - L_1L_2)^{-1} \, (\| u_1 \| + L_2 \| u_2 \| + M_2 + L_2M_1), \\ \| e_2 \| \le (1 - L_1L_2)^{-1} \, (\| u_2 \| + L_1 \| u_1 \| + M_1 + L_1M_2). \end{cases} \tag{5.70}$$

Proof. It follows from

$$e_1 = u_1 - S_2(e_2)$$

that

$$\begin{aligned} \| e_1 \| &\le \| u_1 \| + \| S_2(e_2) \| \\ &\le \| u_1 \| + M_2 + L_2 \| e_2 \|. \end{aligned}$$

Similarly, we have

$$\| e_2 \| \le \| u_2 \| + M_1 + L_1 \| e_1 \|.$$

Combining these two inequalities, we obtain

$$\| e_1 \| \le L_1L_2 \| e_1 \| + \| u_1 \| + L_2 \| u_2 \| + M_2 + L_2M_1,$$

or, using the fact $L_1L_2 < 1$,

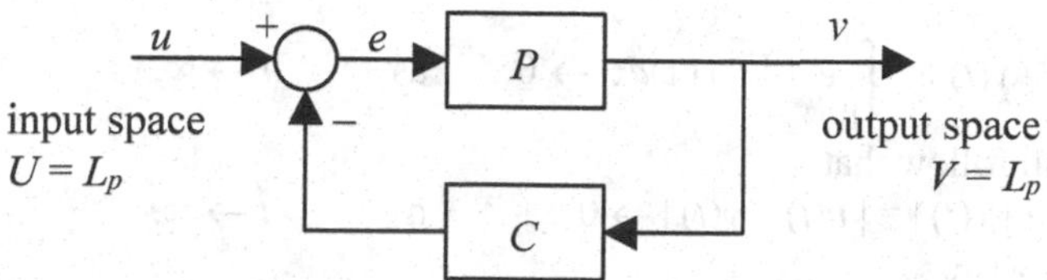

Figure 5.36 A feedback control system.

$$\| e_1 \| \leq (1-L_1L_2)^{-1}(\| u_1 \| + L_2 \| u_2 \| + M_2 + L_2M_1).$$

The rest of the theorem follows immediately.

It is clear that the Small Gain Theorem is applicable to both continuous-time and discrete-time systems, and to both SISO and MIMO systems. Hence, although its statement and proof are quite simple, it is very useful.

We next point out an interesting relation between the BIBO stability and the Lyapunov asymptotic stability. It is clear that the asymptotic stability generally implies the BIBO stability, but the reverse can also be true under some conditions.

Consider a nonlinear system described by the following first-order vector-valued ordinary differential equation:

$$\begin{cases} \dot{x}(t) = Ax(t) - f(x(t),t), \\ x(0) = x_0, \end{cases} \tag{5.71}$$

with an equilibrium solution $\bar{x}(t) = 0$, where A is an $n \times n$ constant matrix whose eigenvalues are assumed to have negative real parts, and $f: R^n \times R^1 \to R^n$ is a real vector-valued integrable nonlinear function of $t \in [0,\infty)$. By adding and then subtracting the term $Ax(t)$, a general nonlinear system can always be written in this form. Let

$$\begin{cases} x(t) = u(t) - \int_0^t e^{(t-\tau)} y(\tau) d\tau, \\ y(x) = f(x(t),t), \end{cases} \tag{5.72}$$

with $u(t) = e^{At}x_0$. Then we can implement system (5.72) by a feedback configuration as depicted in Figure 5.36, where the error signal $e(t) = x(t)$, the plant $P(\cdot)(t) = f(\cdot, t)$, and the compensator $C(\cdot)(t) = \int_0^t e^{(t-\tau)A}(\cdot)(\tau)d\tau$.

Theorem 5.2. Consider the nonlinear system (5.72) and its associate feedback configuration shown in Figure 5.36. Suppose that $U = V = L_p([0,\infty),R^n)$, where $1 \leq p \leq \infty$. Then, if the feedback system shown in Figure 5.36 is BIBO stable, then it is also asymptotically stable.

Proof. Since all eigenvalues of the constant matrix A have negative real parts, we have

$$| e^{tA} x_0 | \leq Me^{-\alpha t}$$

for some constants $0 < \alpha, M < \infty$ for all $t \in [0,\infty)$, so that $|u(t)| = | e^{tA} x_0 | \to 0$ as $t \to \infty$. Hence, in view of the first equation defined above, i.e., $x(t) = u(t) - \int_0^t e^{(t-\tau)A} y(\tau)d\tau$, if we can prove that

$$v(t) := \int_0^t e^{(t-\tau)A} y(\tau) d\tau \to 0 \quad \text{as} \quad t \to \infty,$$

then it will follow that

$$|x(t)| = |u(t) - v(t)| \to 0 \quad \text{as} \quad t \to \infty.$$

To do so, write

$$\begin{aligned} v(t) &= \int_0^{t/2} e^{(t-\tau)A} y(\tau) d\tau + \int_{t/2}^t e^{(t-\tau)A} y(\tau) d\tau \\ &= \int_{t/2}^t e^{\tau A} y(t-\tau) d\tau + \int_{t/2}^t e^{(t-\tau)A} y(\tau) d\tau . \end{aligned}$$

Then, by the Hölder inequality we have

$$\begin{aligned} |v(t)| &\le \left| \int_{t/2}^t e^{\tau A} y(t-\tau) d\tau \right| + \left| \int_{t/2}^t e^{(t-\tau)A} y(\tau) d\tau \right| \\ &\le \left[\int_{t/2}^t | e^{\tau A} |^q \, d\tau \right]^{1/q} \left[\int_{t/2}^t | y(t-\tau) |^p \, d\tau \right]^{1/p} \\ &\quad + \left[\int_{t/2}^t | e^{(t-\tau)A} |^q \, d\tau \right]^{1/q} \left[\int_{t/2}^t | y(\tau) |^p \, d\tau \right]^{1/p} \\ &\le \left[\int_{t/2}^\infty | e^{\tau A} |^q \, d\tau \right]^{1/q} \left[\int_0^\infty | y(t-\tau) |^p \, d\tau \right]^{1/p} \\ &\quad + \left[\int_{t/2}^\infty | e^{(t-\tau)A} |^q \, d\tau \right]^{1/q} \left[\int_0^\infty | y(\tau) |^p \, d\tau \right]^{1/p} . \end{aligned}$$

Since all eigenvalues of A have negative real parts and since the feedback system is BIBO stable from U to V, so that $y \in V = L_p([0,\infty),R^n)$, we have

$$\lim_{t \to \infty} \int_{t/2}^{\infty} | e^{\tau A} |^q \, d\tau = 0$$

and

$$\lim_{t \to \infty} \int_{t/2}^{\infty} | y(\tau) |^p \, d\tau = 0.$$

Therefore, it follows that $| v(t) | \to \infty$ as $t \to \infty$, completing the proof of the theorem.

B. BIBO Stability of Fuzzy PD Control Systems

We are now in a position to study the BIBO stability of the fuzzy PI, PD, and PI+D control systems.

We first discuss in detail the fuzzy PD system. Return to the fuzzy PD control system described in Figure 5.8.

We consider the general case where the system under control is nonlinear, which will be denoted by N. Suppose that the fuzzy control law (5.26)

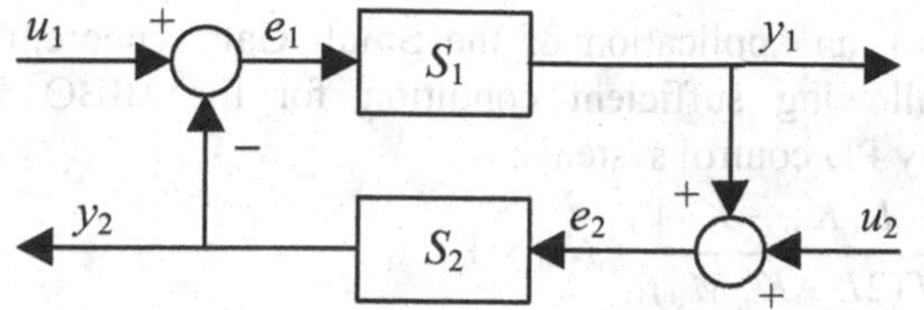

Figure 5.37 An equivalent closed-loop control system.

together with the incremental control formula (5.16) are used, and let the reference signal be $r = r(nT)$ for generality. By defining

$$\begin{cases} e_1(nT) &= e(nT), \\ e_2(nT) &= u(nT), \\ u_1(nT) &= r(nT), \\ u_2(nT) &= -u(nT-T), \\ S_1(e_1(nT)) &= K_u \Delta u(nT), \\ S_2(e_2(nT)) &= N(e_2(nT)), \end{cases} \tag{5.73}$$

it is easy to see that an equivalent closed-loop control system as shown in Figure 5.37 is

$$\begin{cases} u_1(nT) &= r(nT) = e(nT) + N(u(nT)) \\ &= e_1(nT) + S_2(e_2(nT)), \\ u_2(nT) &= -u(nT-T) = u(nT) - K_u \Delta u(nT) \\ &= e_2(nT) + S_1(e_1(nT)). \end{cases} \tag{5.74}$$

Observe that when $e(nT)$ and $r(nT)$ are in the regions IC1, IC2, IC5, IC6, we have formula (5.16), so that

$$\begin{aligned} &\| S_1(e_1(nT)) \| \\ &\leq \frac{K_u L}{2(2L - K_p M_e)} \left[\frac{|K_p - K_d|}{T} |e_1(nT)| + |e(nT-T)| \right] \\ &= \frac{K_u L |K_p - K_d|}{2(2L - K_p M_e)} M_e + \frac{K_u L |K_p - K_d|}{2(2L - K_p M_e)} |e_1(nT)|, \end{aligned} \tag{5.75}$$

and

$$\| S_2(e_2(nT)) \| \leq \| N \| \times | e_2(nT) |, \tag{5.76}$$

where $\| N \|$ is the operator norm of the given $N(\times)$, or the gain of the given nonlinear system, defined as usual by

$$\| N \| := \sup_{v_1 \neq v_2, n \geq 0} \frac{| N(v_1(nT)) - N(v_2(nT)) |}{| v_1(nT) - v_2(nT) |} \tag{5.77}$$

over a set of admissible control signals that have any meaningful function norms, and M_e is defined by

$$M_e := \sup_{n \geq 1} | d(nT) | = \sup_{n \geq 1} \frac{2}{T} | e(nT) |. \tag{5.78}$$

To this end, an application of the Small Gain Theorem (Theorem 5.1) yields the following sufficient condition for the BIBO stability of the nonlinear fuzzy PD control systems:

$$\frac{K_u L \,|\, K_p - K_d \,|}{2T(2L - K_p M_e)} \,\|\, N \,\|\, < 1. \tag{5.79}$$

When $e(nT)$ and $r(nT)$ are in the regions IC3, IC4, IC7, IC8, we can similarly obtain a sufficient stability condition as follows:

$$\frac{K_u L \,|\, K_p - K_d \,|}{2T(2L - K_p M_r)} \,\|\, N \,\|\, < 1, \tag{5.80}$$

where

$$M_r := \sup_{n \geq 1} |\, r(nT) \,| = \sup_{n \geq 1} \frac{1}{T} |\, e(nT) - e(nT - T) \,| \leq M_e. \tag{5.81}$$

When $e(nT)$ and $r(nT)$ are in the rest of the regions, from IC9-IC20, the other incremental control formulas (5.18)-(5.24) are used. In these cases, the stability conditions are found to be

$$\begin{cases} \frac{K_u K_p}{2T} \,\|\, N \,\|\, < 1, \\ \frac{K_u K_d}{2T} \,\|\, N \,\|\, < 1 \\ \|\, N \,\| \text{ is bounded.} \end{cases} \tag{5.82}$$

By combining all the above conditions together, and noting that in IC1-IC8, $K_p M_e \leq L$, and that $K_p > 0$ and $K_d > 0$, we arrive at the following result for the stability of the nonlinear fuzzy PD control systems.

Theorem 5.3. A sufficient condition for the nonlinear fuzzy PD control systems to be BIBO stable is that the given nonlinear system has a bounded norm (gain) $\|\, N \,\| < \infty$ and the parameters of the fuzzy PD controller, K_p, K_d, and K_u, satisfy

$$\frac{\gamma K_m K_u}{2TL} \,\|\, N \,\|\, < 1, \tag{5.83}$$

where

$$\gamma = \max\{\, 1, L \,\} \qquad \text{and} \qquad K_m = \max\{\, K_p, K_d \,\}.$$

We remark that this theorem provides a useful criterion for the design of the nonlinear fuzzy PD controller when a nonlinear process N is given. We may first choose $K_u = T$ and then find a value of K_d (or K_p) for the tracking purpose

$$e(nT) = y(nT) - r(nT) \to 0 \quad \text{as} \quad n \to \infty. \tag{5.84}$$

Finally, we may determine K_d (or K_p) among all possible choices such that the inequality (5.83) is satisfied.

C. BIBO Stability of Fuzzy PI Control Systems

The analysis of the BIBO stability condition for the fuzzy PI control system is similar to that for the fuzzy PD control system discussed in the last subsection.

We again consider the general case where the system under control, N, is nonlinear. We start with the configuration shown in Figure 5.17 with the control law (5.38) and (5.39). By defining

$$\begin{cases} e_1(nT) &=& e(nT), \\ e_2(nT) &=& u(nT), \\ u_1(nT) &=& r(nT), \\ u_2(nT) &=& -u(nT-T), \\ S_1(e_1(nT)) &=& K_u \Delta u(nT), \\ S_2(e_2(nT)) &=& N(e_2(nT)), \end{cases} \tag{5.85}$$

it is easy to see that we obtain an equivalent closed-loop control system as shown in Figure 5.37 where, differing from the fuzzy PD control system, we have

$$\begin{cases} u_1(nT) &=& r(nT) = e(nT) + N(u(nT)) \\ &=& e_1(nT) + S_2(e_2(nT)), \\ u_2(nT) &=& -u(nT-T) = u(nT) - K_u \Delta u(nT) \\ &=& e_2(nT) + S_1(e_1(nT)). \end{cases} \tag{5.86}$$

Observe, moreover, that when $e(nT)$ and $r(nT)$ are in the regions IC1, IC2, IC5, or IC6, we have

$$\begin{aligned} \| S_1(e_1(nT)) \| &= \Bigg\| -\frac{K_u L}{2(2L - K_i \,|\, e_1(nT) \,|)} \times \\ &\quad \left[\left(K_i + \frac{K_p}{T}\right) e_1(nT) - \frac{K_p}{T} e_1(nT-T)\right] \Bigg\| \\ &\le \frac{K_u L}{2(2L - K_i M_e)} \left[\left(K_i + \frac{K_p}{T}\right) | e_1(nT) | - \frac{K_p}{T} M_e\right] \\ &= \frac{K_u K_p M_e L}{2(2L - K_i M_e)} + \frac{K_u K_i + K_p L}{2(2L - K_i M_e)} |e(nT)| \end{aligned} \tag{5.87}$$

and

$$\| S_2(e_2(nT)) \| \le \| N \| \times | e_2(nT) |, \tag{5.88}$$

where $\| N \|$ is the operator norm of the given $N(\times)$, or the gain of the given nonlinear system, defined as in formula (5.77), and M_e is the maximum magnitude of the error signal

$$M_e := \max\{ \sup_{n\ge0} |e(nT)|, \sup_{n\ge1} |e(nT-T)| \}. \tag{5.89}$$

In regions IC1-IC8, $K_i M_e \le L$.

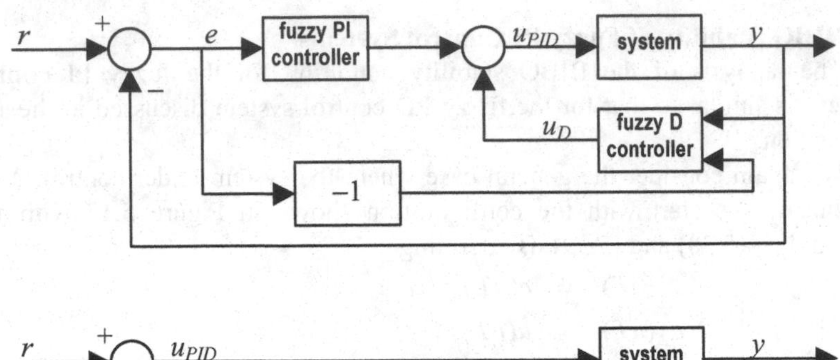

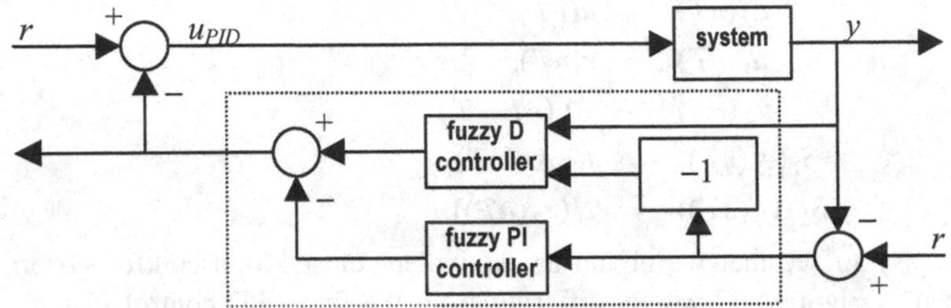

Figure 5.38 Equivalent closed-loop control systems.

To this end, an application of the Small Gain Theorem (Theorem 5.1) produces the following sufficient condition for the BIBO stability of the closed-loop nonlinear fuzzy PI control system:

$$\frac{K_u(TK_i + K_p)}{2T} \| N \| < 1. \tag{5.90}$$

When $e(nT)$ and $r(nT)$ are in the regions IC3, IC4, IC7, or IC8, we can similarly obtain a sufficient stability condition as follows:

$$\frac{K_u(TK_i + K_p)}{2T} \| N \| < 1. \tag{5.91}$$

When $e(nT)$ and $r(nT)$ are in the rest of the regions, from IC9 to IC20, the other control laws are used. In this case, the stability conditions are found to be

$$\begin{cases} \frac{K_u K_i L}{2T} \| N \| < 1, \\ \frac{K_u K_p L}{2T} \| N \| < 1 \\ \| N \| \text{ is bounded.} \end{cases} \tag{5.92}$$

By combining all the above conditions together, we arrive at the following result.

Theorem 5.4. A sufficient condition for the nonlinear fuzzy PI control system shown in Figure 5.17 to be globally BIBO stable is: (1) the given nonlinear system has a bounded norm (gain) $\| N \| < \infty$, and (2) the parameters of the fuzzy PI controller, K_p, K_d, and K_u, satisfy

$$\frac{K_u(\gamma K_i + K_p)}{2T} \| N \| < 1, \tag{5.93}$$

where $\gamma = \max\{L, T\}$.

D. BIBO Stability of Fuzzy PI+D Control Systems

The stability analysis of the fuzzy PI+D control system is in a sense to combine the results obtained individually for the fuzzy PI and PD control systems.

Consider the fuzzy PI+D control system shown in Figure 5.21. First, observe that if we disconnect the fuzzy D controller from Figure 5.21, we have the fuzzy PI control system, exactly the same as Figure 5.17. Hence, all the results obtained in the last subsection for the fuzzy PI control system apply to this situation.

For the fuzzy PI+D control system shown in Figure 5.21, one can easily verify that it is equivalent to either one of the two configurations shown in Figure 5.38. Recall from the Small Gain Theorem (Theorem 5.1) that if we let the system denoted by S_1 and the fuzzy PI+D controller together be denoted by S_2, which is the dashed box in the second picture of Figure 5.38, then we can obtain a sufficient condition for the BIBO stability of the overall closed-loop control system from the bounds

$$\begin{cases} \| S_1(u_{\text{PID}}) \leq M_1 + L_1 \| u_{\text{PID}} \|, \\ \| S_2\left(\begin{bmatrix} y \\ e \end{bmatrix}\right) \leq M_2 + L_2 \| \begin{bmatrix} y \\ e \end{bmatrix} \|, \end{cases} \tag{5.94}$$

where M_1, M_2, L_1, L_2 are constants, with $L_1 L_2 < 1$, as discussed in detail in Theorem 5.1.

Here, we observe that due to the special structure of the D controller, denoted S_D, we actually have

$$\begin{aligned} \left\| S_2\left(\begin{bmatrix} y \\ e \end{bmatrix}\right) \right\| &= \left\| \begin{bmatrix} S_\text{D} & -1 \\ 0 & S_\text{PI} \end{bmatrix} \begin{bmatrix} y \\ e \end{bmatrix} \right\| \\ &\leq \left\| \begin{bmatrix} S_\text{D} & -1 \\ 0 & S_\text{PI} \end{bmatrix} \right\| \times \left\| \begin{bmatrix} y \\ e \end{bmatrix} \right\| \\ &\leq \max\{ \| S_\text{D} \|, \| S_\text{PI} \|, 1 \} \times \max\{ \| y \|, \| e \| \}. \end{aligned}$$

Hence, a sufficient condition for the overall fuzzy PI+D control system to be BIBO stable is the worst one between the fuzzy PI and D control systems. Namely, we may use the larger norms from the right-hand side of the above. Thus, we will have the second equation of inequalities (5.94) in which S_D and S_PI are separated in M_2 and/or L_2, so that the analysis performed in the last two subsections can be repeated here for inequalities (5.94).

Note that in this case we may assume that $\max\{ \|S_\text{D}\|, \|S_\text{PI}\| \} \geq 1$; otherwise, the system will be stable without additional conditions by the contraction mapping principle. Under this inequality, the second condition of (5.94) can be guaranteed.

E. Graphical Stability Analysis of Fuzzy PID Control Systems

We have shown how to analytically analyze the BIBO stability of a fuzzy PID type of control system in the last few sections. Sufficient conditions so obtained are useful for controller design. Although such sufficient conditions are generally conservative, just like many other sufficient conditions derived via different methods for nonlinear systems, they provide useful guidelines for designing "safe" controllers (namely, stabilizing controllers with desirable robustness against system parameter variation and/or external disturbances).

In this section, we introduce a graphical approach to the BIBO stability analysis for PID type of fuzzy control systems. We only discuss the fuzzy PI+D control systems shown in Figure 5.21, but the methodology clearly is applicable to different types of fuzzy PID control systems.

In Figure 5.21, we observe that the control signal $u_{\text{PID}}(nT)$ to the error signal $e(nT)$ can be related implicitly through the closed-loop configuration. In the z-domain, let

$$f(z) = \frac{e(z)}{u_{\text{PID}}(z)} \tag{5.95}$$

and

$$g(z) = \frac{v(z)}{u_{\text{PID}}(z)}, \tag{5.96}$$

where

$$f(z) = f_0 + f_1 z^{-1} + f_2 z^{-2} + \dots$$
$$g(z) = g_0 + g_1 z^{-1} + g_2 z^{-2} + \dots$$

are unknown (not explicitly known) but well-defined, and similarly,

$$e(z) = e_0 + e_1 z^{-1} + e_2 z^{-2} + \dots$$
$$v(z) = v_0 + v_1 z^{-1} + v_2 z^{-2} + \dots$$
$$u_{\text{PID}}(z) = u_0 + u_1 z^{-1} + u_2 z^{-2} + \dots .$$

Let

$$\| F \| = \sum_{i=0}^{\infty} f_i , \qquad \| G \| = \sum_{i=0}^{\infty} g_i ,$$

$$\| E \| = \sum_{i=0}^{\infty} e_i , \qquad \| V \| = \sum_{i=0}^{\infty} v_i , \qquad \| U \| = \sum_{i=0}^{\infty} u_i .$$

For the BIBO stability of the closed-loop system, we must require that both $\| F \|$ and $\| G \|$ be finite. So let

$$\| F \| = \alpha_1,$$
$$\| G \| = \alpha_2,$$

where α_1 and α_2 are constants to be determined. To this end, we have

$$\| E \| \leq \alpha_1 \| U \|, \tag{5.97}$$
$$\| V \| \leq \alpha_2 \| U \|. \tag{5.98}$$

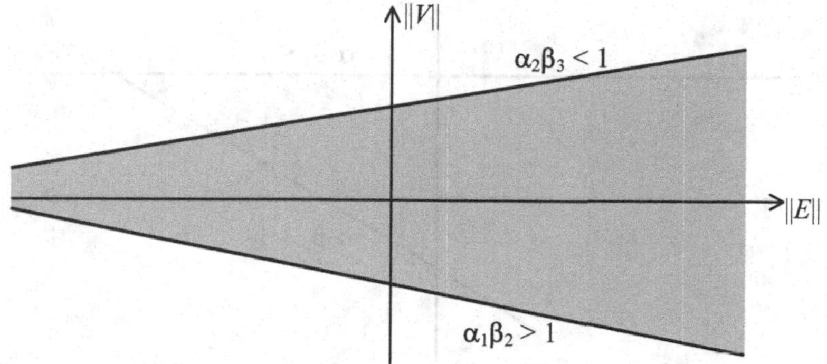

Figure 5.39 An open common region in the error space.

Suppose that

$$y(nT+T) = N(\, y(nT)\,) + H_1(\, u_{\rm PID}(nT)\,) + H_2(\, w(nT)\,), \tag{5.99}$$

where $\{w(nT)\}$ are disturbances ($H_2 = 0$ if it does not exist), N is the closed-loop system operator, and N, H_1, H_2 are bounded nonlinear functions (in operator norm):

$$\| N \| < \infty, \qquad \| H_1 \| < \infty, \qquad \| H_2 \| < \infty.$$

Then, it follows that

$$\begin{aligned} \| H_1(u_{\rm PID}) \| &= \| y(nT+T) - N\,(y(nT)) - H_2(w(nT)) \| \\ &\le \| y(nT+T) \| + \| N \| \times \| y(nT) \| + \| H_2 \| \times \| w \|. \end{aligned}$$

Since $\| H_1(u_{\rm PID}) \| \le \| H_1 \| \times \| u_{\rm PID} \|$, to ensure the left-hand side be bounded, we can require

$$\| H_1 \| \times \| u_{\rm PID} \| \le \| y(nT+T) \| + \| N \| \times \| y(nT) \| + \| H_2 \| \times \| w \|$$

or, even more conservatively,

$$\| u_{\rm PID} \| \le \beta_1 + \beta_2 \| E \| + \beta_3 \| V \|, \tag{5.100}$$

where

$$\beta_1 \le \frac{1}{\| H_1 \|}\,(\, \| y(nT+T) \| + \| N \| \times \| y(nT) \| + \| H_2 \| \times \| w \|\,),$$

$$\beta_2 \le \max\{\, \| K_{\rm P} \|\,,\, 1\,\},$$

$$\beta_3 \le \max\{\, \| K_{\rm I} \|\,,\, \| K_{\rm D} \|\,\}.$$

Substituting (5.100) into (5.97) gives

$$\| E \| \le \alpha_1(\beta_1 + \beta_2 \| E \| + \beta_3 \| V \|\,)$$

which yields (if $\alpha_1\beta_2 < 1$)

$$\| V \| \ge \frac{1}{\alpha_1\beta_3}\,[\,(1 - \alpha_1\beta_2)\, \| E \| - \alpha_1\beta_1\,]. \tag{5.101}$$

Similarly, substituting (5.100) into (5.98) leads to (if $\alpha_2\beta_3 < 1$)

$$\| V \| \le \frac{1}{1 - \alpha_2\beta_3}\,[\, \alpha_2\beta_2 \| E \| + \alpha_2\beta_1\,]. \tag{5.102}$$

It is clear that the two straight lines on the right-hand side of (5.101) and (5.102) will create a common region for $\| V \|$ in the $\| E \|$ - $\| V \|$ plane. If $\alpha_1\beta_2 > 1$ (the first straight line has a negative slope) and $\alpha_2\beta_3 < 1$ (the second

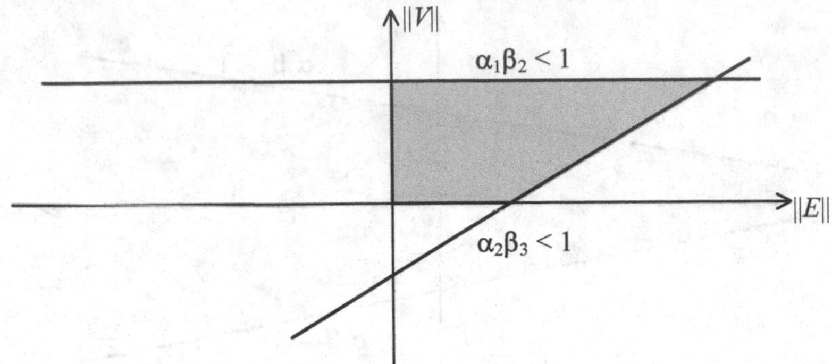

Figure 5.40 A closed common region in the error space.

straight line has a positive slope), then an open region is created as shown in Figure 5.39. In this case, there is no common region in which the errors e and v are guaranteed to be bounded. Therefore, this case should be avoided.

Observe that we are dealing with norms, so only the first quadrant of the $\| E \|$ - $\| V \|$ plane is intersecting. To obtain a closed region on the plane, we can equate (5.101) and (5.102), which gives

$$\frac{1}{\alpha_1\beta_3}[\,(1-\alpha_1\beta_2)\,\| E_r \| - \alpha_1\beta_1\,] = \frac{1}{1-\alpha_2\beta_3}[\,\alpha_2\beta_2\,\| E_r \| + \alpha_2\beta_1\,], \tag{5.103}$$

where $\| E_r \|$ is the value of $\| E \|$ at a point denoted by r on the plane. Solving (5.103) for $\| E_r \|$, we have

$$\| E_r \| = \frac{\alpha_1\beta_1}{1-(\alpha_1\beta_2+\alpha_2\beta_3)}. \tag{5.104}$$

Substituting (5.104) into (5.102) yields the corresponding

$$\| V_r \| = \frac{\alpha_2\beta_1}{1-(\alpha_1\beta_2+\alpha_2\beta_3)}. \tag{5.105}$$

Therefore, the intersection point r of the two straight lines has the coordinates

$$(\,\|E_r\|\,,\,\|V_r\|\,) = \left(\frac{\alpha_1\beta_1}{1-(\alpha_1\beta_2+\alpha_2\beta_3)}\,,\,\frac{\alpha_2\beta_1}{1-(\alpha_1\beta_2+\alpha_2\beta_3)}\right), \tag{5.106}$$

which are finite if $\alpha_1\beta_2 + \alpha_2\beta_3 < 1$. This situation is shown in Figure 5.40. In summary, we have the following sufficient conditions.

Theorem 5.5. A sufficient condition for the BIBO stability of the fuzzy PI+D control system is that the fuzzy control gains K_P, K_I, K_D are chosen such that there are constants α_1, α_2, β_1, β_2, β_3 that together satisfy

(a) $\| E \| \leq \alpha_1 \| U \|$, $\quad \| V \| \leq \alpha_2 \| U \|$,

(b) $\beta_1 \leq \dfrac{1}{\| H_1 \|}[\,(1 + \| N \|)\times\| y \| + \| H_2 \|\times\| w \|\,]$,

(c) $\beta_2 \leq \max\{\,\| K_P \|\,,\,1\,\}$,

(d) $\beta_3 \leq \max\{\,\| K_I \|\,,\,\| K_D \|\,\}$,

(e) $\alpha_1\beta_2 < 1, \quad \alpha_2\beta_3 < 1,$
(f) $\alpha_1\beta_2 + \alpha_2\beta_3 < 1,$

and the closed region shown in Figure 5.40 is not empty. Here,

$$\| y \| = \sup_{n\geq 0} \| y(nT) \| .$$

We finally remark that the norm $\| y \|$ in condition (b) of Theorem 5.5 needs not be finite when verifying the constant β_1; namely, if $\| y \| = \infty$ then condition (b) is trivially satisfied. However, if all the other conditions are satisfied simultaneously, then $\| y \|$ would be finite as a result of the BIBO stability of the overall control system.

PROBLEMS

P5.1 Following the derivation of the digital PD controller shown in Figure 5.7, verify the digital PI controller shown in Figure 5.17.

P5.2 Verify formulas (5.18) – (5.24) for the fuzzy PD controller by showing all the detailed derivation steps.

P5.3 Verify formulas (5.29) – (5.37) for the fuzzy PI controller by showing all the detailed derivation steps.

P5.4 Repeat all numerical simulations shown in Example 5.3, so as to experience how and how well the fuzzy PD controller works.

P5.5 Perform some numerical simulations for the fuzzy PI controller and compare them with the conventional PI controller. You may use the same models as those simulated in Example 5.3.

P5.6 Perform the graphical stability analysis studied in Section 5-III.E on the fuzzy PD and fuzzy PI controller individually.

REFERENCES

[1] C. R. Asfahl. *Robots and Manufacturing Automation*. New York, NY: Wiley (1992).

[2] G. Chen. "Conventional and fuzzy PID controllers: an overview," *Int. J. of Intelligent Control Systems*, Vol. 1, pp 235-246, 1996.

[3] G. Chen and H. Ying. "BIBO stability of nonlinear fuzzy PI control systems," *J. of Intelligent and Fuzzy Systems*, Vol. 5, pp 245-256, 1997.

[4] D. Drianker, H. Hellendoorn, and M. Reinfrank. *An Introduction to Fuzzy Control*. Berlin, Germany: Springer-Verlag (1993).

[5] H. Malki, H. Li, and G. Chen. "New design and stability analysis of fuzzy proportional-derivative control systems," *IEEE Trans. on Fuzzy Systems*, Vol. 2, pp 345-354, 1994.

[6] H. Malki, D. Feigenspan, D. Misir, and G. Chen. "Fuzzy PID control of a flexible-joint robot arm with uncertainties from time-varying loads," *IEEE Trans. on Control Systems Technology*, Vol. 5, pp 371-378, 1997.

[7] D. Misir, H. Malki, and G. Chen. "Design and analysis of a fuzzy proportional-integral-derivative controller," *Int. J. of Fuzzy Sets and Systems*, Vol. 79, pp 297-314, 1996.

[8] P. Sooraksa and G. Chen. "Mathematical modeling and fuzzy control of a flexible-link robot arm," *Mathematical and Computer Modeling*, Vol. 27, pp 73-93, 1998.

[9] K. Tanaka. "Design of model-based fuzzy control using Lyapunov's stability approach and its applications to trajectory stabilization of a model car," *Theoretical Aspects of Fuzzy Control* (ed. H. T. Nguyen, M. Sugeno, R. Tong, and R. R. Yager). pp 31-63, New York, NY: Wiley (1995).

[10] L. X. Wang. *Adaptive Fuzzy Systems and Control*. Englewood Cliffs, NJ: Prentice-Hall (1994).

[11] J. W. Webb and R. A. Reis. *Programmable Logic Controllers: Principles and Applications*. Englewood Cliffs, NJ: Prentice-Hall (1995).

[12] L. A. Zadeh. "The role of fuzzy logic in modeling, identification, and control," *Modeling, Identification, and Control*, Vol. 15, No. 3, pp 191-203 (1994).

CHAPTER 6

Adaptive Fuzzy Control

Adaptive control is a method of designing a controller with some adjustable parameters and an embedded mechanism for adjusting these parameters. Adaptive controllers have been used mainly to improve the controller's performance online. For each control cycle, the adaptive algorithm is normally implemented in three basic steps: (i) observable data are collected to calculate the controller's performance, (ii) the controller's performance is used as a guidance to calculate the adjustment to a set of controller parameters, and (iii) the controller's parameters are then adjusted to improve the performance of the controller in the next cycle.

Normally, an adaptive controller is designed based on one of the available techniques. Each technique is originally designed for a specific class of dynamic systems. The controller is then adjusted as data are collected during run time to extend its effectiveness to control a larger class of dynamic systems.

A typical application of adaptive control was to calibrate a system at startup. In this case, a controller is also designed for a specific class of dynamic systems. However, the parameters that characterize the dynamic behavior of a particular system might not be known in advance. A controller is then designed and arbitrary values are assigned to initialize these parameters. After a few control cycles, parameters are adjusted to converge to the actual parameters of the system. This approach is often used for cases in which a system is designed to handle a variable payload. The payload is different each time, e.g., a crane is used to pick up a sizeable object. The payload will alter the basic dynamic behavior of a dynamic system. Adaptive control is normally used to calibrate these parameters that characterize such dynamic behavior.

Traditionally, there are four basic approaches for adaptive control: (i) gain scheduling, (ii) model reference adaptive system, (iii) self-tuning regulator, and (iv) dual control. Gain scheduling is a method of adjusting the control signal based on a known look-up table describing changes of a dynamic system. The model reference adaptive system is a method of comparing the performance of the actual system against an assumed mathematical model that describes the actual system, and designing control input to drive this comparison error to zero. Self-tuning regulator is a method of updating the parameters of a model that describes the plant based on observed data, and channeling the updated information into the controller that is designed based on these parameters. Dual control is a method of extending adaptive control to stochastic model dealing with uncertainties.

This chapter seeks to extend the principle of adaptive control to the fuzzy control techniques developed in earlier chapters. In this setting, a dynamic system is controlled to ensure tracking and stability. A controller is designed based on a fuzzy representation of that system, and is therefore called an adaptive fuzzy controller. This fuzzy controller is parameterized by a set of constants. These constants are adjusted in each instance of the adaptation to reflect improvement in effectiveness of the controller. Four traditional adaptive approaches (gain scheduling, model reference adaptive system, self-tuning regulator, and dual control) are applied to fuzzy controllers in this chapter.

In addition, an alternative approach to adaptive control is included: a sub-optimal fuzzy control. In this approach, a fuzzy controller is optimally designed for use only during a short period of time. Then, as new data are collected to reflect more knowledge of the system, the optimal fuzzy control problem is adjusted and solved for a new solution.

Simple applications are given for each method mainly to serve as illustrative examples. Practical and industrial examples are deferred to the next chapter to show how fuzzy logic can be applied to real-life control systems.

I. FUNDAMENTAL ADAPTIVE FUZZY CONTROL CONCEPT

Adaptive fuzzy control is an extension of fuzzy control theory to allow the fuzzy controller extending its applicability, either to a wider class of uncertain systems or to fine-tune the parameters of a system to accuracy. In this scheme, a fuzzy controller is designed based on knowledge of a dynamic system. This fuzzy controller is characterized by a set of parameters. These parameters are either the controller constants or functions of a model's constants.

A. Operational Concepts

A controller is designed based on an assumed mathematical model representing a real system. It must be understood that the mathematical model does not completely match the real system to be controlled. Rather, the mathematical model is seen as an approximation of the real system. A controller designed based on this model is assumed to work effectively with the real system if the error (difference) between the actual system and its mathematical representation is relatively insignificant. However, there exists a threshold constant that sets a boundary for the effectiveness of a controller. An error (difference) above this threshold will render the controller ineffective toward the real system.

An adaptive controller is set up to take advantage of additional data collected at run time for better effectiveness. At run time, data are collected periodically at the beginning of each constant time interval, $t_n = t_{n-1} + \Delta t$, where Δt is a constant measurement of time, and $[t_n, t_{n+1})$ is a duration between data collection. Let D_n be a set of data collected at time $t = t_n$. It is assumed that at any particular time, $t = t_n$, a history of data $\{D_0, D_1, ..., D_n\}$ is always

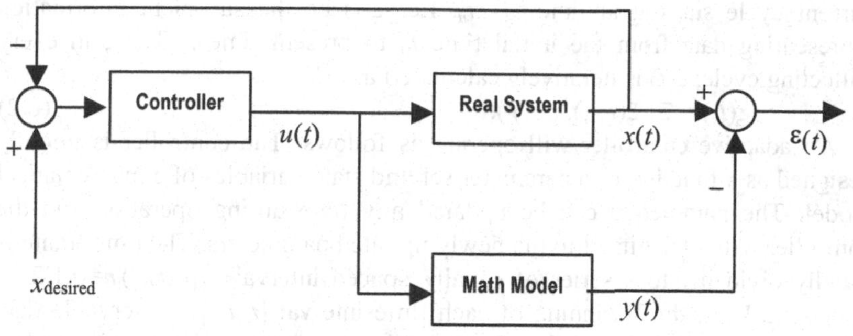

Figure 6.1 Tracking the error function between outputs of a real system and mathematical model.

available. The more data available, the more accurate the approximation of the system will become.

At run time, the control input is fed into both the real system and the mathematical model representing that system. The output of the real system and the output of that mathematical model are collected and an error representing the difference between these two outputs are calculated. Let $x(t)$ be the output of the real system, and $y(t)$ the output of the mathematical model. The error $\varepsilon(t)$ is defined as

$$\varepsilon(t) = x(t) - y(t). \tag{6.1}$$

Figure 6.1 depicts this tracking of the difference between the mathematical model and the real system it represents.

Notice that in the context of this chapter, the mathematical model will be in the form of fuzzy IF-THEN rules. In this case, the output of the mathematical model is still crisp numerical data because of the defuzzification process that converts a fuzzy representation to a numerical one.

An adaptive controller will be adjusted based on the error function $\varepsilon(t)$. This calculated data will be fed into either the mathematical model or the controller for adjustment. Since the error function $\varepsilon(t)$ is available only at run time, an adjusting mechanism must be designed to accept this error as it becomes available, i.e., it must evolve with the accumulation of data in time. At any time, $t = t_n$, the set of calculated data in the form of a time series $\{\varepsilon(t_0),\varepsilon(t_1),\varepsilon(t_2),\ldots,\varepsilon(t_n)\}$ is available and must be used by the adjusting mechanism to update appropriate parameters.

In normal practice, instead of doing re-calculation based on a lengthy set of data, the adjusting algorithm is reformulated to be based on two entities: (i) sufficient information, and (ii) newly collected data. The sufficient information is a numerical variable representing the set of data $\{\varepsilon(t_0),\varepsilon(t_1),\varepsilon(t_2),\ldots,\varepsilon(t_{n-1})\}$ collected from the initial time t_0 to the previous collecting cycle starting at time $t = t_{n-1}$. The new datum $\varepsilon(t_n)$ is collected in the current cycle starting at time $t = t_n$. Let $\xi(t)$ be the sufficient information representing data from the initial time t_0 to present time t. Then, in every collecting cycle, $\xi(t)$ is iteratively calculated as

$$\xi(t_n) = \Xi(\ \xi(t_{n-1}), \varepsilon(t_n)\ ). \tag{6.2}$$

An adaptive controller will operate as follows. The controller is initially designed as a function of a parameter set and state variables of a mathematical model. The parameters can be updated any time during operation and the controller will adjust itself to the newly updated parameters. The time frame is usually divided into a series of equally spaced intervals $\{[t_n,t_{n+1})|n=0,1,2,...;\ t_{n+1}= t_n+\Delta t\}$. At the beginning of each time interval $[t_n,t_{n+1})$, observable data are collected and the error function $\varepsilon(t_n)$ is calculated. This error is used to calculate the adjustment in the parameters of the controller. New control input $u(t_n)$ for the time interval $[t_n,t_{n+1})$ is then calculated based on the newly calculated parameters and fed into both the real dynamic system under control and the mathematical model upon which the controller is designed. This completes one control cycle. The next control cycle will consist of the same steps repeated for the next time interval $[t_{n+1},t_{n+2})$, and so on.

B. System Parameterization

As discussed in Chapter 3, a dynamic system can be described by a set of fuzzy IF-THEN statements that correlate the input and the output. These statements include a set of parameters that is used to uniquely calculate the estimated output of a system given the inputs and current states of the system.

A dynamic system is mathematically modeled as a set of IF-THEN rules:

$$\begin{aligned}
R^1:\ & \text{IF} && x_1(t_n) \in X_{11} \text{ AND ... AND } x_i(t_n) \in X_{1i}\\
& \text{THEN} && y_1(t_{n+1}) = \alpha_{11}x_1(t_n) + \alpha_{12}x_2(t_n) + ... + \alpha_{1i}x_i(t_n)\\
& && \qquad + \beta_{11}u_1(t_n) + \beta_{12}u_2(t_n) + ... + \beta_{1i}u_j(t_n),\\
R^2:\ & \text{IF} && x_1(t_n) \in X_{21} \text{ AND ... AND } x_i(t_n) \in X_{2i}\\
& \text{THEN} && y_2(t_{n+1}) = \alpha_{21}x_1(t_n) + \alpha_{22}x_2(t_n) + ... + \alpha_{2i}x_i(t_n)\\
& && \qquad + \beta_{21}u_1(t_n) + \beta_{22}u_2(t_n) + ... + \beta_{2i}u_j(t_n),\\
\vdots\\
R^k:\ & \text{IF} && x_1(t_n) \in X_{k1} \text{ AND ... AND } x_i(t_n) \in X_{ki}\\
& \text{THEN} && y_k(t_{n+1}) = \alpha_{k1}x_1(t_n) + \alpha_{k2}x_2(t_n) + ... + \alpha_{ki}x_i(t_n)\\
& && \qquad + \beta_{k1}u_1(t_n) + \beta_{k2}u_2(t_n) + ... + \beta_{ki}u_j(t_n),
\end{aligned}$$

where x_1, x_2, ..., x_i are the observable state variables of the real system, y_1, y_2, ..., y_k the calculated state variables of the mathematical model, and u_1, u_2, ..., u_j the control inputs to both the real system and the mathematical model. In addition, a set of fuzzy membership functions $\{\mu_{X_1}(\times), \mu_{X_2}(\times), ..., \mu_{X_i}(\times)\}$ is given.

The above fuzzy mathematical model is characterized by a set of parameters $\{\alpha_{k'i'},\beta_{k'j'}|k'=1,2,...,k;i'=1,2,...,i;j'=1,2,...,j\}$. These parameters will determine the behavior of the mathematical model. Theoretically, the model will approximate the real system. The more accurately the model approximates the real dynamic system, the better the controller designed based on this model behaves.

Generally, there can be more than one set of fuzzy rules describing a system. This is a special feature of fuzzy modeling: different (and sometimes contradicting) rules for a same set of conditions can co-exist. In case of having more than one rule applicable to a particular condition, the membership functions will be used to determine a unique result. This process is defuzzification, which resolves difference in a compromising fashion. The most popular defuzzification technique is the weighted average formula which assigns a weight for each result according to a given fuzzy membership function.

Similarly, a fuzzy controller, if designed based on the fuzzy mathematical model above, will have the following form of fuzzy rules:

$$
\begin{aligned}
R^1:\ & \text{IF} && x_1(t_n) \in X_{11} \text{ AND } \ldots \text{ AND } x_i(t_n) \in X_{1i} \\
& \text{THEN} && u_1(t_n) = \kappa_{11}x_1(t_n) + \kappa_{12}x_2(t_n) + \ldots + \kappa_{1i}x_i(t_n) \\
R^2:\ & \text{IF} && x_1(t_n) \in X_{21} \text{ AND } \ldots \text{ AND } x_i(t_n) \in X_{2i} \\
& \text{THEN} && u_2(t_n) = \kappa_{21}x_1(t_n) + \kappa_{22}x_2(t_n) + \ldots + \kappa_{2i}x_i(t_n) \\
& \vdots && \\
R^j:\ & \text{IF} && x_1(t_n) \in X_{k1} \text{ AND } \ldots \text{ AND } x_i(t_n) \in X_{ki} \\
& \text{THEN} && u_j(t_n) = \kappa_{j1}x_1(t_n) + \kappa_{j2}x_2(t_n) + \ldots + \kappa_{ji}x_i(t_n).
\end{aligned}
$$

In this case, the set of parameters $\{\kappa_{j'i'} \mid i'=1,2,\ldots,i; j'=1,2,\ldots,j\}$ characterizes the controller. It can be understood intuitively that the parameter set $\{\kappa_{j'i'}\}$ is designed as a function of the parameter set $\{\alpha_{k'i'},\beta_{k'j'}\}$, so that the state variables of the fuzzy mathematical model are driven to a target point with stability.

Therefore, it is natural for an engineer who designs the controller to either adjust the system parameters $\{\alpha_{k'i'},\beta_{k'j'}\}$, which will lead to the adjustment of the control parameters $\{\kappa_{j'i'}\}$, or directly adjust the controller parameters $\{\kappa_{j'i'}\}$ as a scheme to implement adaptive strategies. The adjustment of the system parameters is sometimes called *system identification*. This system identification process is often applied to a system that can be described by a time-varying parameter set in the fuzzy model. Sometimes the identification process can be used to establish the parameters of a dynamically changing system, such as a flying jet that continuously changes its mass due to fuel consumption.

C. Adjusting Mechanism

Adjusting the system parameters is normally carried out to improve the performance of the overall closed-loop system, or to improve accuracy of the assumed model upon which the controller is designed. The process of calculating the adjustment to the parameters is the adjusting mechanism mentioned as part of the adaptive control problem.

Let $\theta^{(n)}$ be a set of parameters to be adjusted at time $t = t_n$. The adjustment can be the recalculation of the parameters,

$$\theta^{(n)} = \Theta(D_0,D_1,\ldots,D_n),$$

where D_m is the data collected at time $t = t_m$. For numerical efficiency, this formulation is normally reformulated by the update of the parameter set based on its previous value, namely,

$$\theta^{(n)} = \Phi(\theta^{(n-1)}, D_n).$$

A special case of such an update formulation is the separation of the function Φ into a recursive form:

$$\theta^{(n)} = \theta^{(n-1)} + \Lambda(D_n, \Lambda(D_{n-1})), \tag{6.3}$$

where $\Lambda(D_n)$ is known as the *update* (or adjustment) of the parameter set. This quantity is sometimes referred to as *sufficient information*. This formulation improves clarity and efficiency. It is easy to see the new parameters as the sum of the old parameters and the changes in these parameters. It is also computationally efficient not having to repeat the same type of calculation based on n sets of data.

When accuracy of system identification is considered, the problem of calculating the adjustment to the parameters can be formulated as an optimization:

$$\min_{\theta} \|\varepsilon(t)\|, \tag{6.4}$$

where $\varepsilon(t)$ is the error function representing the difference between the real system and its approximated model. The real system is given in terms of observable data collected over time. The approximated model is given in terms of a mathematical expression. The expression $\| \times \|$ represents an adequate norm function of a vector. For numerical simplicity, it is convenient to use an l_2-norm defined as

$$\| \varepsilon(t) \|_2 = \sqrt{\varepsilon^2(t_1) + \varepsilon^2(t_2) + \ldots + \varepsilon^2(t_n)} \,.$$

The optimization problem in this l_2-normed space is often known as the least-squares problem. Given a set of linear constraints, this problem will lead to a standard linear solution (see Chapter 3). This solution can be reformulated to become an efficient calculation of the update of the parameters in the form of (6.3).

When the performance of the overall closed-loop system is considered, the problem of calculating the adjustment to the parameters can be formulated as an optimization problem minimizing the rate of convergence to a desired stable point:

$$\min_{\theta} \|x(t) - x_f\|. \tag{6.5}$$

There are normally variations of the objective function to be optimized. It is common to select a norm formulation that will lead directly to a closed-form solution. This closed-form solution is important because it will allow rapid calculation of the updated parameters in a short time interval, $[t_n, t_n+\Delta t)$, before the next cycle begins. In this time interval, the algorithm to compute the control input $u(t_n)$ is also performed, increasing the demand for numerical efficiency. The capability to calculate everything required in a small time

Table 6.1 Criteria for Selecting an Adaptive Fuzzy Controller

Adaptive Type	System Description
Fuzzy STR	Time invariant with unknown parameters
Fuzzy STR, MRAFS	Time varying
Sub-optimal Fuzzy Control	Dynamic environment is changing with no prior knowledge
Dual Control	Highly nonlinear
Fuzzy Gain Scheduling	Only modeled accurately piecewise and represented by multiple models each for a specific range of time

interval (with a predefined duration Δt) is sometimes called the real-time processing capability.

D. Guidelines for Selecting an Adaptive Fuzzy Controller

It should be noted that an adaptive fuzzy controller, while providing increasing accuracy and effectiveness, should be carefully selected according to the criteria it was designed for. Otherwise, the increase in calculation workload for implementation of an adaptive fuzzy controller might not be offset by the gain in accuracy or effectiveness.

In order to determine whether to apply adaptive fuzzy control to a system, one must understand the nature of that system and the mathematical fuzzy model used to design a controller. To demonstrate an understanding of a system, one may try to answer the following questions:

1. Can the system be approximated entirely by a fuzzy model?
2. If a system can be approximated entirely by a fuzzy model, are the parameters of this fuzzy model readily available or must they be determined online?
3. If a system cannot be approximated entirely by a fuzzy model, can it be approximated piecewise by a set of fuzzy models?
4. If a system can be approximated by a set of fuzzy models, are these models having the same format with different parameters or are they having different formats?
5. If a system can be approximated by a set of fuzzy models having the same format, each with a different set of parameters, are these parameter sets readily available or must they be determined online?

Satisfactorily answering the above questions will help determine whether an adaptive controller is necessary and, if so, which type of adaptive controller is appropriate. Table 6.1 summarizes some criteria to determine whether an adaptive control design is necessary and, if so, which type of adaptive control method is appropriate. In the table, STR stands for self-tuning regulator and MRAFS is a model-reference adaptive fuzzy system.

If a system can be approximated piecewise by a set of fuzzy models, with the parameters of each model readily known, then it is best to implement a fuzzy gain scheduling control algorithm. This algorithm is similar to the fuzzy PD/PI/PI+D controllers presented in Chapter 5.

If a system can be approximated entirely by a set of fuzzy models, but the parameters of these models are not readily available, then it is best to implement a STR fuzzy controller based on these parameters. These parameters will be constantly evaluated and updated, consequently changing the controller parameters.

If a system can be approximated piecewise by a set of fuzzy models, with all models having the same format but different parameter sets, then it is best to implement a fuzzy sub-optimal controller. In this approach, as soon as the controller moves the closed-loop system into a particular area in the control space, an optimal fuzzy control problem is solved only for a short time duration. As soon as control action is pushing the system into another area in the control space, the sub-optimal fuzzy control problem is solved again with a different set of parameters.

II. GAIN SCHEDULING

Gain scheduling is a method of combining several controllers together to control a particular dynamic system. In this approach, each controller is the most effective in a particular region. Joining them together will require a condition checking procedure to determine which controller is activated.

In order to design a gain scheduling fuzzy controller, one needs to design two sets of rules: (i) a set of several fuzzy controllers that are applicable to a dynamic system, each effective under certain conditions, and (ii) a set of meta rules to determine when to activate which controller for an appropriate action.

This type of fuzzy controller is normally useful for applying rule-based expert systems to control a complicated dynamic system. The expert system will contain fuzzy rules on how to control that dynamic system. The fuzzy rules are usually extracted from several human operators. In this setting, the fuzzy system will contain several rules: either each will be consistently applicable to a scenario or some might contradict one another for a scenario. In both cases, a meta rule set is required, either to determine when to activate a rule for the case of having many exclusively effective controllers, or to provide a compromised solution for the case of having contradicting rules.

Another application of interest is the case of a piecewise linear system that behaves linearly under a unique set of parameters in each region. A piecewise linear mathematical model is normally used to approximate a nonlinear system. An example of this type of application is the dynamic behavior of a moving object in a gravity field generated by a large planetary body such as the earth. Near the surface of this planetary body, the gravity field can be approximated as a cause that results in a constant downward acceleration of 9.8 m/sec^2.

A fuzzy gain scheduling consists of a series of rule sets, each consisting of several IF-THEN fuzzy logic rules. Each rule set is in the form of

$$\text{Rule Set } k\text{: } R^{(k1)}: \bigwedge_{i=1}^{N} (x_i \in X_{k1i}) \Rightarrow y_1 = \sum_{i=1}^{N} a_{k1i} x_i + \sum_{i=1}^{M} b_{k1i} u_i$$

$$\vdots$$

$$R^{(kN)}: \bigwedge_{i=1}^{N} (x_i \in X_{kNi}) \Rightarrow y_N = \sum_{i=1}^{N} a_{kNi} x_i + \sum_{i=1}^{M} b_{kNi} u_i$$

for $k = 1,2,\ldots,K$. Fuzzy membership functions $\mu_{X_{kni}}(x)$ for appropriate values of k, n, i are available either in closed-form or in look-up tables.

In order to determine when to activate a rule set, one must implcment a set of meta rules that provides selection conditions. The meta rules should be of the form:

$$M^{(j)}: \bigvee_{k=1}^{K} [\bigwedge_{i=1}^{N} (x_i \in X_{jki})] \Rightarrow \text{Rule Set } j \qquad \text{for } j = 1,2,\ldots,J.$$

The above meta rules are cumbersome and detailing. One might prefer to simplify the above meta rule by approximating it with

$$\widetilde{M}^{(j)}: \bigwedge_{i=1}^{N} (x_i \in Y_{ji}) \Rightarrow \text{Rule Set } j \qquad \text{for } j = 1,2,\ldots,J,$$

where Y_{ki} is defined as the union of the sets X_{kji}:

$$Y_{ji} = \bigcap_{k=1}^{K} X_{jki},$$

and the fuzzy membership function associated with Y_{ji} is

$$\mu_{Y_{ji}}(x) = \max\{\mu_{X_{j1i}}(x),\ldots, \mu_{X_{jKi}}(x)\}.$$

There are two difficulties associated with a fuzzy gain scheduling scheme: (i) it is often time-consuming to define a set of meta rules determining when to activate which set of fuzzy rules, and (ii) the meta rule set is used in a feed-forward fashion without feedback.

There is no formal method of defining a set of meta rules. As a result, one has to design the rules heuristically, most of the time exhaustively. The meta rules are used to select a set of fuzzy rules to activate. This selection does not include feedback to confirm whether the right set of rules has been used.

Example 6.1. Controller for a moving body is normally based on a set of equations of motion. These equations include dynamic description of an object moving in a gravitational field, accelerating toward the ground at a constant rate called gravity effect. From Newtonian physics law, this gravitational effect is described as an interactional force between two bodies, one of which is the earth:

$$F = \frac{GM}{R^2} m,$$

where G is the gravitational constant, M the mass of the earth, m the mass of the moving object, and R the distance between the object and the center of the

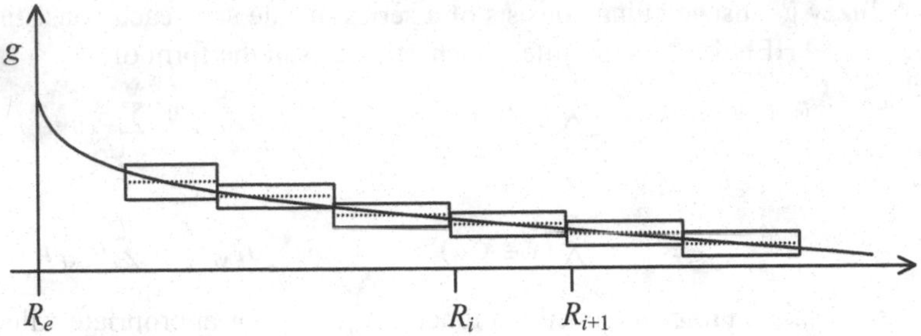

Figure 6.2 Approximation of the constant g.

earth. Near the surface of the earth, where $R = R_E$, the radius of the earth, the force is estimated as

$$F = \frac{GM}{R_E^2} m \approx m\,g,$$

where g is the gravity on earth, often estimated as GM / R_E^2, which yields a constant value of 9.8 m/sec^2.

An object with mass m falling near the surface of the earth will obey the equation of motion which can be described by the following fuzzy rules:

$$\text{R}^{(1)}\text{:}\quad \text{IF } x_2(t) \in S_k \text{ THEN } \dot{x}_1(t) \in S_k,$$

$$\text{R}^{(2)}\text{:}\quad \text{IF } u(t) \in S_f \text{ THEN } \dot{x}_2(t) \in S_{f/m+g},$$

where $x_1(t)$ is the displacement, $x_2(t)$ the velocity, and $u(t)$ the external force acting on the object. The fuzzy set S_k is defined as

$$S_k = [(k-1)\sigma, (k+1)\sigma),$$

$$\mu_{S_k}(x) = \begin{cases} \dfrac{x}{\sigma} - k + 1 & \text{for } (k-1)\sigma \le x \le k\sigma \\ k + 1 - \dfrac{x}{\sigma} & \text{for } k\sigma \le x < (k+1)\sigma, \end{cases}$$

where σ is half the width of the fuzzy set S_k. Notice that a real line is partitioned into a series of fuzzy sets, each indexed by an integer k, where $k = \ldots,-2,-1,0,1,2,\ldots$ (see Chapter 3).

A typical fuzzy controller for the system of two rules representing the equations of motion for a moving object near earth's surface is

$$\text{C}^{(1)}\text{:}\quad \text{IF } x_1(t) \in S_i \text{ AND } x_2(t) \in S_j \text{ THEN } u(t) \in S_{-m(6i+5j+g)}.$$

This controller will yield a stable closed-loop system described by the following rules:

$$\text{R}^{(1)}\text{:}\quad \text{IF } x_1(t) \in S_i \text{ AND } x_2(t) \in S_j \text{ THEN } \dot{x}_1(t) \in S_j,$$

$$\text{R}^{(2)}\text{:}\quad \text{IF } x_1(t) \in S_i \text{ AND } x_2(t) \in S_j \text{ THEN } \dot{x}_2(t) \in S_{-6i-5j},$$

which can describe any system whose output is bounded from above and from below by the two bounds, as illustrated by Figure 6.2.

These two bounds characterize the approximation accuracy, implying that all systems bounded by them will also have the same error bounds of model approximation.

However, the controller is working only with an accurate estimate of the gravitational effect *g*. Near earth's surface, this value *g* is estimated at 9.8 m/sec^2. For planetary spacecraft, the value of *g* is dependent on the distance of the spacecraft from earth, to be given in terms of GM/R^2. The further the craft from earth, the smaller the value of *g*.

A simple solution to the control problem with variation in gravity constant is the use of gain scheduling. The controller is still effective provided an accurate value of *g* is used. Therefore, one can program a set of possible values of *g* that the spacecraft will likely encounter in its operating range. For example, a spacecraft can be designed to be functional in many orbital levels, each defined at a specific altitude. In this case, a value *g* is pre-determined for each orbit. As soon as sensor indicates the altitude $x_1(t)$, a set of meta rules can be used to determine which value of *g* to use, thus triggering a new set of control gains to apply to the system. This meta rule set can either be crisp or fuzzy, determined by the designer. For a set of crisp meta rules, the pre-calculated *g* is used for the controller. For a set of fuzzy meta rules, the final value of *g* is determined from the defuzzification of the results based on membership values of each individual result.

Figure 6.2 shows the actual value of *g* versus the altitude of an object. Instead of recalculation of the value of *g* as a function of an altitude and increasing the complexity of the mathematical model used to design the controller, one can use a constant value of *g* for each pre-defined range (shown in the rectangular box in the figure).

Let *G* be the set of possible values of *g*, $G = \{g_1, g_2, \ldots, g_P\}$. Each value of g_i is applicable only for a range $[R_i, R_{i+1}]$. The meta rules for selecting different values of g can be listed as the following rules:

$M^{(1)}$: IF $R_1 \le x_1(t) \le R_2$ THEN $g = g_1$

$M^{(2)}$: IF $R_2 \le x_1(t) \le R_3$ THEN $g = g_2$

⋮

$M^{(P)}$: IF $R_P \le x_1(t) \le R_{P+1}$ THEN $g = g_P$.

Example 6.2 (Cash flow analysis of potential of a project). This example discusses the use of fuzzy logic in a set of expert systems. Each expert system will operate according to a set of fuzzy logic rules. It calculates the confidence level of each rule using fuzzy logic processing. A meta rule set will be used to select which expert system (or set of rules) is to be used throughout the entire process. This is the gain-scheduling technique applied to the problem of decision making on investment. Currently, there are several models that a financial manager can use to analyze cash inflow and outflow of a project before making a commitment decision. Each model has its own unique implication and calculation rules.

A financial manager must have an ability to analyze the cash flow of a project before committing to it. This analysis is based on the profit returned from the investment. There are several factors involved in this analysis, e.g., predicted inflation rate, periodic dividends, initial investment, length of commitment, etc. Using these factors, a financial manager computes some index that shows profits or losses. If the index shows profits, the manager can comfortably commit to the project.

When a financial manager has to select one investment project among several, the analysis above can be modified to become the selection of a project with the highest index of profit or lowest index of loss. However, there are several techniques of analyzing a project. All such techniques are consistent with each other in determining the expected profit (or expected loss) situation. Yet, when one ranks projects using these techniques, the rankings are not consistent. Depending on the criteria that each method is designed for, the use of each method will yield different ranking.

To illustrate the inconsistencies in ranking of profit index using different techniques of analyzing a cash flow, we summarize the five indicators commonly used to analyze cash flows: profitablity index, internal rate of return, net present value, payback period, and modified internal rate of return.

Profitability Index (PI) is an indicator that determines the ratio of a net present value of all cash inflows to net present value of all cash outflows of a project. This indicator has a profit threshold value of one. If the PI of a project is greater than one, then the project returns profit; if it is less than one, then the project shows losses; if it is equal to one, then the project breaks even. In case of a single-project analysis, if the PI is greater than one, a financial manager can commit to the project. The profitability index is calculated as

$$\mathrm{PI} = \frac{\displaystyle\sum_{t=0}^{n} \frac{C_{in}(t)}{(1+k)^t}}{\displaystyle\sum_{t=0}^{n} \frac{C_{out}(t)}{(1+k)^t}},$$

where $C_{in}(t)$ is the cash inflow, $C_{out}(t)$ the outflow, k the index of inflation, and n the number of time units.

The fuzzy rules determining if a project is commitable are

$R^{(1)}$: if (PI – 1) is positive then accept,
$R^{(2)}$: if (PI – 1) is negative then reject.

Net Present Value (NPV) is the computation converting all cash flows at different times into a single quantity equivalent to the value at presence. This method is based on the expected inflation rate in the future, the amount of cash received periodically, and the initial investment. The total of these amounts after indexed for inflation rate is the expected profit. If the NPV of a project is positive, the project shows profits and is commitable; if it is negative, the project shows losses and is not commitable; if it is zero, the project breaks even and the commitment shows no effects. The NPV is calculated as

$$\text{NPV} = \sum_{t=0}^{n} \frac{[C_{in}(t) - C_{out}(t)]}{(1+k)^t},$$

with notation as above.

The fuzzy rules determining if a project is commitable are

$R^{(1)}$: if NPV is positive then accept,

$R^{(2)}$: if NPV is negative then reject.

Internal Rate of Return (IRR) is the minimum rate of return that allows break-even in the project. This is the yield rate that, if used for calculating net present value, shows a total of zero net present value of cash flows. This rate is compared to the current cost of capital, i.e., the interest rate available in the market. If the IRR is greater than the cost of capital, then it is profitable to commit investment to the project; if it is less than the cost of capital, then the project is not profitable and should not be committed to. The IRR is calculated as

$$\text{IRR} = r \quad \text{such that} \quad \sum_{t=0}^{n} \frac{[C_{in}(t) - C_{out}(t)]}{(1+r)^t} = 0,$$

with notation as above.

The fuzzy rules determining if a project is commitable are

$R^{(1)}$: if $(\text{IRR}-k_{\text{inflation}})$ is positive then accept,

$R^{(2)}$: if $(\text{IRR}-k_{\text{inflation}})$ is negative then reject.

Payback Period (PP) is the period required to collect back all the initial investments. Normally, a company has a threshold for payback period. This threshold is the maximum length of a period that the company is willing to wait before showing profit. If the PP of a project is less than this threshold, then the project is commitable; if it is greater than this threshold, then the project should not be invested. This method has been popular for quick calculation inside an executive's head. Recently, it has been modified to reflect the inflation rate. This rate is used to calculate the net present value of each cash flow. These net-present-valued cash flows are used to calculate the payback period:

$$\text{PP} = n \quad \text{such that} \quad \sum_{t=0}^{n} \frac{[C_{in}(t) - C_{out}(t)]}{(1+k)^t} = 0,$$

with notation as above.

The fuzzy rules determining if a project is commitable are

$R^{(1)}$: if (PP–threshold) is positive then accept,

$R^{(2)}$: if (PP–threshold) is negative then reject.

Modified Internal Rate of Return (MIRR) is a slight variation of the IRR. Here, the terminal value, i.e., future value at the time of the final payment of all the cash inflows, is computed. Then, this amount is adjusted to reflect the net present value. The rate that returns such a zero net present value is the

MIRR. The analysis of this indicator is similar to that of the IRR. The main advantage of using this indicator instead of the IRR is the claim that the MIRR provides an alternative solution of simply investing the company's fund to saving institutions where the MIRR is smaller than the rate of return offered by such saving institutions. The MIRR is calculated as

$$\text{MIRR} = r \quad \text{such that} \quad \sum_{t=0}^{n} \frac{C_{out}(t)}{(1+k)^t} = \frac{\sum_{t=0}^{n} C_{in}(t)(1+k)^{n-t}}{(1+r)^n},$$

with notation as above.

The fuzzy rules determining if a project is committable are

$R^{(1)}$: if (MIRR–$k_{\text{inflation}}$) is positive then accept,

$R^{(2)}$: if (MIRR–$k_{\text{inflation}}$) is negative then reject.

The meta rules used to select which value to use is as follows. If profits are the main determining factor, then the values PI, IRR, and MIRR will be emphasized. If the principle of the investment is protected, then the value NPV should be used. One can vary the emphasis on each value and combine them together using the weighted average formula, which is equivalent to resolving results of different fuzzy rules, each with a different membership value.

III. FUZZY SELF-TUNING REGULATOR

Fuzzy Self-Tuning Regulator (FSTR) is the extension of traditional self-tuning regulator, which is an adaptive control method to handle systems modeled by fuzzy logic rules. In this approach, a set of fuzzy logic rules is used to describe the dynamics of a system. A fuzzy controller, therefore, is designed based on the system parameters and is characterized by its own controller parameters. The design of a fuzzy controller can be set up so that the controller's fuzzy parameters are dependent on the system's fuzzy parameters. In this setup, the fuzzy self-tuning regulator approach is utilized to update the system's parameters at every control cycle. Then, the controller's parameters will be updated as a consequence. The new controller (with newly computed parameters) will then calculate the control signal to be sent to the system to control it for tracking with stability. Figure 6.3 illustrates this FSTR concept, where $u_c(t)$ is the reference input, $u(t)$ and $y(t)$ the input and output of the system, respectively, $\{a_{ij}\}$ and $\{b_{mn}\}$ are system parameters, and $\{K_{pq}\}$ controller parameters (including the control gains).

Notice that in the FSTR approach, the problem is partitioned into two parts: (i) system identification and (ii) adjusting the control gain. The first part, system identification, is an estimation of the parameters in the fuzzy model. The adjusting of the control gains is the recalculation of the control gains based on new parameters of the system model. In this setup, inputs and outputs of the system to be controlled are monitored and profiled at the

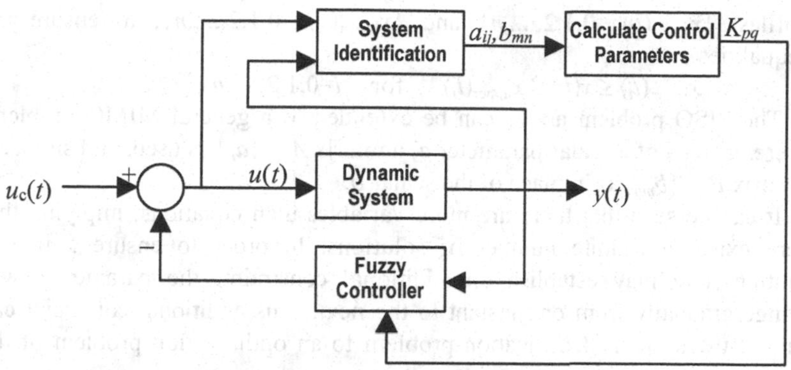

Figure 6.3 Fuzzy self-tuning regulator.

beginning of every control cycle, $[t_n, t_n+\Delta t]$. These two profiles will contain data from the initial time, $t = t_0$, to the current time, $t = t_n$. Based on the input-output data, a relation can be established in terms of a fuzzy model that will map the input into the output.

Consider an SISO fuzzy model in the fuzzy IF-THEN form (see Chapter 3):

$$R^{(i)}: \quad \text{IF } x \in S_n \text{ AND } u \in S_m \text{ THEN } \dot{x} \in S_{an+bm}, \quad i=1,\ldots,l, \quad n,m=0,\pm1,\pm2,\ldots .$$

Under specific conditions that $S_n = [-\sigma(n-1), \sigma(n+1))$, and that the associated membership function of this set S_n is triangular shape of the form

$$\mu_{S_n}(x) = \begin{cases} 1-|x-n\sigma|/\sigma & \text{for} \quad x \in [(n-1)\sigma, (n+1)\sigma), \\ 0 & \text{else,} \end{cases}$$

it can be shown that $x(t)$ is bounded from both above and below by the bounds

$$\dot{x}_{\text{upper}}(t) = a\, x_{\text{upper}}(t) + b\, u(t) + [\, |a| + |b|\,]\, \sigma,$$
$$\dot{x}_{\text{lower}}(t) = a\, x_{\text{lower}}(t) + b\, u(t) - [\, |a| + |b|\,]\, \sigma.$$

A typical fuzzy controller for this model is given in terms of the following rules:

$$C^{(i)}: \quad \text{IF } x \in S_n \text{ THEN } u = kx, \quad i=1,\ldots,l, \quad n=0,\pm1,\pm2,\ldots,$$

where the constant k is designed so that

$$a + bk < 0,$$

in order to have a stable closed-loop system. For purpose of illustration, k can be selected as

$$k = -2\ \text{sign}(b)\frac{|a|}{|b|}.$$

For the first part of system identification, given two sets of data profiles $\{u(t_i)|i=0,1,2,\ldots,n\}$ and $\{x(t_i)|i=0,1,2,\ldots,n\}$, the problem is to find the two constants a and b in the IF-THEN fuzzy model that will produce the two profiles, $\{x_{\text{upper}}(t_i)|i=0,1,2,....,n\}$ and $\{x_{\text{lower}}(t_i)|i=0,1,2,....,n\}$, to ensure the inequalities

$$x_{\text{lower}}(t_i) \le x(t_i) \le x_{\text{upper}}(t_i) \quad \text{for} \quad i=0,1,2,....,n.$$

The SISO problem above can be extended to a general MIMO problem, where, instead of a scalar parameter a, a matrix $A = \{a_{ij}\}$ is used, and similarly a matrix $B = \{b_{mn}\}$ is in place of the scalar b.

It can be seen that there are more variables than equations, implying that there exist an infinite number of solutions. In order to ensure a unique solution, one may establish an additional constraint: the parameters will change gradually from one instant to the next. This additional constraint can convert the system identification problem to an optimization problem of the form

$$\min_{a_{ij},b_{mn}} \left\{ \sum_{i=1}^{K}\sum_{j=1}^{K}[a_{ij}(t_k)-a_{ij}(t_{k-1})]^2 + \sum_{m=1}^{K}\sum_{n=1}^{J}[b_{mn}(t_k)-b_{mn}(t_{k-1})]^2 \right\}$$

subject to

$$\dot{x}_r(t_k) = \sum_{n=1}^{K} a_{rn}x_n + \sum_{m=1}^{J} b_{rm}u_m .$$

This optimization problem would minimize the deviation of the parameters from the previous set of parameters. This minimization provides the smooth change in the values of the parameters, and the constraint would put the estimated transition right in the middle of the two bounds.

The control parameters will be updated based on the new values of the model parameters calculated from the above optimization problem. There are several techniques to design a controller that can control the two bounds of the fuzzy model to ensure convergence and stability.

The solution to the first part of the FSTR problem is calculated in three steps as follows.

Step 1. Estimate the transition of the state variables: In this calculation, the following simple finite difference may be used:

$$\dot{x}_n(t_k) = \frac{x_n(t_k)-x_n(t_{k-1})}{t_k - t_{k-1}}.$$

Step 2. Estimate the parameters of the fuzzy model: The solution to the optimization problem is solved by Calculus in closed form as follows:

$$a_{ij}(t_k) = a_{ij}(t_{k-1}) + \frac{x_j(t_k)\dot{x}_i(t_k)}{\sum_{m=1}^{K}x_m^2(t_k)+\sum_{n=1}^{J}u_n^2(t_k)}$$

$$- \frac{x_j(t_k)\sum_{m=1}^{K}x_m(t_k)a_{im}(t_{k-1})+\sum_{n=1}^{J}u_n(t_k)b_{in}(t_{k-1})}{\sum_{m=1}^{K}x_m^2(t_k)+\sum_{n=1}^{J}u_n^2(t_k)},$$

$$b_{mn}(t_k) = b_{mn}(t_{k-1}) + \frac{u_n(t_k)\dot{x}_m(t_k)}{\sum_{p=1}^{K} x_p^2(t_k) + \sum_{q=1}^{J} u_q^2(t_k)}$$

$$- \frac{u_n(t_k)\sum_{p=1}^{K} x_p(t_k)a_{mp}(t_{k-1}) + \sum_{q=1}^{J} u_q(t_k)b_{mq}(t_{k-1})}{\sum_{p=1}^{K} x_p^2(t_k) + \sum_{q=1}^{J} u_q^2(t_k)}.$$

Step 3. Update the control parameters: The control parameters are designed to stabilize the system in terms of the constants a and b for the SISO case. A typical solution was given earlier, for which the control parameter k can be updated as

$$k(t_k) = -2\,\mathrm{sign}(b(t_k))\frac{|a(t_k)|}{|b(t_k)|}.$$

The adaptation of the system parameters to input data can be reformulated into a matrix form, as

$$A_k = A_{k-1} + \frac{\dot{x}_k x_k^T}{\|x_k\|^2 + \|u_k\|^2} - \frac{A_{k-1}x_k x_k^T}{\|x_k\|^2 + \|u_k\|^2} - \frac{B_{k-1}u_k x_k^T}{\|x_k\|^2 + \|u_k\|^2},$$

$$B_k = B_{k-1} + \frac{\dot{x}_k u_k^T}{\|x_k\|^2 + \|u_k\|^2} - \frac{A_{k-1}x_k u_k^T}{\|x_k\|^2 + \|u_k\|^2} - \frac{B_{k-1}u_k u_k^T}{\|x_k\|^2 + \|u_k\|^2},$$

which can be implemented efficiently on a digital computer in the updating scheme.

IV. MODEL REFERENCE ADAPTIVE FUZZY SYSTEMS

Model Reference Adaptive Fuzzy Systems (MRAFS) is a method of adjusting the gain(s) to a system so that the output of that system tracks the output of a model having the same control input. In this setting, a control signal is being fed into both the model and the system. The model, being fuzzy in this case, is designed to be stable. It is the reference that the system is controlled to match. Figure 6.4 illustrates the concept of MRAFS. Associated with the MRAFS approach are different numerical techniques to calculate the adjusting gain(s). This approach basically establishes a desired dynamic behavior of a system through its model. Then, the system is controlled so that its output matches that of the model.

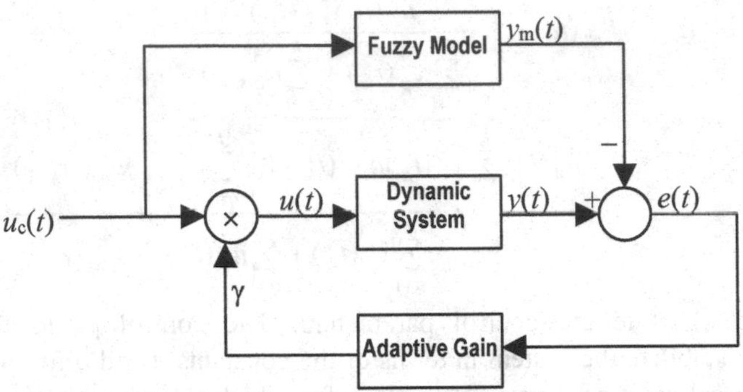

Figure 6.4 Model Reference Adaptive Fuzzy Systems (MRAFS).

Even though there is no consideration about stability of the overall system in the discussion of the MRAFS approach below, it is understood that if the model is designed to be stable, i.e., with internal feedback loop that brings its output to a desired set-point in a stable manner, then matching the output of a system with this reference model will ensure the system to be stable as well. Therefore, an important criterion in this design is the requirement that the reference model be stable.

In fuzzy logic applications, it is assumed that the model being used is represented by a set of fuzzy logic rules. The adaptive gain is also given in terms of fuzzy logic rules.

A basic guideline to calculate the gain constant is the minimization of the error function defined as the difference between the output of the reference model and the output of the actual system. In this setup, the error function $e(t)$ is defined as

$$e(t) = y(t) - y_m(t),$$

where $y(t)$ is the output of the system, $y_m(t)$ the output of the reference model.

If the reference model is given in terms of fuzzy logic rules of the form

$$\text{IF } x(t) \in S_n \text{ AND } u(t) \in S_n \text{ THEN } \dot{x}(t) \in S_{an+bm}, \tag{6.6}$$

it describes a system that has an output bounded from both above and below by two analytical systems (see Chapter 3):

$$\dot{\Xi}(t) = a\,\Xi(t) + b\,u(t) + [\,|a| + |b|\,]\sigma,$$
$$\dot{\xi}(t) = a\,\xi(t) + b\,u(t) - [\,|a| + |b|\,]\sigma.$$

The task of guiding the dynamic system (to be controlled) to have the output bounded by these two bounds is the goal of the MRAFS. In this case, the error function is slightly modified to reflect these bounds, as the average of the error with respect to the upper bound and the error with respect to the lower bound:

$$e(t) = \frac{1}{2}\,[\,y(t) - y_{\text{upper}}(t)] + \frac{1}{2}\,[\,y(t) - y_{\text{lower}}(t)].$$

Let $J(e)$ be the objective function for designing the MRAFS gain. This function $J(e)$ is normally the magnitude of the error function $e(t)$. Two popular magnitude functions are the l_2-norm

$$J_2(e) = e^{\mathrm{T}}(t)\, e(t)$$

and the l_1-norm

$$J_1(e) = |\, e(t)\, |.$$

Each formulation will lead to a specific closed-form solution. The l_2 solution is common because it normally yields a linear solution. It can be shown that the l_1 solution is similar to the traditional bang-bang control.

Let θ be the parameter used to control the system so that it tracks the reference model. The approach here is to optimize the objective function $J(e)$ with respect to this parameter θ. The objective function is gradually reduced when the parameter θ is adjusted in the negative gradient of J, i.e.,

$$\frac{d\theta}{dt} = -\gamma \frac{\partial J}{\partial \theta},$$

where γ is a scalar which determines the convergence rate in general. In this approach, the parameter θ is assumed to be continuously adjusted. Depending on the definition of J and the nature of the reference model, the formulation to calculate the adjustment to the parameter θ is derived on a case by case basis.

V. DUAL CONTROL

The traditional concept of dual control was originally designed to handle stochastic control systems. It consists of a controller and an estimator for the statistical properties of the closed-loop system. The estimator is used to provide statistical information to the controller that will adjust itself accordingly. This concept is usually difficult to implement on an actual system.

The same dual control concept can be extended to fuzzy control systems. In this setting, the fuzzy setup consisting of definitions for fuzzy sets and its associated membership functions is re-evaluated during the operational process of the system. These definitions are continuously adjusted and, thus, the controller is continuously adjusted accordingly.

The main difficulty of implementing the traditional dual control is also translated into the fuzzy dual control concept. The process of adjusting the fuzzy definitions involves changing the membership functions. This change does not permit the analytical process of simplifying the rules to be converted into two analytical bounds during the design phase. It is therefore difficult to implement this simplification method on a digital computer and operate it on line in synchronization with the processing rate at which data are collected.

The second difficulty in the fuzzy dual control concept is in the change of the boundaries of the fuzzy sets that lead to possible widening of the coverage of the area of applicability.

Basically, the concept of fuzzy dual control is often proposed for application in fuzzy clustering where input and output are analyzed to determine an effective set of fuzzy rules that correlate the two. It shows promising future in mimicking human's intelligence in observing surrounding environment and making intelligent decisions, but generally not in control applications.

VI. SUB-OPTIMAL FUZZY CONTROL

An optimal controller is a controller designed to optimize its performance criteria. Normally, the criteria can be either the tracking performance (various specifications), the cost of operation (amount of control input required), or a combination of the two. An optimal controller is usually designed for its operative time duration from the initial time t_i to some specific ending time t_f. Sometimes the ending time t_f can be infinite. The optimal controller is a solution to the following optimization problem:

$$\min_{u(t)} \int_{t_i}^{t_f} J(x(t),u(t),x_f)dt$$

subject to

$$\dot{x}(t) = f(x(t),u(t)),$$

where the function $J(\cdot,\cdot)$ is the criterion function, x_f the desired set point, and $u(t)$ the controller. When the mathematical model is given in terms of fuzzy logic rules, the resulting controller is referred to as the optimal fuzzy controller.

There are cases in which the desired set point x_f is changing with time and cannot be profiled in advance. This scenario is often seen in the case of a system trying to synchronously track another dynamic system, e.g., a spacecraft docking with another moving spacecraft in orbit. In this case, a prediction scheme can be implemented to provide the moving set point x_f. However, the prediction method will result in an error that normally increases with time. It is impractical to design an optimal controller with operative time duration $[t_i,t_f)$ for a large t_f based on this predictive scheme for the moving set point. In this case, it is reasonable to take an alternate approach, called sub-optimal fuzzy control.

The sub-optimal fuzzy control method is a strategy of optimally solving a control problem only over a short time interval when prediction of dynamic environment is still accurate. At the beginning of a new interval, data are collected and assessment of the dynamic environment is revised to give a new prediction. Another optimal control problem is then set up and solved. In this adaptive way, the same problem format is repeatedly used, with only parameters being revised for better accuracy in the prediction of the changing environment.

A. SISO Control Systems

The optimal fuzzy control problem for an SISO dynamic system can be formulated as follows:

$$\min_{u(t)} \{ \int_{t_{initial}}^{t_{final}} [R_1\Xi^2(t)+R_2\xi^2(t)+R_3u^2(t)]dt + R_4\Xi^2(t_{\text{final}}) + R_5\xi^2(t_{\text{final}})\}$$

subject to the constraint(s):

$$\text{IF } x(t) \in S_n \text{ AND } u(t) \in S_n \text{ THEN } \dot{x}(t) \in S_{an+bm},$$

where $\Xi(t)$ and $\xi(t)$ are the upper and lower bounds of $x(t)$. Again, this setup refers to the thorough study given in Chapter 3. Since $x(t)$ is not given precisely, one can try minimizing the upper and lower bounds of the errors between the controlled trajectories and the target. As a result, any state variable that is between these two bounds will also be controlled to the target. The quadratic terms of control input $u(t)$ indicate a requirement for a bounded minimum energy solution. Additional constraints such as time minimization can be added to the problem and the solution can be derived in a similar manner.

Theorem 6.1. The SISO fuzzy model used as a constraint in the optimal fuzzy control problem posed above is bounded from above by $x_{\text{upper}}(t) = \Xi(t)$ and from below by $x_{\text{lower}}(t) = \xi(t)$, where

$$\dot{\Xi}(t) = a\,\Xi(t) + b\,u(t) + [\,|a| + |b|\,]\sigma,$$
$$\dot{\xi}(t) = a\,\xi(t) + b\,u(t) - [\,|a| + |b|\,]\sigma,$$

with the initial conditions

$$\Xi(t_{\text{initial}}) = \xi(t_{\text{initial}}) = x(t_{\text{initial}}).$$

Proof. Given a positive constant σ, for any numbers $x(t)$ and $u(t)$ at any instant t, there exist two unique integers, n and m, such that

$$(n)\sigma \le x \le (n+1)\sigma,$$
$$(m)\sigma \le u \le (m+1)\sigma.$$

The rule in (6.6) will have four possible implications:

(i) $x \in S_n \wedge u \in S_m \Rightarrow \dot{x} \in S_{an+bm}$,

(ii) $x \in S_{n+1} \wedge u \in S_m \Rightarrow \dot{x} \in S_{an+bm+a}$,

(iii) $x \in S_n \wedge u \in S_{m+1} \Rightarrow \dot{x} \in S_{an+bm+b}$,

(iv) $x \in S_{n+1} \wedge u \in S_{m+1} \Rightarrow \dot{x} \in S_{an+bm+a+b}$.

These four implications yield four different results for $\dot{x}$. The weighted average formulation is used to calculate the final value for $\dot{x}$. The membership values for each of the implications above are calculated using the triangular membership functions. There are four possible cases. The final result by cases and the derivation steps can be derived mathematically to show that all four cases yield the same results given in the theorem.

Theorem 6.2. The solution for the SISO optimal fuzzy control problem defined above can be obtained as a direct consequence of Theorem 6.4 below on an MIMO system:

$$u^*(t) = -\frac{1}{R_3}[\,b \quad b\,]\,P(t)\begin{bmatrix}\Xi(t)\\ \xi(t)\end{bmatrix},$$

where $P(t)$ is the solution to the matrix Riccati equation

$$-\dot{P}(t) = \begin{bmatrix} R_1 & 0 \\ 0 & R_2 \end{bmatrix} - \frac{b^2}{R_3} P^2(t) + 2a\,P(t)$$

with the boundary condition $P(t_{\text{final}}) = \begin{bmatrix} R_4 & 0 \\ 0 & R_5 \end{bmatrix}$.

Proof. This is a special case of the MIMO case presented next.

B. MIMO Control Systems

The optimal fuzzy control problem for an MIMO dynamic system (see Chapter 3 for system setup) can be formulated as follows:

$$\min_{u(t)} \Big\{ \int_{t_{initial}}^{t_{final}} [\Xi^{\mathrm{T}}(t)R_1\Xi(t) + \xi^{\mathrm{T}}(t)R_2\xi(t) + u^{\mathrm{T}}(t)R_3u(t)]dt$$
$$+ \Xi^{\mathrm{T}}(t_{\text{final}})\mathrm{R}_4\Xi(t_{\text{final}}) + \xi^{\mathrm{T}}(t_{\text{final}})\mathrm{R}_5\xi(t_{\text{final}}) \Big\}$$

subject to the constraint(s)

$R^{(i)}$: IF $(x_1(t) \in S_{n_1})$ AND ... AND $(x_p(t) \in S_{n_p})$ AND
$(u_1(t) \in S_{m_1})$ AND ... AND $(u_q(t) \in S_{m_q})$
THEN $\dot{x}_i(t) \in \mathrm{S}_{a_{i1}n_1+\ldots+a_{ip}n_p+b_{i1}m_1+\ldots+b_{iq}m_q}$.

An MIMO system described above, under some specific definitions of fuzzy sets S_n and membership functions $\mu_{S_n}(x)$, will yield closed-form upper and lower bounds. These bounds will then lead to a specific objective function in the optimal fuzzy control problem, which finally yield a unique closed-form solution.

Theorem 6.3. The MIMO fuzzy model used as constraint in the optimal fuzzy control problem defined above yields the upper bound $\Xi_i(t)$ and the lower bound $\xi_i(t)$ as follows:

$$\dot{\Xi}_i(\mathrm{t}) = a_{i1}\Xi_1 + \ldots + a_{ip}\Xi_p + b_{i1}u_1 + \ldots + b_{iq}u_q +$$
$$[\,|a_{i1}| + \ldots + |a_{ip}| + |b_{i1}| + \ldots + |b_{iq}|\,]\sigma,$$

$$\dot{\xi}_i(\mathrm{t}) = a_{i1}\xi_1 + \ldots + a_{ip}\xi_p + b_{i1}u_1 + \ldots + b_{iq}u_q -$$
$$[\,|a_{i1}| + \ldots + |a_{ip}| + |b_{i1}| + \ldots + |b_{iq}|\,]\sigma.$$

Proof. This theorem can be proved by induction as follows. First, for the induction base, i.e., for $p = 1$ and $q = 1$, which is the SISO case, the result was established by Theorem 6.1 presented earlier. For the induction assumption, we assume that for any two arbitrary integers p and q, the two transition equations above hold. Then, the induction steps for the cases of $p+1$ and/or $q+1$ can be derived by direct but tedious computation.

Theorem 6.4. The solution for the MIMO optimal fuzzy control problem, with fixed t_{final}, is

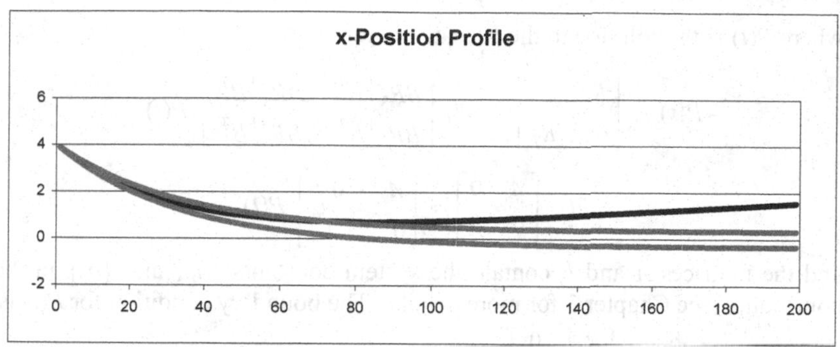

Figure 6.5 Solution trajectory of a fuzzy model without optimization goes beyond the upper boundary.

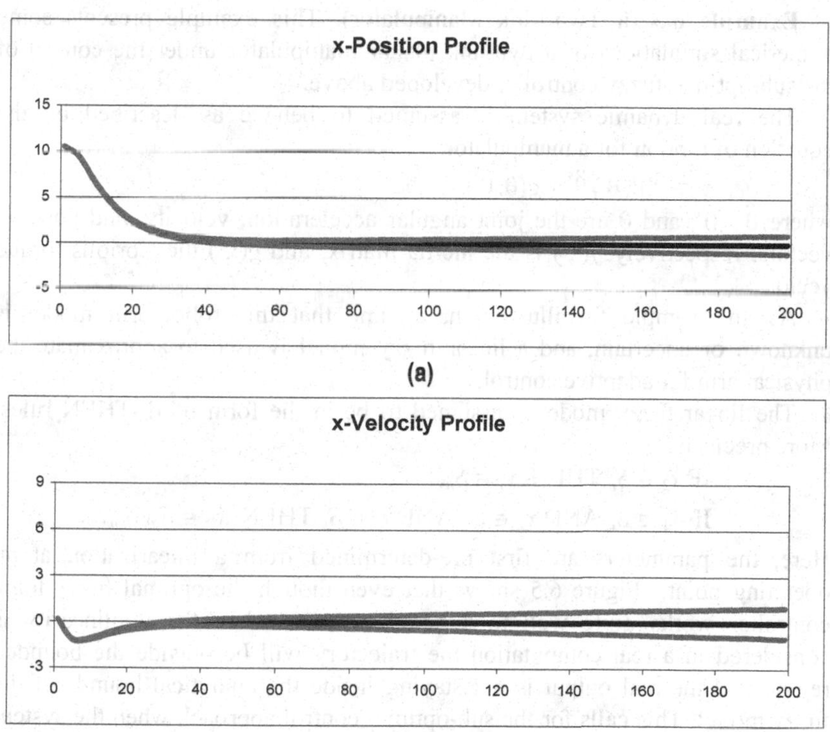

Figure 6.6 Solution trajectory of the same fuzzy model subject to sub-optimal control.

$$u^*(t) = -\ R_3^{-1}\ [\,B^{\mathrm{T}} \quad B^{\mathrm{T}}]\ P(t) \begin{bmatrix} \Xi(t) \\ \xi(t) \end{bmatrix},$$

where $P(t)$ is the solution to the Riccati equation:

$$-\dot{P}(t) = \begin{bmatrix} R_1 & 0 \\ 0 & R_2 \end{bmatrix} - P(t)\begin{bmatrix} BR_3^{-1}B^T & BR_3^{-1}B^T \\ BR_3^{-1}B^T & BR_3^{-1}B^T \end{bmatrix} P(t)$$

$$+ P(t)\begin{bmatrix} A & 0 \\ 0 & A \end{bmatrix} + \begin{bmatrix} A^T & 0 \\ 0 & A^T \end{bmatrix} P(t)$$

and the matrices A and B contain the system constants $\{a_{ij}\}$ and $\{b_{ij}\}$ in the constraints (see Chapter 3 for more detail). The boundary condition for $P(t)$ is

$$P(t_{\text{final}}) = \begin{bmatrix} R_4 & 0 \\ 0 & R_5 \end{bmatrix}.$$

Proof. See [3], pp 201-219.

Example 6.3 (a Two-Link Manipulator). This example presents some numerical simulations of a two-link planar manipulator under the control of the sub-optimal fuzzy controller developed above.

The real dynamic system is assumed to behave as described by the equation of motion for a manipulator:

$$\tau = f(\theta, \dot{\theta})\ddot{\theta} + g(\theta, \dot{\theta}),$$

where $\ddot{\theta}$, $\dot{\theta}$, and θ are the joint angular acceleration, velocity, and position vectors, respectively, $f(\cdot,\cdot)$ is the inertia matrix, and $g(\cdot,\cdot)$ the Coriolis torque vector.

As an example for illustration, assume that this robot arm model is unknown or uncertain, and a linear fuzzy model is used to approximate the physical arm for adaptive control.

The linear fuzzy model is assumed to be in the form of IF-THEN rules. More precisely,

$$\text{IF } x_2 \in S_n \text{ THEN } \dot{x}_1 \in S_n,$$

$$\text{IF } x_1 \in S_n \text{ AND } x_2 \in S_m \text{ AND } u \in S_p \text{ THEN } \dot{x}_2 \in S_{an+bm+cp}.$$

Here, the parameters are first pre-determined from a linearization at an operating point. Figure 6.5 shows that even though the optimal fuzzy logic controller works quite well to a certain extent, when the nonlinearity is considered in a real computation the trajectory will be outside the bounded region and the real output is not staying inside the analytical bounds of the fuzzy model. This calls for the sub-optimal control approach when the system is moving outside the accuracy range of linearization. In this case, the parameters must be re-established and the control problem is solved again with these new parameters.

The result is now improved: the end-effector is stabilized within some small bounds as shown in Figure 6.6(a) and (b).

Example 6.4 (Docking a Boat on a 2-D Surface). The problem of docking a boat can be defined as follows: Given a description of the dynamics of a boat, its current state, and the desired state (docking position), design a control input that drives the boat from its current state to stop at its docking position.

In this example, a fuzzy logic model is used to solve the docking problem in an optimal sense. This is the problem of finding a controller that performs the best for a fuzzy model under a set of pre-defined criteria. The solution is also given as a set of fuzzy logic rules. A fuzzy description of the closed-loop system is derived to establish the stability of the controlled system.

First, a fuzzy model for the system is derived. This model will be used as a set of constraints in the optimal control problem. Then, a set of criteria is defined as the objective for the optimal control problem. Finally, a solution optimizing this objective under the fuzzy constraints is obtained. This solution is given in terms of parameters for the fuzzy logic model of the system. The desired destination can change and, therefore, the objective function will change as a result. It is impractical to derive an optimal solution with some assumed knowledge of a destination. In general, the destination is located by external sensors and data of its position are continuously provided to the controller. The controller has to adjust the parameters to the optimal control problem and a new solution is recalculated based on the new data. The process is adaptively carried out until the docking is completed.

The state of the boat is characterized by its current location, orientation (the direction of the vector from its stern to its bow), and speed. It is controlled by a pedal that generates acceleration either forward or backward, and by a rudder that turns the boat. When the rudder is set at zero, the boat will move in a straight line, either forward or backward. When the rudder is set to one side or the other, it exerts a lateral force that causes the boat to follow a circular arc. In this example, for simplicity, it is assumed that there are no sideways slippage, no water currents, no vibrational dynamics, no drag, and no pitching, rolling, or tipping of the boat. The boat will move according to the control input with no outside interference.

The fuzzy model for the boat can be described in terms of five fuzzy rules such as:

$R^{(1)}$: IF $\varphi_4 \in S_n$ THEN $\dot{\varphi}_1 \in S_n$,

$R^{(2)}$: IF $\varphi_5 \in S_n$ THEN $\dot{\varphi}_2 \in S_n$,

$R^{(3)}$: IF $\varphi_4 \in S_n$ AND $\varphi_5 \in S_m$ AND $u_2 \in S_q$
THEN $\dot{\varphi}_3 \in S_{n+m+q}$,

$R^{(4)}$: IF $\varphi_3 \in S_n$ AND $\varphi_4 \in S_m$ AND $\varphi_5 \in S_k$ AND
$u_1 \in S_p$ AND $u_2 \in S_q$
THEN $\dot{\varphi}_4 \in S_{n+m+k+p+q}$,

$R^{(5)}$: IF $\varphi_3 \in S_n$ AND $\varphi_4 \in S_m$ AND $\varphi_5 \in S_k$ AND
$u_1 \in S_p$ AND $u_2 \in S_q$
THEN $\dot{\varphi}_5 \in S_{n+m+k+p-q}$.

Here, φ_1 is the x-coordinate, φ_2 is the y-coordinate of the position of the boat, φ_3 is the displacement of the steering wheel, φ_4 is the x-velocity component, and φ_5 is the y-velocity component of the boat, and $n, m, p, q, k = 0, \pm 1, \pm 2, \ldots$.

The objective function consists of two terms: the difference between the state of the system and the desired state, and a measure of the energy of the control input, as discussed before, is often used in optimal control problems. The optimization of the error term seeks to bring the system to the desired state. The optimization of the energy term seeks to minimize the control energy. The solution that optimizes the objective function yields a compromise between convergence and minimal effort.

The solution to this problem takes the form of the following set of fuzzy rules:

$$\begin{aligned}
L^{(1)}:\ & \text{IF } \varphi_1 \in S_{n_1} \text{ AND } \varphi_2 \in S_{n_2} \text{ AND } \varphi_3 \in S_{n_3} \text{ AND} \\
& \qquad \varphi_4 \in S_{n_4} \text{ AND } \varphi_5 \in S_{n_5} \\
& \text{THEN } u_1 \in S_{k_1}, \\
L^{(2)}:\ & \text{IF } \varphi_1 \in S_{n_1} \text{ AND } \varphi_2 \in S_{n_2} \text{ AND } \varphi_3 \in S_{n_3} \text{ AND} \\
& \qquad \varphi_4 \in S_{n_4} \text{ AND } \varphi_5 \in S_{n_5} \\
& \text{THEN } u_2 \in S_{k_2},
\end{aligned}$$

where k_1 and k_2 are functions of $n_1, \ldots, n_5$, determined based on the Riccati equation, as discussed before.

An exact mathematical model is used as the "actual system" in order to test and validate the optimal fuzzy control solution. This model corresponds to the exact response of the boat to any control input. Note that the derivation of the control input discussed in previous sections does not depend on knowledge of this mathematical model. By validating the approach with a fully specified system, one reinforces the confidence needed to deal with the general case where the precise equations of motion are not known to the person developing the controller. A specified system for this purpose is

$$\begin{aligned}
\dot{\varphi}_1 &= \varphi_4, \\
\dot{\varphi}_2 &= \varphi_5, \\
\dot{\varphi}_3 &= c\,[\varphi_4 \cos(\varphi_3) + \varphi_5 \sin(\varphi_3)\,]\, u_2, \\
\dot{\varphi}_4 &= u_1 \cos(\varphi_3) - c\,[\,\varphi_4^2 + \varphi_5^2\,] \sin(\varphi_3)\, u_2, \\
\dot{\varphi}_5 &= u_1 \sin(\varphi_3) + c\,[\,\varphi_4^2 + \varphi_5^2\,] \cos(\varphi_3)\, u_2.
\end{aligned}$$

The constant c is a system parameter specifying the ratio of steering wheel angle to the curvature of the path that the boat will make while turning (the inverse of the turning radius). In this simulation, the value $c = 1.0$ was used. Figure 6.7 shows how the first two state variables of this system behave when the system starts at the origin, oriented 45° from the x-axis, with initial x- and y-velocities of 1.0, and the inputs are constant: $u_1 = 0$, and $u_2 = 0.3$. The sinusoidal patterns for x and y correspond to a counterclockwise circular path of diameter 6.66, centered at (–2.3570, –2.3570), with a constant speed of 1.412.

A theoretical analysis of the boat model can be carried out to confirm that this trajectory agrees with the assumptions of the model.

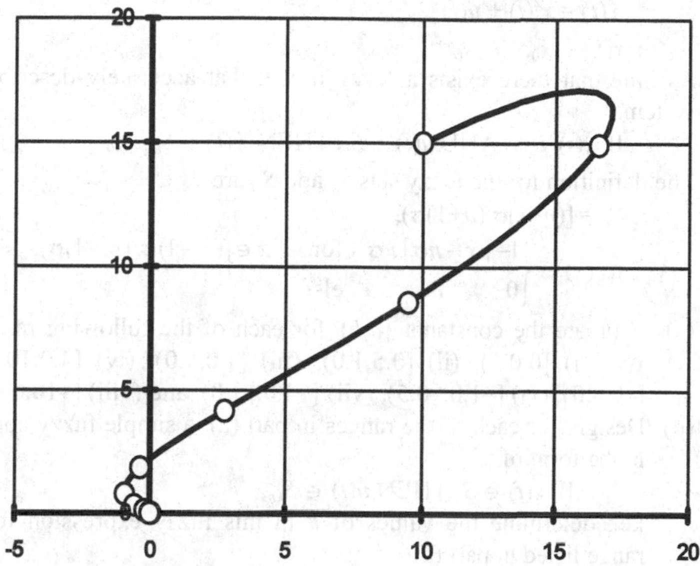

Figure 6.7 Simulation of fuzzy sub-optimal control for a boat: x-y plot of the trajectory.

PROBLEMS

P6.1 Consider an SISO dynamic system of the form

$$\dot{x}(t) = x^2(t) + u(t)$$
$$x(t_0) = x_0.$$

Assume that there exists a fuzzy model that accurately describes this system:

IF $x(t) \in S_n$ AND $u(t) \in S_m$ THEN $\dot{x}(t) \in S_{an+bm}$.

The definition for the fuzzy sets S_n and S_m are

$$S_n = [(n-1)\sigma,(n+1)\sigma),$$
$$\mu_{S_n}(x) = \begin{cases} 1-|x-n\sigma|/\sigma & \text{for} \quad x \in [(n-1)\sigma,(n+1)\sigma), \\ 0 & \text{else.} \end{cases}$$

(a) Estimate the constants $\{a,b\}$ for each of the following ranges for $x(t)$: (i) [0,0.5), (ii) [0.5,1.0), (iii) [1.0,4.0), (iv) [4.0,10.0), (v) [–0.5,0), (vi) [–1.0,–0.5), (vii) [–4.0,–1.0), and (viii) [–10.0,–4.0).

(b) Design, for each of the ranges in part (a), a simple fuzzy controller in the form of

IF $x(t) \in S_n$ THEN $u(t) \in S_{kn}$,

i.e., determine the values of k in this fuzzy expression for each range listed in part (a).

(c) Write a simulation program that combines the actual system, the fuzzy controllers in part (b), along with condition checking, to determine when $x(t)$ falls into a particular range to activate a specific controller designed for that range. Use the following conditions x_0 = 3.0 and Δt = 0.02 sec. Plot the output $x(t)$ for $t \in [0.0,5.0)$ in seconds.

P6.2 Consider an SISO dynamic system of the form

$$\dot{x}(t) = \sin(x(t)) + u(t)$$
$$x(t_0) = x_0.$$

Assume that a set of data is collected for output $x(t)$ at the beginning of every time interval $[t_n,t_n+\Delta t)$. Denote this data set $D_{x,n}$ with

$$D_{x,n} = \{x(t_0), x(t_1), x(t_2), \ldots, x(t_n)\}.$$

Similarly, assume that a set of data $D_{u,n}$ is collected for the input $u(t)$ that drives the system, with

$$D_{u,n} = \{u(t_0), u(t_1), u(t_2), \ldots, u(t_n)\}.$$

(a) Set up recursive formulas to update the parameters a and b in the corresponding fuzzy model

IF $x(t) \in S_n$ AND $u(t) \in S_m$ THEN $\dot{x}(t) \in S_{an+bm}$,

with fuzzy sets S_n and S_m defined as

$$S_n = [(n-1)\sigma,(n+1)\sigma),$$

$$\mu_{S_n}(x) = \begin{cases} 1-|x-n\sigma|/\sigma & \text{for} \quad x \in [(n-1)\sigma,(n+1)\sigma), \\ 0 & \text{else,} \end{cases}$$

for the data sets $D_{x,n}$ and $D_{u,n}$.

(b) For an input signal, $u(t) = \sin(\pi t)$, write a computer program to generate input and output profiles $D_{u,n}$ and $D_{x,n}$ for the system at time $t_0 = 0.0$ to $t_n = 10.0$ sec. with $\Delta t = 0.02$ sec. Plot the input $u(t)$ and output $x(t)$ versus time t.

(c) Write a computer program to update the parameters a and b in the fuzzy model used to approximate the given nonlinear system. Plot the parameters $a(t)$ and $b(t)$ versus time t.

(d) Design a simple fuzzy controller in the form of

IF $x(t) \in S_n$ THEN $u(t) \in S_{kn}$

and determine the values of k in the above fuzzy expression as functions of parameters a and b.

(e) Write a computer program to combine the dynamic system, the update of the parameters a and b for the fuzzy model, the update of the parameter k for the fuzzy controller, and the input/output for the closed-loop system with initial condition $x(0) = 3.0$. Plot the input $u(t)$ and output $x(t)$ versus time t. How long does it take for the system to converge to final position $x_f = 0$? Does the output system oscillate at this value? Explain.

P6.3 Consider an SISO dynamic system of the form

$$\dot{x}(t) = \sin(t)\, x(t) + \cos(t)\, u(t)$$
$$x(t_0) = x_0.$$

Assume that a set of data is collected for output $x(t)$ at the beginning of every time interval $[t_n, t_n+\Delta t)$. Denote this data set $D_{x,n}$ with

$$D_{x,n} = \{x(t_0), x(t_1), x(t_2), \ldots, x(t_n)\}.$$

Similarly, assume that a set of data $D_{u,n}$ is collected for the input $u(t)$ that drives the system, where

$$D_{u,n} = \{u(t_0), u(t_1), u(t_2), \ldots, u(t_n)\}.$$

Repeat the steps from (a) through (e) in Problem ***P6.2*** for the present case.

P6.4 Consider an SISO dynamic system of the form

$$\dot{x}(t) = 3\, x(t) + 2\, u(t)$$
$$x(t_0) = x_0.$$

(a) Design a Model Reference Adaptive Fuzzy System so that

$$u(t) = \gamma(t)\, x(t)$$

will drive the closed-loop system to track the fuzzy model

IF $x(t) \in S_n$ AND $u(t) \in S_m$ THEN $\dot{x}(t) \in S_{6n+4n}$.

(b) Write a computer program that plots the outputs and inputs of both the bounds of the fuzzy model and the dynamic system. Compare the results. Is the output of the closed-loop dynamic system tracking the fuzzy model, i.e., does it fall in between the two bounds?

P6.5 Consider an SISO dynamic system of the form

$$\dot{x}(t) = x^2(t) + u(t)$$

$$x(t_0) = x_0.$$

(a) Design a Model Reference Adaptive Fuzzy System, so that

$$u(t) = \gamma(t)\, x(t)$$

will drive the closed-loop system to track the fuzzy model

$$\text{IF } x(t) \in S_n \text{ AND } u(t) \in S_m \text{ THEN } \dot{x}(t) \in S_{2n+n}.$$

(b) Write a computer program that plots the outputs and inputs of both the bounds of the fuzzy model and the dynamic system. Compare the results. Is the output of the closed-loop dynamic system tracking the fuzzy model, i.e., does it fall in between the two bounds?

REFERENCES

[1] K. J. Åstrom and B. Wittenmark. *Adaptive Control.* Reading, MA: Addison Wesley Publishing Company (1989).

[2] G. Chen, T. T. Pham, and J. J. Weiss. "Fuzzy Modeling of Control Systems," *IEEE Transactions on Aerospace and Electrical Systems*, Vol. 30, pp 414-429, 1995.

[3] D. E. Kirk. *Optimal Control Theory, An Introduction.* Englewood Cliffs, NJ: Prentice-Hall (1970)

[4] H. Kwakernaak and R. Sivan. *Linear Optimal Control Systems.* New York, NY: Wiley Interscience (1972).

[5] J. W. Polderman. *Adaptive Control & Identification: Conflict or Conflux?* Amsterdam, the Netherlands: Centrum voor Wiskunde en Informatica (1989).

[6] S. Sastry and M. Bodson. *Adaptive Control: Stability, Convergence, and Robustness.* Englewood Cliffs, NJ: Prentice-Hall (1984).

[7] M. Sugeno. "An Introductory Survey of Fuzzy Control," *Information Sciences*, Vol. 36, pp 59-83, 1985.

[8] M. Sugeno and T. Yasukawa. "A Fuzzy Logic Based Approach to Qualitative Modeling," *IEEE Transactions on Fuzzy Systems*, Vol. 1, pp 7-31, 1993.

[9] T. Takagi and M. Sugeno. "Fuzzy Identification of Systems and Its Applications to Modeling and Control," *IEEE Transaction on Systems, Man, and Cybernetics,* Vol. 15. pp 116-132, 1985.

[10] R. M. Tong. "A Control Engineering Review of Fuzzy Systems," *Automatica*, Vol. 13, pp 559-569, 1977.

[11] R. M. Tong. "An Annotated Bibliography of Fuzzy Control." *Industrial Applications of Fuzzy Control* (edited by M. Sugeno). Amsterdam, the Netherlands: Elsevier Science Publishers (1985), pp 249-269.

CHAPTER 7

Some Applications of Fuzzy Control

This chapter introduces some engineering applications of fuzzy logic and fuzzy control systems. The purpose of this introduction is mainly to illustrate workability and applicability of fuzzy logic and fuzzy control in real-life situations, based on the fuzzy systems theories developed in the previous chapters of the book.

Three main applications are introduced in the following order: a fuzzy rule-based expert system for a health care diagnostic system monitoring vital signs of a human patient, a fuzzy control system for an autofocus camera lens system, and a fuzzy modeling and design of a fuzzy controller for a servo system. For each application, a description of the functionality of the system is given, followed by the fuzzy logic rules and computer simulation results.

I. HEALTH MONITORING FUZZY DIAGNOSTIC SYSTEMS

Diagnostic systems are used to monitor the behavior of a process and identify certain pre-defined patterns that associate with well-known problems. These problems, once identified, imply suggestions for specific treatment. Most diagnostic systems are in the form of a rule-based expert system: a set of rules is used to describe certain patterns. Observed data are collected and used to evaluate these rules. If the rules are logically satisfied, the pattern is identified, and a problem associated with that pattern is suggested. Each particular problem might imply a specific treatment. In general, the diagnostic systems are used for consultation rather than replacement of human expert. Therefore, the final decision is still with the human expert to determine the cause and to prescribe the treatment.

Most current health monitoring systems only check the body's temperature, blood pressure, and heart rate against individual upper and lower limits and start an audible alarm should each signal move out of its predefined range (either above the upper limit or below the lower limit). Then, human experts (nurses or physicians) will have to examine the patient and probe the patient's body further for additional data that lead to proper diagnosis and its corresponding treatment.

Other more complicated systems normally involve more sensors that provide more data but still follow the same pattern of independently checking individual sets of data against some upper and lower limits. The warning alarm from these systems only carries a meaning that there is something wrong with the patient. Thus, attending staff would have to wait for the physician to make a diagnostic examination before they could properly prepare necessary equipment for a corresponding treatment. In a life-threatening situation, reducing the physician's reaction time (the time between the warning and the time proper treatment is given to the patient) by preparing proper equipment for specific treatment in advance would significantly increase the patient's chance of surviving.

In the field of exploration where a team of humans is sent to a distant and isolated location, it is important to have a health monitoring system that can give early diagnostic data of each explorer's health status to prepare for the continuation of the mission. In space exploration missions, an astronaut who suddenly gets sick will be diagnosed and the data reviewed by the onboard physician as well as the physician team on earth to determine if a proper treatment onboard is possible (for the continuation of the mission) or if specific treatment on earth is required (for the abortion of the mission). Therefore, the health monitoring system is playing two roles in every space mission: to monitor the health of the astronaut, and to aid in determining if a mission should be continued or aborted (due to serious illness).

This section presents a simple implementation of a health monitoring expert system utilizing fuzzy rules for its rule base. This health monitoring expert system consists of a set of sensors monitoring three vital signs of a patient: body temperature, blood pressure, and heart rate.

A. Fuzzy Rule-Based Health Monitoring Expert Systems

In this system, we will implement a fuzzy rule-based expert health monitoring system with three basic sensors: body temperature, heart rate, and blood pressure. Note that the blood pressure is measured in two readings: systolic pressure (the maximum pressure that the blood exerts on the blood vessel, i.e., the aorta, when the pumping chamber of the heart contracts), and diastolic pressure (the lowest pressure that remains in the small blood vessel when the pumping chamber of the heart relaxes). For simplicity of discussion, only diastolic pressure is used, with the understanding that an additional reading can be easily added to the system and the number of diagnostic cases will increase accordingly. The expert system will check for combinations of data instead of individual data and thus will identify twenty-seven different scenarios instead of three in the conventional system.

Individual sensors can identify three isolated cases: (i) high body temperature indicates high fever normally associated with the body fighting against some infectious virus or bacteria, some hormone disorder such as hyperthyroidism, some autoimmune disorder such as rheumatoid arthritis, or some damage to the hypothalamus; (ii) high blood pressure indicates hypertension normally associated with some kidney disease, hormonal disorder such as hyperaldoteronism, or acute lead poisoning; and (iii) high heart rate indicates rapid heart beat normally associated with an increase in adrenaline (a hormone produced by the adrenal glands) production. In addition, three more cases can be identified: (iv) low body temperature indicates hypothermia; (v) low blood pressure normally associates with excess bleeding, muscle damage, heart valve disorder, or excessive sweating/urination; and (vi) low heart rate normally associates with abnormal pacemaker or with the blockage between the pacemaker and the atria where pacemaker signal is received to stimulate heartbeats.

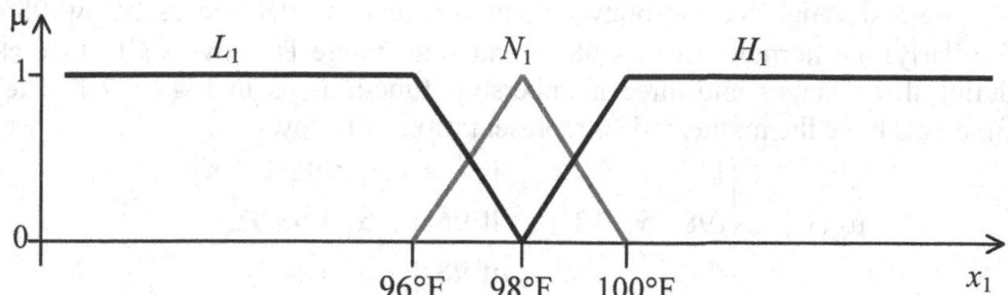

Figure 7.1 Definitions and membership functions of three different ranges for body temperatures.

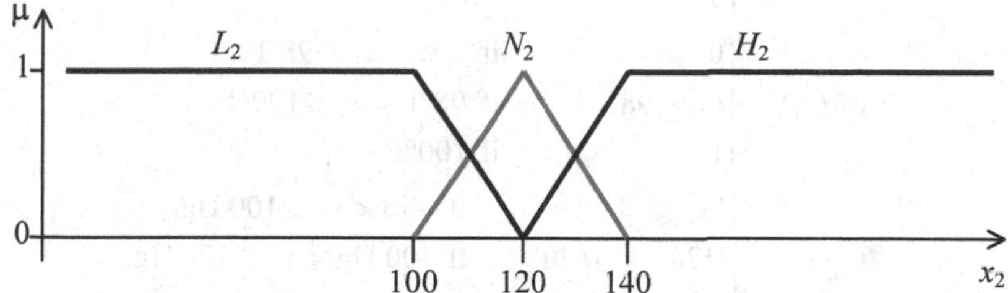

Figure 7.2 Definitions and membership functions of three different ranges for blood pressure.

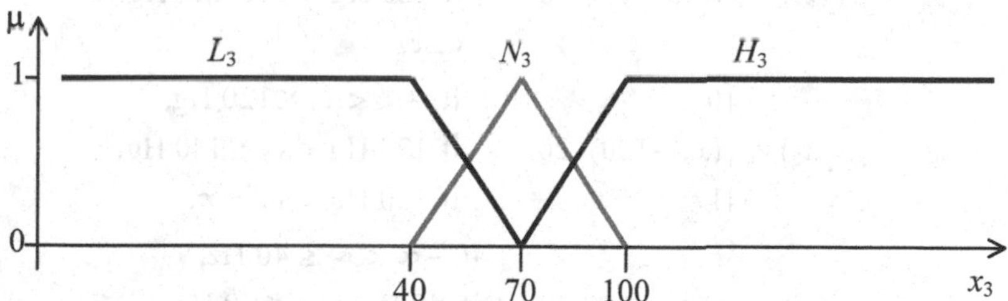

Figure 7.3 Definitions and membership functions of three different ranges for heart rate.

However, the three individual sensors, each with three settings (normal, high, and low), can be combined to give 27 different scenarios. With the perfectly normal case (where all three give normal readings) and the six cases defined above, there are additionally 20 more cases where combinations of abnormal readings can be observed. For example, low blood pressure and high body temperature might indicate severe exposure to heat, a severe loss of blood, or lack of body fluid.

Let x_1 be the body temperature, x_2 the (systolic) blood pressure, x_3 the heart rate, and y the diagnostic statement. Let L_i, N_i, and H_i represent the three sets of low range, normal range, and high range for sensor data x_i, where $i = 1, 2$, or 3. Furthermore, let $C_0, C_1, C_2, \ldots, C_{26}$ be the individual scenarios that could happen for each combination of the different data sets.

The low range for the body temperature can be defined as below 98°F. Similarly, the normal range is 98°F, and high range is above 98°F. One can define three ranges and three membership functions as in Figure 7.1. These functions have the mathematical representation as follows:

$$\mu_{L_1}(x_1) = \begin{cases} 1 & \text{if } -\infty < x_1 \le 96°\text{F}, \\ (98 - x_1)/2 & \text{if } 96°\text{F} < x_1 \le 98°\text{F}, \\ 0 & \text{if } 98°\text{F} < x_1 < \infty, \end{cases}$$

$$\mu_{N_1}(x_1) = \begin{cases} (x_1 - 96)/2 & \text{if } 96°\text{F} < x_1 \le 98°\text{F}, \\ (100 - x_1)/2 & \text{if } 98°\text{F} < x_1 \le 100°\text{F}, \\ 0 & \text{else}, \end{cases}$$

$$\mu_{H_1}(x_1) = \begin{cases} 0 & \text{if } -\infty < x_1 \le 98°\text{F}, \\ (x_1 - 98)/2 & \text{if } 98°\text{F} < x_1 \le 100°\text{F}, \\ 1 & \text{if } 100°\text{F} < x_1 < \infty, \end{cases}$$

$$\mu_{L_2}(x_2) = \begin{cases} 1 & \text{if } -\infty < x_2 \le 100 \text{ Hg}, \\ (120 - x_2)/20 & \text{if } 100 \text{ Hg} < x_2 \le 120 \text{ Hg}, \\ 0 & \text{if } 120 \text{ Hg} < x_2 < \infty, \end{cases}$$

$$\mu_{N_2}(x_2) = \begin{cases} (x_2 - 100)/20 & \text{if } 100 \text{ Hg} < x_2 \le 120 \text{ Hg}, \\ (140 - x_2)/20 & \text{if } 120 \text{ Hg} < x_2 \le 140 \text{ Hg}, \\ 0 & \text{else}, \end{cases}$$

$$\mu_{H_2}(x_2) = \begin{cases} 0 & \text{if } -\infty < x_2 \le 120 \text{ Hg}, \\ (x_2 - 120)/20 & \text{if } 120 \text{ Hg} < x_2 \le 140 \text{ Hg}, \\ 1 & \text{if } 140 \text{ Hg} < x_2 < \infty, \end{cases}$$

$$\mu_{L_3}(x_3) = \begin{cases} 1 & \text{if } -\infty < x_3 \le 40 \text{ Hz}, \\ (70 - x_3)/30 & \text{if } 40 \text{ Hz} < x_3 \le 70 \text{ Hz}, \\ 0 & \text{if } 70 \text{ Hz} < x_3 < \infty, \end{cases}$$

$$\mu_{N_3}(x_3) = \begin{cases} (x_3 - 40)/30 & \text{if } 40 \text{ Hz} < x_3 \le 70 \text{ Hz}, \\ (100 - x_3)/30 & \text{if } 70 \text{ Hz} < x_3 \le 100 \text{ Hz}, \\ 0 & \text{else}, \end{cases}$$

$$\mu_{H_3}(x_3) = \begin{cases} 0 & \text{if } -\infty < x_3 \le 70 \text{ Hz}, \\ (x_1 - 70)/30 & \text{if } 70 \text{ Hz} < x_3 \le 100 \text{ Hz}, \\ 1 & \text{if } 100 \text{ Hz} < x_3 < \infty. \end{cases}$$

From this knowledge, four basic rules can be defined as follows:

$R^{(0)}$: IF x_1 is N_1 AND x_2 is N_2 AND x_3 is N_3 THEN y is C_0
$R^{(1)}$: IF x_1 is H_1 AND x_2 is N_2 AND x_3 is N_3 THEN y is C_1
$R^{(2)}$: IF x_1 is N_1 AND x_2 is H_2 AND x_3 is N_3 THEN y is C_2
$R^{(3)}$: IF x_1 is N_1 AND x_2 is N_2 AND x_3 is H_3 THEN y is C_3.

In addition, three scenarios can be identified when only one sensor provides data that are lower than the lower limit:

$R^{(4)}$: IF x_1 is L_1 AND x_2 is N_2 AND x_3 is N_3 THEN y is C_4
$R^{(5)}$: IF x_1 is N_1 AND x_2 is L_2 AND x_3 is N_3 THEN y is C_5
$R^{(6)}$: IF x_1 is N_1 AND x_2 is N_2 AND x_3 is L_3 THEN y is C_6.

It can be seen that C_0 corresponds to normal condition, C_1 high fever, C_2 hypertension, C_3 rapid heart rate, C_4 hypothermia, C_5 low blood pressure, and C_6 low heart rate. The remaining 20 cases can be defined as follows:

$R^{(7)}$: IF x_1 is L_1 AND x_2 is L_2 AND x_3 is L_3 THEN y is C_7
$R^{(8)}$: IF x_1 is L_1 AND x_2 is L_2 AND x_3 is N_3 THEN y is C_8
$R^{(9)}$: IF x_1 is L_1 AND x_2 is L_2 AND x_3 is H_3 THEN y is C_9
$R^{(10)}$: IF x_1 is L_1 AND x_2 is N_2 AND x_3 is L_3 THEN y is C_{10}
$R^{(11)}$: IF x_1 is L_1 AND x_2 is N_2 AND x_3 is H_3 THEN y is C_{11}
$R^{(12)}$: IF x_1 is L_1 AND x_2 is H_2 AND x_3 is L_3 THEN y is C_{12}
$R^{(13)}$: IF x_1 is L_1 AND x_2 is H_2 AND x_3 is N_3 THEN y is C_{13}
$R^{(14)}$: IF x_1 is L_1 AND x_2 is H_2 AND x_3 is H_3 THEN y is C_{14}
$R^{(15)}$: IF x_1 is N_1 AND x_2 is L_2 AND x_3 is L_3 THEN y is C_{15}
$R^{(16)}$: IF x_1 is N_1 AND x_2 is L_2 AND x_3 is H_3 THEN y is C_{16}
$R^{(17)}$: IF x_1 is N_1 AND x_2 is H_2 AND x_3 is L_3 THEN y is C_{17}
$R^{(18)}$: IF x_1 is N_1 AND x_2 is H_2 AND x_3 is H_3 THEN y is C_{18}
$R^{(19)}$: IF x_1 is H_1 AND x_2 is L_2 AND x_3 is L_3 THEN y is C_{19}
$R^{(20)}$: IF x_1 is H_1 AND x_2 is L_2 AND x_3 is N_3 THEN y is C_{20}
$R^{(21)}$: IF x_1 is H_1 AND x_2 is L_2 AND x_3 is H_3 THEN y is C_{21}
$R^{(22)}$: IF x_1 is H_1 AND x_2 is N_2 AND x_3 is L_3 THEN y is C_{22}
$R^{(23)}$: IF x_1 is H_1 AND x_2 is N_2 AND x_3 is H_3 THEN y is C_{23}
$R^{(24)}$: IF x_1 is H_1 AND x_2 is H_2 AND x_3 is L_3 THEN y is C_{24}
$R^{(25)}$: IF x_1 is H_1 AND x_2 is H_2 AND x_3 is N_3 THEN y is C_{25}
$R^{(26)}$: IF x_1 is H_1 AND x_2 is H_2 AND x_3 is H_3 THEN y is C_{26}.

Physicians can provide their knowledge in medical science to label individual cases C_i for i = 7, 8, ..., 26. For example, the condition C_{26} with high body temperature, high blood pressure, and high heart rate might indicate hyperthyroidism (a condition in which excess thyroid hormone is produced); the condition C_{25} with high body temperature, high blood pressure, and normal heart rate might indicate pheochromocytoma (a tumor in the adrenal gland causing overproduction of a hormone that triggers high blood pressure and raises body temperature), or the condition C_{20} with high body temperature, low blood pressure, and normal heart rate might indicate heat stroke (a condition associated with prolonged exposure to heat).

The membership function for a rule is calculated as a minimum of the memberships of individual conditions, i.e., the membership functions for rules $R^{(0)}$, $R^{(1)}$, $R^{(2)}$, and $R^{(3)}$ are:

$$\mu_{R^{(0)}}(x_1,x_2,x_3) = \min\{\, \mu_{N_1}(x_1),\ \mu_{N_2}(x_2),\ \mu_{N_3}(x_3) \,\},$$
$$\mu_{R^{(1)}}(x_1,x_2,x_3) = \min\{\, \mu_{H_1}(x_1),\ \mu_{N_2}(x_2),\ \mu_{N_3}(x_3) \,\},$$
$$\mu_{R^{(2)}}(x_1,x_2,x_3) = \min\{\, \mu_{N_1}(x_1),\ \mu_{H_2}(x_2),\ \mu_{N_3}(x_3) \,\},$$
$$\mu_{R^{(3)}}(x_1,x_2,x_3) = \min\{\, \mu_{N_1}(x_1),\ \mu_{N_2}(x_2),\ \mu_{H_3}(x_3) \,\}.$$

The membership functions for rules $R^{(4)}$ through $R^{(26)}$ can be similarly established according to their corresponding rules given earlier.

B. Computer Simulations

In this section, numerical simulations are created to demonstrate the workability of the 27 rules given in the previous section representing a health monitoring expert system.

Example 7.1. In this example, the rules $R^{(1)}$ for high fever, $R^{(2)}$ for hypertension, and $R^{(3)}$ for rapid heart rate are tested. The temperature profile $x_1(t)$, blood pressure profile $x_2(t)$, and heart rate profile $x_3(t)$ are generated according to the following formulas:

$$x_1(t) = 98 + 4\sin(\frac{\pi}{20}t)\, e(t) + n(t),$$

$$x_2(t) = 120 + 40\sin(\frac{\pi}{20}t)\, e(t-20) + n(t),$$

$$x_3(t) = 70 + 60\sin(\frac{\pi}{20}t)\, e(t-40) + n(t),$$

where the function $n(t)$ is random white noise and $e(t)$ the rectangular envelop function of the form:

$$e(t) = \begin{cases} 1 & \text{if } 0 \leq x \leq 20, \\ 0 & \text{else.} \end{cases}$$

These three formulas for $x_1(t)$, $x_2(t)$, and $x_3(t)$ create three separate time intervals, $I_1 \approx [7,13)$, $I_2 \approx [27,33)$, and $I_3 \approx [47,53)$. These data profiles will result in rule $R^{(1)}$ being valid in I_1, rule $R^{(2)}$ valid in I_2, and rule $R^{(3)}$ valid in I_3. Thus, the expected membership function $\mu_{R^{(1)}}(x_1,x_2,x_3)$ should yield the value 1 for $t \in I_1$ and the value 0 for everywhere else. Similarly, the expected membership function $\mu_{R^{(2)}}(x_1,x_2,x_3)$ should yield the value 1 for $t \in I_2$ and the value 0 for everywhere else, and the membership function $\mu_{R^{(3)}}(x_1,x_2,x_3)$ should yield the value 1 for $t \in I_3$ and the value 0 for everywhere else.

Example 7.2. In this example, the rule $R^{(20)}$ for heat stroke condition (high body temperature, low blood pressure, and normal heart rate) is tested. The temperature profile $x_1(t)$, blood pressure profile $x_2(t)$, and heart rate profile $x_3(t)$ are generated according to the following formulas:

$$x_1(t) = 98 + 4\sin(\frac{\pi}{40}t)\, e(t) + n(t),$$

$$x_2(t) = 120 - 40\sin(\frac{\pi}{40}t)\, e(t) + n(t),$$

$$x_3(t) = 70 + n(t),$$

where the function $n(t)$ is random white noise and $e(t)$ the rectangular envelop function described in the previous example. These three formulas for $x_1(t)$, $x_2(t)$, and $x_3(t)$ create an interval $I_4 \approx [10,30)$ where the rule $R^{(20)}$ will be valid, i.e., the membership function $\mu_{R^{(20)}}(x_1,x_2,x_3)$ should yield the value 1 for $t \in I_4$ and the value 0 for everywhere else.

Example 7.3. In this example, the rule $R^{(26)}$ for hyperthyroidism (high body temperature, high blood pressure, and rapid heart rate) is tested. The temperature profile $x_1(t)$, blood pressure profile $x_2(t)$, and heart rate profile $x_3(t)$ are generated according to the following formulas:

$$x_1(t) = 98 + 4\sin(\frac{\pi}{40}t)\, e(t) + n(t),$$

$$x_2(t) = 120 + 40\sin(\frac{\pi}{40}t)\, e(t) + n(t),$$

$$x_3(t) = 70 + 60\sin(\frac{\pi}{40}t)\, e(t) + n(t),$$

where the function $n(t)$ is random white noise and $e(t)$ the rectangular envelop function described in Example 7.1. These three formulas for $x_1(t)$, $x_2(t)$, and $x_3(t)$ create an interval $I_5 \approx [10,30)$ where the rule $R^{(26)}$ will be valid, i.e., the membership function $\mu_{R^{(26)}}(x_1,x_2,x_3)$ should yield the value 1 for $t \in I_5$ and the value 0 for everywhere else.

C. Numerical Results

In this section, computer simulations are created for the three examples listed in the previous section. Data are generated at the rate of one sample per second for the duration of 75 seconds. For each example, three sets of data, $X_1 = \{x_1(n) \mid n = 0, 1, 2, \ldots, 74\}$, $X_2 = \{x_2(n) \mid n = 0, 1, 2, \ldots, 74\}$, and $X_3 = \{x_3(n) \mid n = 0, 1, 2, \ldots, 74\}$, are generated. In Example 7.1, the membership functions $\mu_{R^{(1)}}(x_1(n),x_2(n),x_3(n))$, $\mu_{R^{(2)}}(x_1(n),x_2(n),x_3(n))$, $\mu_{R^{(3)}}(x_1(n),x_2(n),x_3(n))$ are generated for $n = 0, 1, 2, \ldots, 74$. In Examples 7.2 and 7.3, the membership functions $\mu_{R^{(20)}}(x_1(n),x_2(n),x_3(n))$ and $\mu_{R^{(26)}}(x_1(n),x_2(n),x_3(n))$ are also generated for $n = 0, 1, 2, \ldots, 74$.

Figures 7.4, 7.5, and 7.6 show the plot of the data $x_1(t)$ for body temperature, $x_2(t)$ for blood pressure, and $x_3(t)$ for heart rate generated for Example 7.1. Figure 7.8 shows the resulting membership functions of the three conditions C_1, C_2, and C_3 tested in Example 7.1. The data are generated at the beginning of each time interval of 1 second, for 75 seconds. In the first 20 seconds, $x_2(t)$ and $x_3(t)$ are in normal range, and $x_1(t)$ is gradually moving toward the high range, causing the membership function of condition C_1 to rise toward 1. Similarly, during the time interval [20,40], $x_1(t)$ and $x_3(t)$ are in normal range, and $x_2(t)$ is gradually moving toward the high range, causing the membership function of condition C_2 to rise toward 1; and during the time interval [40,60], $x_1(t)$ and $x_2(t)$ are in normal range, and $x_3(t)$ is gradually moving toward the high range, causing the membership function of condition C_3 to rise toward 1. After that, all three signals go back to normal, causing the membership function for condition C_0 to rise toward 1.

Figures 7.8, 7.9, and 7.10 show the plot of the data $x_1(t)$, $x_2(t)$, and $x_3(t)$ generated for Example 7.2. Figure 7.11 shows the resulting membership function for condition C_{20}. Here, conditions for $x_1(t)$ to be high, $x_2(t)$ to be low, and $x_3(t)$ to be normal are simulated. The corresponding membership for condition C_{20} is shown to confirm that the rule $R^{(20)}$ is valid.

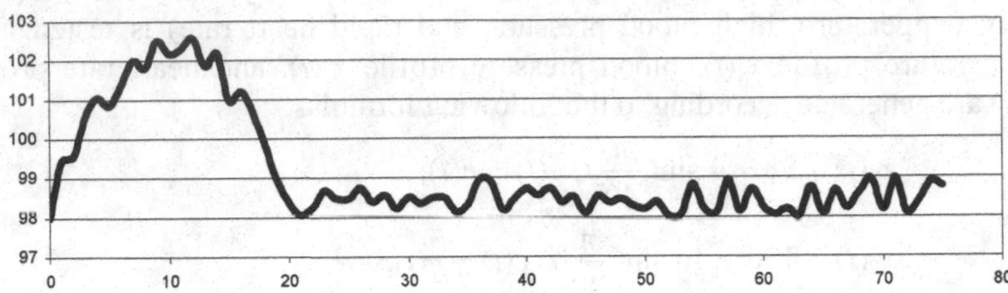

Figure 7.4 Body temperature profile for Example 7.1.

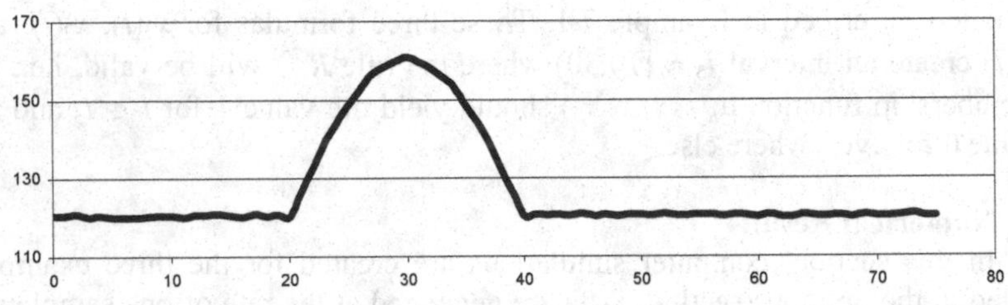

Figure 7.5 Blood pressure profile for Example 7.1.

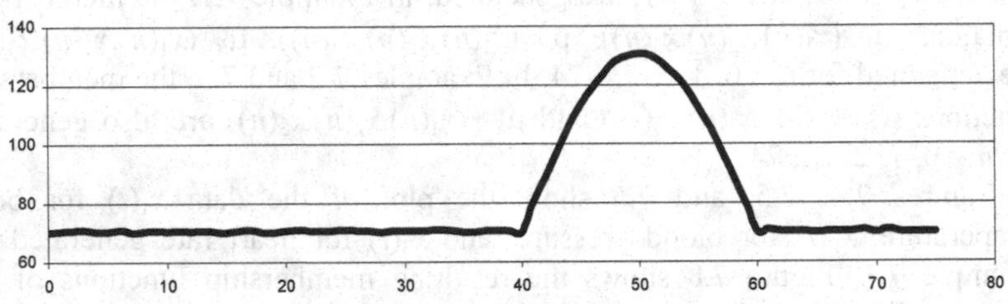

Figure 7.6 Heart rate profile for Example 7.1.

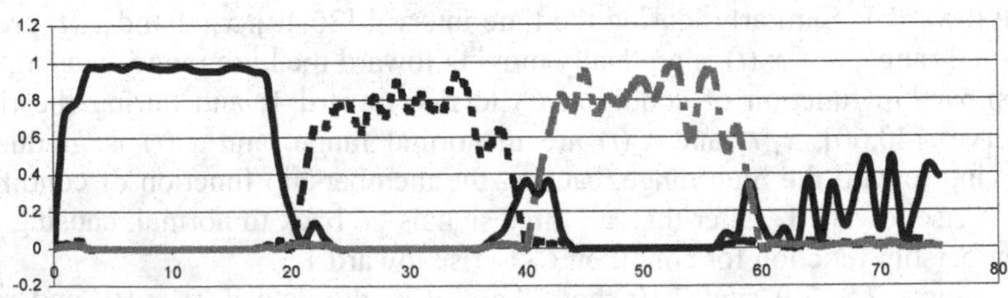

Figure 7.7 Membership functions of conditions C_1, C_2, and C_3 for Example 7.1.

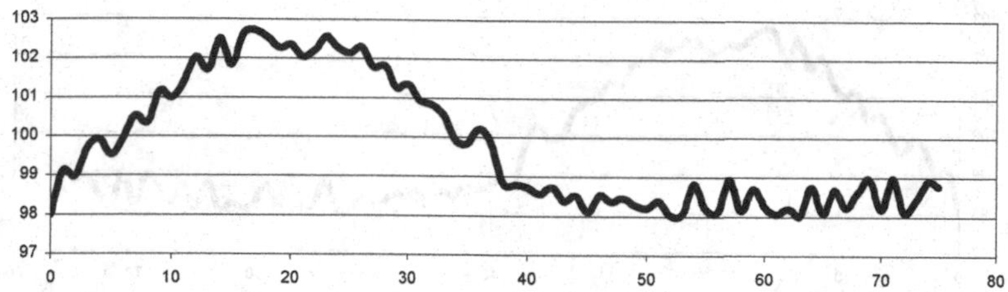

Figure 7.8 Body temperature profile for Example 7.2.

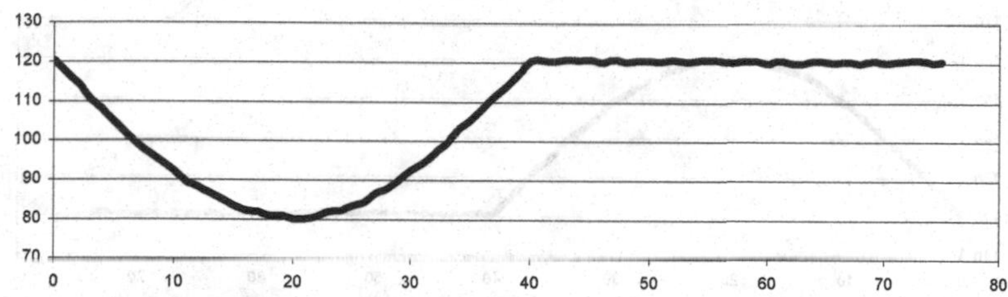

Figure 7.9 Blood pressure profile for Example 7.2.

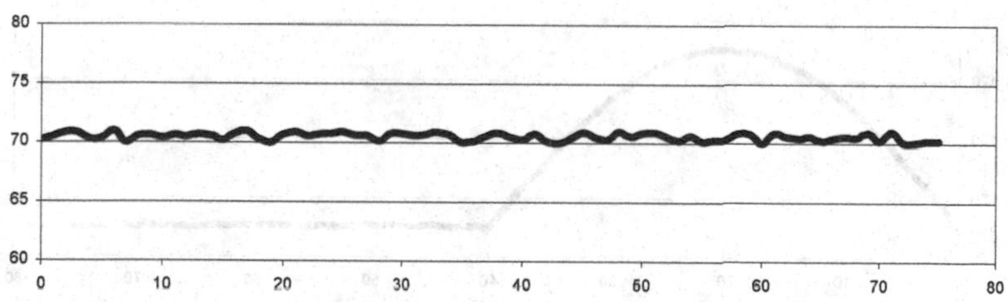

Figure 7.10 Heart rate profile for Example 7.2.

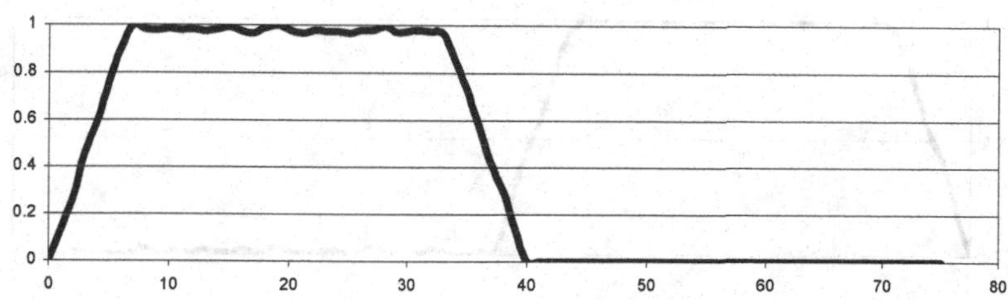

Figure 7.11 Membership function of condition C_{20} for Example 7.2.

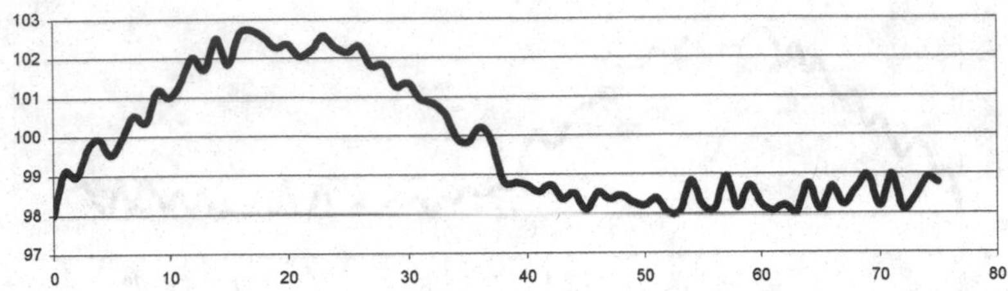

Figure 7.12 Body temperature profile for Example 7.3.

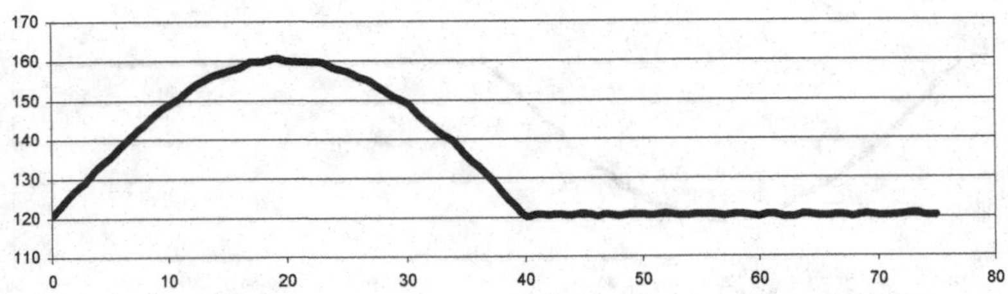

Figure 7.13 Blood pressure profile for Example 7.3.

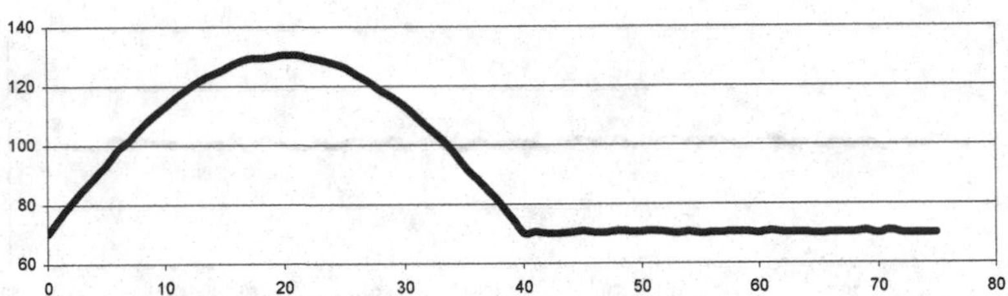

Figure 7.14 Heart rate profile for Example 7.3.

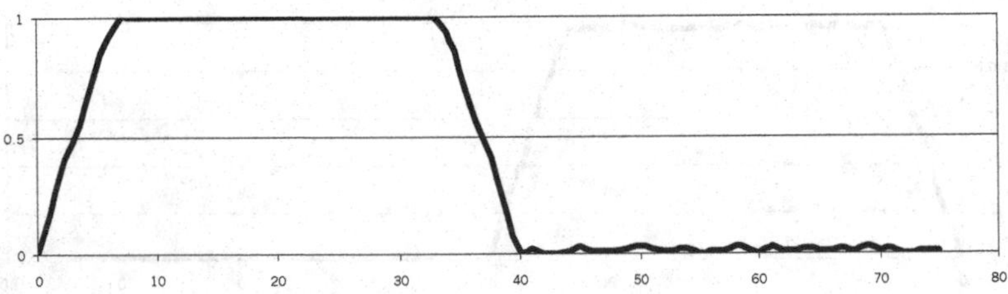

Figure 7.15 Membership function of condition C_{26} for Example 7.3.

Figures 7.12, 7.13, and 7.14 show the plot of the data $x_1(t)$, $x_2(t)$, and $x_3(t)$ generated for Example 7.3. Here, conditions for $x_1(t)$ to be high, $x_2(t)$ to be high, and $x_3(t)$ to be high are simulated. The corresponding membership for condition C_{26} is shown in Figure 7.15 to confirm that the rule $R^{(26)}$ is valid.

II. FUZZY CONTROL OF IMAGE SHARPNESS FOR AUTOFOCUS CAMERAS

Recent implementation of fuzzy logic in popular household items has increased the public awareness of fuzzy logic. One of such applications is the use of fuzzy logic rules to control a focus ring of a camera to allow automatic focusing for a sharp image.

A camera is an imaging system consisting of a set of lens that projects an image of a scene onto a sensor (either a film plate for conventional cameras or a set of CCD sensors for digital cameras). Traditional cameras require humans to manually adjust the focus control ring to obtain a sharp picture, i.e., the focusing position where the film plate (or sensor plate) coincides with the projected image. If the film plate is not at the position the image is projected, the captured picture on the film plate (or sensor plate) appears to be blurry; this phenomenon is called "out-of-focus."

Image-processing techniques have been developed to help determine the quality of a digital picture. The quality of a picture can be the color balance, the smoothness of spatial picture elements, the sharpness of the image, etc. For the purpose of controlling the focusing ring, the sharpness of an image is used as the determining factor. This quality determination technique for a digital image is used to help control the focusing ring of the lens system, so that the film plate is placed precisely at the distance an image is projected and a sharp picture is thus obtained. It is understood that the image is in-focus when its sharpness is at maximum level.

From a control standpoint, a servomotor is connected to the internal gear that controls the focusing ring to adjust the focal point and projected image point. The position of this focusing ring determines the position of each lens element to put the image (of an object) at the film plate located at a fixed distance behind the lens system. This automatic control system constitutes an autofocus lens of a camera.

In this section, the use of fuzzy logic to control the focusing ring in an autofocus lens system is outlined. This control scheme is a combination of position fuzzy control rules and the calculated sharpness of a picture as feedback. The calculation effort involved in this procedure is relatively simple; therefore, it is easy to be implemented on an inexpensive system with little computing power. The operating cycle can be summarized in four steps: (i) a digital image is captured, (ii) the sharpness index is calculated, (iii) the new position of the focusing ring is calculated and sent to the mechanical servo, and (iv) the servo controls the ring to the desired position.

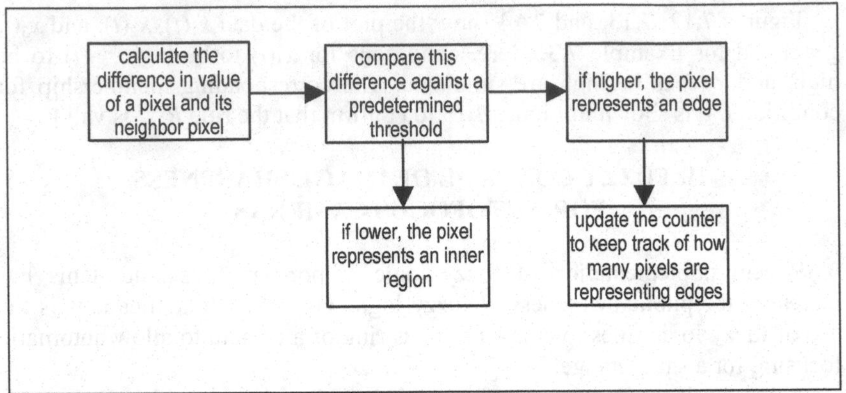

Figure 7.16 Algorithm for calculating sharpness index of a digital picture.

This section describes a procedure to calculate the sharpness of a digital image, and the fuzzy control algorithm that uses this sharpness index as feedback to control the focusing ring to maximize the sharpness index.

A. Basic Image Processing Techniques

A digital picture consists of a two-dimensional array of picture elements (or pixels). Each pixel represents light intensity at a spatial location. Together, the pixels represent the image of an object (or a scene of several objects).

There are several numerical techniques to determine if a digital picture is sharp (or in focus). This section describes a basic technique based on the edge-detection algorithm. The edge-detection algorithm will determine the edge of the object. It is assumed that a sharp picture will yield clearly defined edges around the objects. Therefore, a sharp picture will yield more edges than a blurred picture (of the same scenario or of the same object). This edge-detection technique is introduced here because it is simple to understand, computationally efficient for real-time implementation, and effective enough for determining the sharpness of a digital picture.

Edge-detection algorithm is a numerical method performed on the spatial relations between adjacent pixels to determine the boundary between two distinct regions on a two-dimensional space. The method consists of two steps: (i) calculation of the difference between two adjacent pixels, and (ii) applying a threshold filter to determine if this difference represents an edge of an object.

Let $i(n,m)$ represents a gray-scaled digital picture of dimension $N{\times}M$. The edge-detection algorithm will calculate a binary image $e(n,m)$ (with pixels of values of either zero or one) representing the edges of the objects in the digital picture $i(n,m)$. The edge-detection algorithm can be represented by the formula:

$$e(n,m) = \begin{cases} 1 & \text{if } |\, i(n,m) - i(n+1,m)\,| \geq \theta \\ 0 & \text{else,} \end{cases} \tag{7.1}$$

```
int sharpnessIndex( int image[][], int Ndim, intMdim, int t )
{
  int sIndex = 0;
  for( int n = 1; n <= Ndim; n++ )
    for( int m = 1; m <= Mdim; m++ )
      if( abs(image[n][m]-image[n+1][m]) < t ) sIndex++
  return( sIndex );
}
```

Figure 7.17 Implementation of sharpness index using C-programming language.

where θ is a pre-determined numerical constant representing a threshold that distinguishes between edge and non-edge. A digital picture is considered sharp if the edges are distinctive, i.e., the number of pixels representing the edges calculated in (7.1) is at its maximum. A blurred picture will not produce so many distinct edges. Therefore, a scalar variable s can be formulated as follows to determine the sharpness of a digital picture:

$$s = \sum_{n=1}^{N} \sum_{m=1}^{M} e(n,m)\,. \tag{7.2}$$

The variable s is sometimes referred to as a sharpness index. The higher the sharpness index, the more focused the picture becomes. This quantity can also be normalized so that its range is standardized at between the value of zero and one. The normalized sharpness is calculated as

$$s = \frac{1}{NM} \sum_{n=1}^{N} \sum_{m=1}^{M} e(n,m)\,.$$

The higher the value of s, the more contrast is the image. It is reasoned that for the same scene, the image which contains more contrast is assumed to be the one correctly focused. When incorrectly focused, the image becomes blurred. This effect is similar to that of a smooth filter. Therefore, the contrast is less for images that are off-focused.

Figure 7.16 depicts an algorithm for calculating the sharpness index of a digital picture. A typical implementation of this calculation using C-programming language is listed in Figure 7.17 to illustrate the development efficiency of the algorithm.

B. Fuzzy Control Model

The fuzzy control model is relatively simple for a position control. The rules can be intuitively derived with minor adjustment to improve the performance.

1. Fixed Rate. This is the simplest form of position control. For each control cycle, the position of the focusing ring is moved by a fixed constant in either direction. The focusing ring can also stay where it was. Thus, there are only three choices: (i) move the focusing ring counterclockwise by an angle Δθ, (ii) move the focusing ring clockwise by an angle Δθ, or (iii) do not move

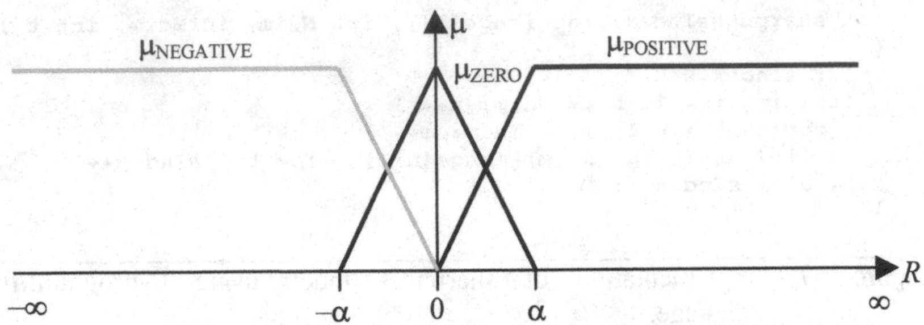

Figure 7.18 Membership functions $\mu_{POSITIVE}(x)$, $\mu_{NEGATIVE}(x)$, and $\mu_{ZERO}(x)$.

the focusing ring at all. These selections are similar to that of a bang-bang control, where the fixed control measure is either positive or negative. When the system is in a deadband, the control measure is zero.

Basically, the three rules above can be applied to the three situations: when the focusing is moving in the right direction, i.e., the sharpness index is increasing, keep moving in that direction; when the focusing ring is moving in the wrong direction, i.e., the sharpness index is decreasing, start moving in the opposite direction; and when there is no improvement in the sharpness index, stop.

Let σ_{now} be the current sharpness index, σ_{prev} the previous sharpness index, θ_{now} the current position of the focusing ring, θ_{now} the previous position of the focusing ring, D_{prev} (which can be either +1 or −1) the previous direction of the ring movement, and $\Delta\theta$ the fixed step that the focusing ring can move for each control cycle. Then these three rules can be given as follows:

$R^{(1)}$: IF $(\sigma_{now} - \sigma_{prev})$ is POSITIVE
THEN $D_{now} = D_{prev}$ AND $\theta_{now} = \theta_{prev} + D_{now}\Delta\theta$,

$R^{(2)}$: IF $(\sigma_{now} - \sigma_{prev})$ is NEGATIVE
THEN $D_{now} = -D_{prev}$ AND $\theta_{now} = \theta_{prev} + D_{now}\Delta\theta$,

$R^{(3)}$: IF $(\sigma_{now} - \sigma_{prev})$ is ZERO
THEN $D_{now} = D_{prev}$ AND $\theta_{now} = \theta_{prev}$.

For each description POSITIVE, NEGATIVE, and ZERO, there associate three ranges $R_{POSITIVE} = [x_1,\infty)$, $R_{NEGATIVE} = [-x_1,-\infty)$, and $R_{ZERO} = [-x_2,x_2]$, and three membership functions $\mu_{POSITIVE}(x)$, $\mu_{NEGATIVE}(x)$, and $\mu_{ZERO}(x)$. These ranges and membership functions will determine the final numerical angle that the focusing ring should be rotated to.

Figure 7.18 shows a typical setup for the three ranges $R_{POSITIVE}$, $R_{NEGATIVE}$, and R_{ZERO}, and their corresponding membership functions $\mu_{POSITIVE}(x)$, $\mu_{NEGATIVE}(x)$, and $\mu_{ZERO}(x)$. In this setup, the constant x_1 is set to be equal to x_2, i.e., $x_1 = x_2 = \alpha$. The membership functions in Figure 7.18 can be expressed mathematically as:

$$\mu_{\text{NEGATIVE}}(x) = \begin{cases} 1 & \text{if } -\infty < x \le \alpha, \\ -x/\alpha & \text{if } \alpha < x \le 0, \\ 0 & \text{else,} \end{cases} \tag{7.3}$$

$$\mu_{\text{ZERO}}(x) = \begin{cases} (x+\alpha)/\alpha & \text{if } -\alpha < x \le 0, \\ (\alpha - x)/\alpha & \text{if } 0 < x \le \alpha, \\ 0 & \text{else,} \end{cases} \tag{7.4}$$

$$\mu_{\text{POSITIVE}}(x) = \begin{cases} x/\alpha & \text{if } 0 \le x < \alpha, \\ 1 & \text{if } \alpha \le x < \infty, \\ 0 & \text{else.} \end{cases} \tag{7.5}$$

These specific definitions, along with the weighted average method of defuzzification, lead to the exact formulation of D_{now} and θ_{now} as follows:

$$D_{now} = \begin{cases} D_{prev} & \text{if } (\sigma_{now} - \sigma_{prev}) < 0, \\ -D_{prev} & \text{if } (\sigma_{now} - \sigma_{prev}) \ge 0, \end{cases} \tag{7.6}$$

$$\theta_{now} = \begin{cases} \theta_{prev} + D_{now}\Delta\theta & \text{if } (\sigma_{now} - \sigma_{prev}) < -\alpha, \\ \theta_{prev} - \beta D_{now}\Delta\theta & \text{if } -\alpha \le (\sigma_{now} - \sigma_{prev}) < 0, \\ \theta_{prev} + \beta D_{now}\Delta\theta & \text{if } 0 \le (\sigma_{now} - \sigma_{prev}) < \alpha, \\ \theta_{prev} + D_{now}\Delta\theta & \text{if } \alpha \le (\sigma_{now} - \sigma_{prev}), \end{cases} \tag{7.7}$$

where

$$\beta = \frac{\sigma_{now} - \sigma_{prev}}{\alpha}. \tag{7.8}$$

2. Adjustable Control Rate. It was reasoned that the fixed rate control above might be sped up faster if step $\Delta\theta$ is initially set at a large value and thus moves the focusing ring closer and faster to the right setting. As it moves past the right setting, the focusing ring will be moved back in the opposite direction, similar to the situation of overshooting. Here, when this change of direction occurs, step $\Delta\theta$ will be adjusted to be smaller, thus providing a finer control at the range closer to the right setting. The same three rules in the fixed rate control above can be modified to reflect this variable rate strategy:

$R^{(1)}$: IF $(\sigma_{now} - \sigma_{prev})$ is POSITIVE
THEN $D_{now} = D_{prev}$ AND $\theta_{now} = \theta_{prev} + D_{prev}\Delta\theta$,

$R^{(2)}$: IF $(\sigma_{now} - \sigma_{prev})$ is NEGATIVE
THEN $D_{now} = -D_{prev}$ AND $\theta_{now} = \theta_{prev} + \gamma D_{now}\Delta\theta$,

$R^{(3)}$: IF $(\sigma_{now} - \sigma_{prev})$ is ZERO
THEN $D_{now} = D_{prev}$ AND $\theta_{now} = \theta_{prev}$.

Again, for each description POSITIVE, NEGATIVE, and ZERO, there associate three ranges $R_{\text{POSITIVE}} = [x_1, \infty)$, $R_{\text{NEGATIVE}} = [-x_1, -\infty)$, and $R_{\text{ZERO}} = [-x_2, x_2]$, and three membership functions $\mu_{\text{POSITIVE}}(x)$, $\mu_{\text{NEGATIVE}}(x)$, and $\mu_{\text{ZERO}}(x)$. These ranges and membership functions will determine the final numerical angle that the focusing ring should be rotated to.

The constant γ in Rule $R^{(2)}$ is between zero and one (or $\gamma \in (0,1]$). When one selects $\gamma = 1$, the rules converge to the fixed rate control presented earlier. When one selects any arbitrary $0 < \gamma < 1$, the steps are refined smaller each time the focusing ring goes past its targeted setting. One can also select a scheme to calculate different values for the parameter γ for every control cycle according to the adaptive scheme discussed in Chapter 6.

If the definitions for R_{POSITIVE}, R_{NEGATIVE}, and R_{ZERO}, and their corresponding membership functions $\mu_{\text{POSITIVE}}(x)$, $\mu_{\text{NEGATIVE}}(x)$, and $\mu_{\text{ZERO}}(x)$ defined in Sections B.1, B.2, and B.3 of Chapter 2 are used, the defuzzification process of the three fuzzy rules $R^{(1)}$, $R^{(2)}$, and $R^{(3)}$ will lead directly to the following closed-form formulas:

$$D_{\text{now}} = \begin{cases} D_{\text{prev}} & \text{if } (\sigma_{\text{now}} - \sigma_{\text{prev}}) < 0, \\ -D_{\text{prev}} & \text{if } (\sigma_{\text{now}} - \sigma_{\text{prev}}) \geq 0, \end{cases} \tag{7.9}$$

$$\theta_{\text{now}} = \begin{cases} \theta_{\text{prev}} + D_{\text{now}}\Delta\theta & \text{if } (\sigma_{\text{now}} - \sigma_{\text{prev}}) < -\alpha, \\ \theta_{\text{prev}} - \beta D_{\text{now}}\Delta\theta & \text{if } -\alpha \leq (\sigma_{\text{now}} - \sigma_{\text{prev}}) < 0, \\ \theta_{\text{prev}} + \beta D_{\text{now}}\Delta\theta & \text{if } 0 \leq (\sigma_{\text{now}} - \sigma_{\text{prev}}) < \alpha, \\ \theta_{\text{prev}} + D_{\text{now}}\Delta\theta & \text{if } \alpha \leq (\sigma_{\text{now}} - \sigma_{\text{prev}}), \end{cases} \tag{7.10}$$

where

$$\beta = \frac{\sigma_{\text{now}} - \sigma_{\text{prev}}}{\alpha}. \tag{7.11}$$

C. Computer Simulation Results

In this computer simulation, simulation of a lens system is developed to show that a correctly focused image blurs away as the focusing ring is rotated away from the right position. A fuzzy control system is implemented to control the position of the focusing ring on the lens system. Sharpness of an image is used as feedback to the fuzzy control system.

1. Focusing System. A focusing system is simulated so that when the focusing ring is not in the correct position, in this case, the image is blurred. The more the focusing ring is off target, the more the image is blurred. In order to create the blurring effect, one can use the smoothing filter to apply to the original image (the one in focus). Let the in-focus image be $I_{\text{in-focus}}(n,m)$; the off-focus image $I_{\text{off-focus}}(n,m)$ can be generated according to the difference δ between the correct distance and the estimated distance as follows:

$$I_{\text{off-focus}}(n,m) = \frac{1}{(2\delta+1)^2} \sum_{j=-\delta}^{\delta} \sum_{k=-\delta}^{\delta} I_{\text{in-focus}}(n+j, m+k). \tag{7.12}$$

The formula above is the smoothing filter base on the average value of the pixels inside the square area around a pixel of indices (n,m). The larger the size of this square area, the more blurred the resulting image. Therefore, to create the blurring effect, we use the difference δ between the correct distance and the estimated distance to determine the size of the square area used in the smoothing filter in (7.12).

2. Fuzzy Control System. The fuzzy control system will use the fuzzy rules described earlier, with the index of sharpness calculated from the formulation for edge detection used as feedback to control the focusing ring of the camera.

3. Computer Simulation Results. Three experiments are performed in the simulation. The first experiment demonstrates the fixed rate control algorithm. The second and third experiments demonstrate the adjustable control algorithm. The results are shown in Figures 7.19, 7.20, and 7.21. The corresponding control data are listed in Tables 7.1, 7.2, and 7.3.

Experiment 1. An object (the fireworks) is at a large distance (infinity setting) from the lens. The control algorithm initializes the system to start at the closest distance allowable on the lens system. This distance represents the macro setting. From this distance, the lens focusing ring is moved one step at a time at a fixed rate of 10° for each control cycle. The time duration between two consecutive control cycles is 100 milliseconds.

Experiment 2. An object (the astronaut) is at a moderate distance (between the macro setting and infinity setting) from the lens. The control algorithm also initializes the system to start at the macro setting. From this distance, the lens focusing ring is moved one step at a time at an initial rate of 20° for each control cycle. The time duration between two consecutive control cycles is 100 milliseconds. At the first overshoot, the control algorithm reverses the direction of the focusing ring and reduces the control step to the rate of 10° for each control cycle. This reduction corresponds to the selection of the constant $\gamma = 0.5$ in the adjustable rate algorithm.

Experiment 3. This is an extension of automatic focusing to a microscope. This application is particularly useful for the space program where biological experiments are often conducted in the space shuttle (or in the future space station) with little supervision. In this scenario, data are collected automatically and sent back to Earth for analysis by humans.

The object for this experiment is a slide of plant cells. This is a biological experiment in which microscopic lives are observed unattended through a microscope. The condition for growth is dependent on the nature of the experiment. The microscope is used in such a way that cells' lives are preserved. This life-preservation requirement is normally satisfied by using the phase-contrast microscope (a special kind of microscope that uses the difference in the refraction indices instead of the intensity of the reflected light). The phase-contrast microscope only records images in black-and-white data, which is the format used in this example.

For color images, one can convert it into gray-level by applying the average filter that calculates the average of the three main colors (namely, red, green, and blue) representing each color pixel into a single scalar quantity representing a gray-level pixel.

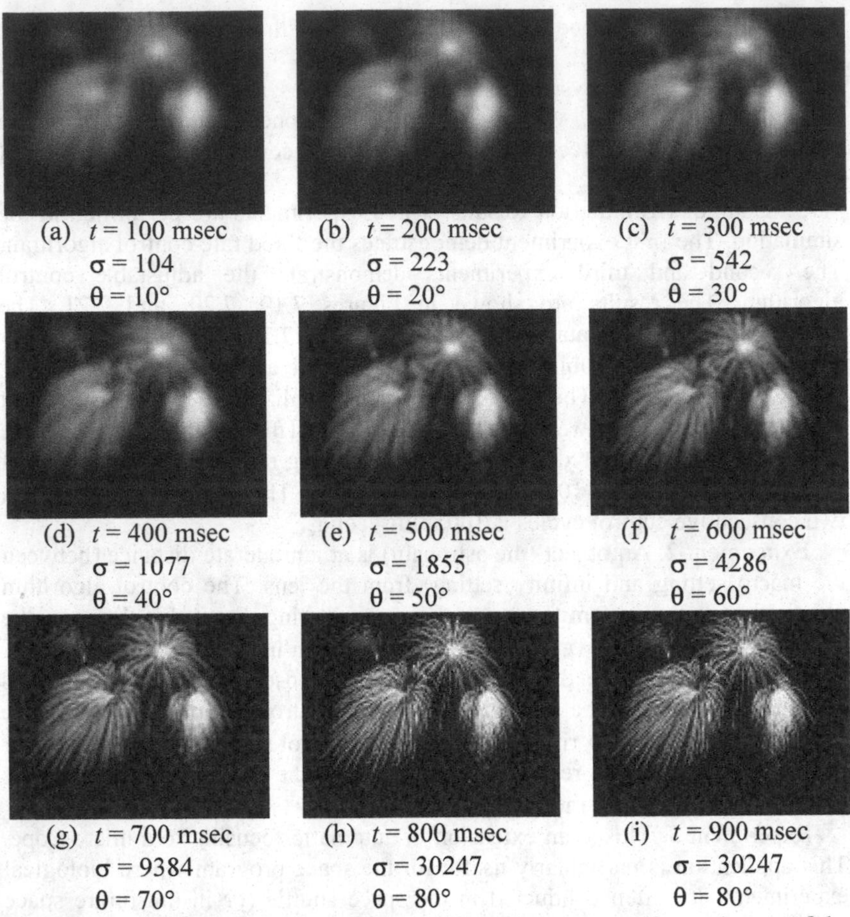

Figure 7.19 Automatic control of an autofocus lens system for an image of the fireworks display at distance of infinity (from the lens). The control ring is starting at the macro end, i.e., at the smallest distance allowed. The angle is adjusted by 10° in either direction every cycle.

Table 7.1 Numerical Data and Results for Experiment 1

time t	sharpness σ	position θ	rule $R^{(1)}$	rule $R^{(2)}$	rule $R^{(3)}$
100 msec	104	10°	1	0	0
200 msec	223	20°	1	0	0
300 msec	542	30°	1	0	0
400 msec	1077	40°	1	0	0
500 msec	1855	50°	1	0	0
600 msec	4266	60°	1	0	0
700 msec	9384	70°	1	0	0
800 msec	30247	80°	1	0	0
900 msec	30247	80°	0	0	1

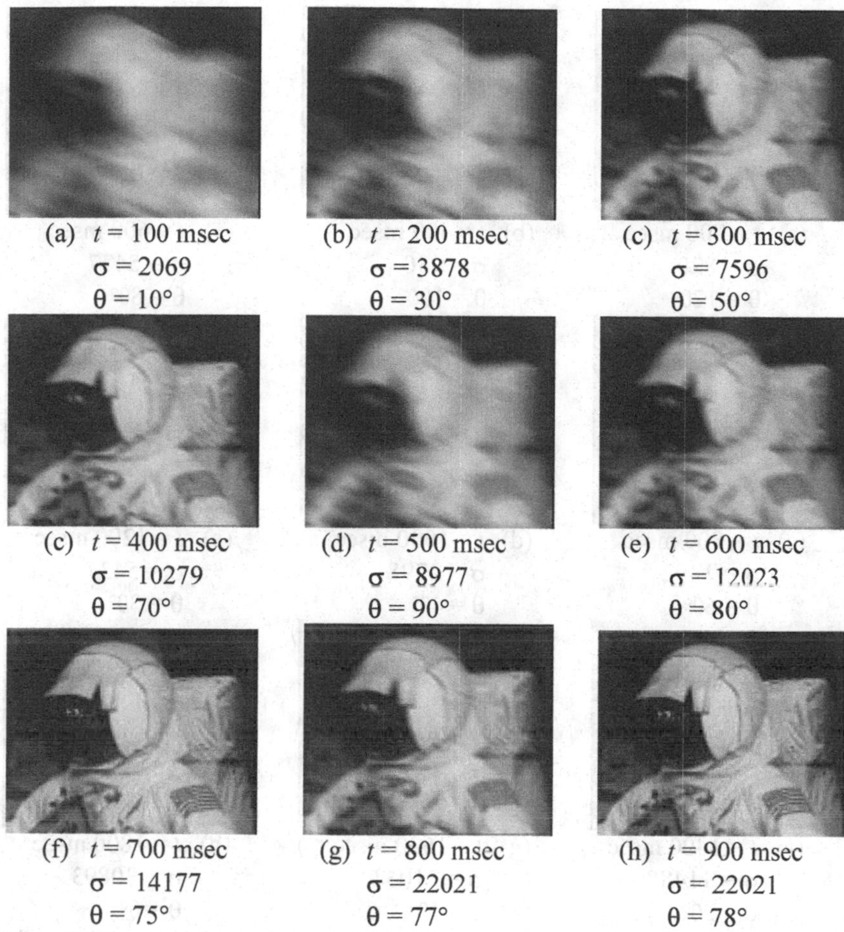

(a) $t = 100$ msec, $\sigma = 2069$, $\theta = 10°$

(b) $t = 200$ msec, $\sigma = 3878$, $\theta = 30°$

(c) $t = 300$ msec, $\sigma = 7596$, $\theta = 50°$

(c) $t = 400$ msec, $\sigma = 10279$, $\theta = 70°$

(d) $t = 500$ msec, $\sigma = 8977$, $\theta = 90°$

(e) $t = 600$ msec, $\sigma = 12023$, $\theta = 80°$

(f) $t = 700$ msec, $\sigma = 14177$, $\theta = 75°$

(g) $t = 800$ msec, $\sigma = 22021$, $\theta = 77°$

(h) $t = 900$ msec, $\sigma = 22021$, $\theta = 78°$

Figure 7.20 Automatic control of an autofocus lens system for an image of an astronaut.

Table 7.2 Numerical Data and Results for Experiment 2

time t	**sharpness** σ	**position** θ	**rule** $R^{(1)}$	**rule** $R^{(2)}$	**rule** $R^{(3)}$
100 msec	104	10°	1	0	0
200 msec	223	30°	1	0	0
300 msec	542	50°	1	0	0
400 msec	1077	70°	1	0	0
500 msec	1855	90°	1	0	0
600 msec	4266	80°	0	1	0
700 msec	9384	75°	0	1	0
800 msec	30247	77°	0	1	0
900 msec	30247	78°	0	0	1

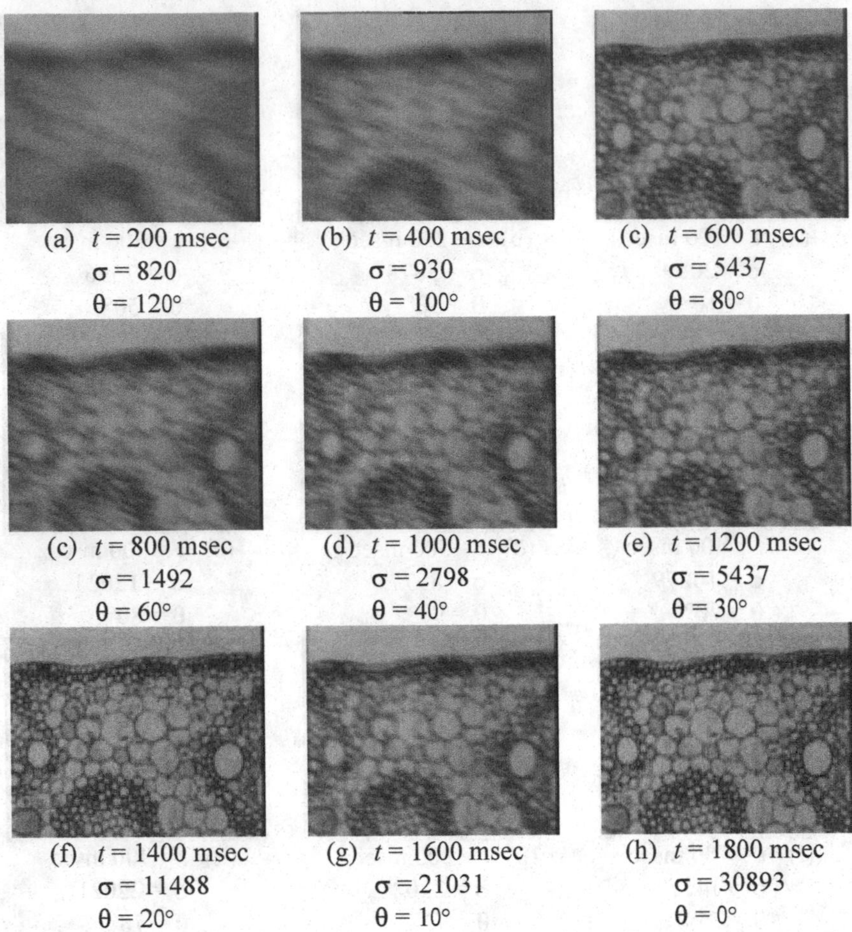

Figure 7.21 Automatic control of an autofocus lens system for an image of a set of plant cells on a microscope.

Table 7.3 Numerical Data and Results for Experiment 3

time t	sharpness σ	position θ	rule $R^{(1)}$	rule $R^{(2)}$	rule $R^{(3)}$
200 msec	820	120°	1	0	0
400 msec	930	100°	1	0	0
600 msec	5437	80°	1	0	0
800 msec	1492	60°	1	0	0
1000 msec	2798	40°	0	1	0
1200 msec	5437	30°	0	1	0
1400 msec	11488	20°	0	1	0
1600 msec	21031	10°	0	1	0
1800 msec	30893	0°	0	0	1

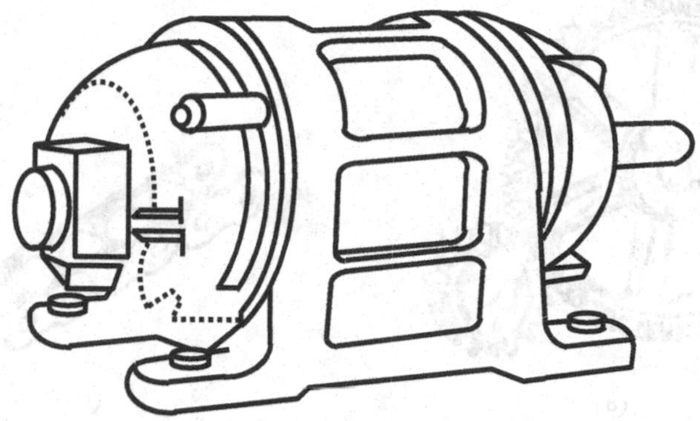

Figure 7.22 A typical servo motor.

III. FUZZY CONTROL FOR SERVO MECHANIC SYSTEMS

Servo mechanic systems are electromechanical machines that drive mechanical platforms to achieve rotational movements. A servo mechanic system can be used on a miniature scale to drive the focusing ring of an automatic focusing lens system of a camera. It can be used on a larger scale to drive the joint of each link of a robotic manipulator. It can also be used to drive electric automobiles. Figure 7.22 shows the exterior appearance of a typical industrial servo motor.

Traditionally, servo mechanic systems were treated with linear control systems. The servo mechanic system was modeled as a linear system and a corresponding controller was designed based on the model. This approach was popular partly because of its simple design that is easy for implementation, and partly because of the accuracy of the linear model that approximates the behavior of an actual servo mechanic system in its operating range. Outside this operating range, the error will magnify and require different treatments such as a piecewise linear approach or an adaptive approach.

A servo mechanic system can be characterized by a linear system of the form

$$J\frac{d^2}{dt^2}\theta(t) + K\frac{d}{dt}\theta(t) + B\,\theta(t) = \tau(t),$$

where B is the coefficient representing damping characteristic, K the stiffness characteristic, J the inertia characteristic, $\theta(t)$ the state variable representing an angle displacement, and $\tau(t)$ the applied torque. A more complicated model will include the electrical current used as the control input to deliver necessary torque $\tau(t)$. Normally, this model can accurately approximate the real servo motor in a certain operating range. Outside this operating range, the model will result in severe error and the general practice is to use another different set of constant $\{J,K,B\}$.

Figure 7.23 Gears and accessory: (a) direct connection and (b) a segment of chain used to drive belt-driven gear.

A complicated model of a servo motor often includes a gear box. This is an additional mechanical device containing some gears connecting to the rotational axis of the servo motor. The result is an increase in angle displacement by a constant factor. Correspondingly, the angular velocity and angular acceleration will also be increased by the same constant factor. Figure 7.23 shows an example of two sets of mechanical gears representing a gear box. Mathematically, a gear box can be modeled by a multiplication factor. This is the simplest model. A more complicated model of a gear box normally includes a time delay factor, slippage (for belt-driven gears), and backlash (for loosely connected gears).

This section presents application of fuzzy control to a general class of servo mechanic systems. Computer simulations are developed and numerical results are used to illustrate the concepts.

A. Fuzzy Modeling of a Servo Mechanic System

A dynamic system can be modeled by a set of fuzzy logic rules in the form of IF-THEN statements, where the conditional IF-clause consists of logical conditions concerning the values of its state variables, and the consequence THEN-clause consists of values of its transition states. A servo mechanic system with two state variables, $x_1(t)$ representing the angle displacement $\theta(t)$ and $x_2(t)$ representing the angular velocity $\dot{\theta}(t)$, and one control input, $u(t)$ representing the torque $\tau(t)$, can be represented with the fuzzy logic rules as follows:

$R^{(1)}$: IF $x_1 \in S_n$ AND $x_2 \in S_m$ AND $u \in S_p$ THEN $\dot{x}_1 \in S_m$,

$R^{(2)}$: IF $x_1 \in S_n$ AND $x_2 \in S_m$ AND $u \in S_p$ THEN $\dot{x}_2 \in S_{an+bm+cp}$,

where $a = -K/J$, $b = -B/J$, and $c = 1/J$, and K is the stiffness coefficient, B the damping coefficient, and J the moment of inertia of the servo mechanic system.

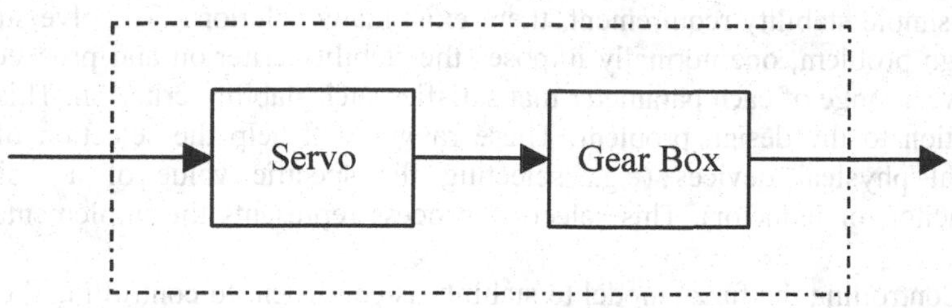

Figure 7.24 A servo motor and a gear box sequentially put together.

It has been shown that under specific definitions for each fuzzy set indexed by an arbitrary integer n as follows,

$$S_n = [(n-1)\sigma,(n+1)\sigma],$$

$$\mu_{S_m}(x) = \begin{cases} 1-|x/\sigma - n| & (n-1)\sigma \le x \le (n+1)\sigma, \\ 0 & \text{else,} \end{cases}$$

the two fuzzy rules above represent a class of dynamic systems with state variables behaving within two sets of upper and lower bounds:

$$\zeta_1(t) \le x_1(t) \le \xi_1(t) \quad \text{and} \quad \zeta_2(t) \le x_2(t) \le \xi_2(t),$$

where

$$\begin{bmatrix} \dot{\zeta}_1 \\ \dot{\zeta}_2 \end{bmatrix} = \begin{bmatrix} 0 & 1 \\ -K/J & -B/J \end{bmatrix} \begin{bmatrix} \zeta_1 \\ \zeta_2 \end{bmatrix} + \begin{bmatrix} 0 \\ 1/J \end{bmatrix} u(t) - \begin{bmatrix} \delta \\ \delta \end{bmatrix},$$

$$\begin{bmatrix} \dot{\xi}_1 \\ \dot{\xi}_2 \end{bmatrix} = \begin{bmatrix} 0 & 1 \\ -K/J & -B/J \end{bmatrix} \begin{bmatrix} \xi_1 \\ \xi_2 \end{bmatrix} + \begin{bmatrix} 0 \\ 1/J \end{bmatrix} u(t) + \begin{bmatrix} \delta \\ \delta \end{bmatrix},$$

and

$$\delta = (1 + |K/J| + |B/J| + |1/J|)\sigma.$$

The selection of the constant σ will determine the width between the upper and lower bounds, i.e., the flexibility in the fuzzy description that will cover a larger class of systems.

A gear box can be treated as a separated system sequentially connected to the servo motor (Figure 7.24). This way, a mathematical model of the gear box can be developed independently and separately (from the servo motor). The overall model contained within the dashed line in Figure 7.23 can be easily derived. A typical (and simple) fuzzy model of a gear box is:

$$R^{(1)}: \quad \text{IF } x_1 \in S_n \text{ THEN } y_1 \in S_{\chi n},$$

where the constant χ represents the gear box ratio (between the input and the output).

B. Fuzzy Controller of a Servo Mechanic System

A controller can be designed to drive the fuzzy system to stability. With this simple stability requirement, there exist many solutions. To solve such a design problem, one normally imposes the stability criterion and proceeds to derive a range of each parameter that satisfies such stability criterion. This is a solution to the design problem. These ranges will help the selection of the actual physical devices (e.g., selecting the specific value of a resistor, capacitor, or inductor). This selection process represents the implementation step.

Controlling the fuzzy model to stability is equivalent to controlling the two bounds to stability. Let a fuzzy control rule be of the form

$$R_C^{(1)}:\ \text{IF } x_1 \in S_n \text{ AND } x_2 \in S_m \text{ THEN } u \in S_{-\kappa n - \nu m}.$$

Then the closed loop system will take the form

$$R_{CL}^{(1)}:\ \text{IF } x_1 \in S_n \text{ AND } x_2 \in S_m \text{ THEN } \dot{x}_1 \in S_m,$$

$$R_{CL}^{(2)}:\ \text{IF } x_1 \in S_n \text{ AND } x_2 \in S_m \text{ THEN } \dot{x}_2 \in S_{an+bm-c(\kappa n+\lambda m)}.$$

These fuzzy rules describe any system that has the state variable bounded from above and below by

$$\begin{bmatrix} \dot{\zeta}_1 \\ \dot{\zeta}_2 \end{bmatrix} = \begin{bmatrix} 0 & 1 \\ -K/J-\kappa/J & -B/J-\lambda/J \end{bmatrix} \begin{bmatrix} \zeta_1 \\ \zeta_2 \end{bmatrix} - \begin{bmatrix} \delta \\ \delta \end{bmatrix},$$

$$\begin{bmatrix} \dot{\xi}_1 \\ \dot{\xi}_2 \end{bmatrix} = \begin{bmatrix} 0 & 1 \\ -K/J-\kappa/J & -B/J-\lambda/J \end{bmatrix} \begin{bmatrix} \xi_1 \\ \xi_2 \end{bmatrix} + \begin{bmatrix} \delta \\ \delta \end{bmatrix}.$$

Thus the 2×2 matrix in the equations above must be negative definite, which yields the following conditions:

$$\frac{K}{J}+\frac{\kappa}{J}>0 \quad \text{and} \quad \frac{B}{J}+\frac{\lambda}{J}>0.$$

Since the moment of inertia J is always positive, the two conditions above are reduced to

$$K+\kappa>0 \quad \text{and} \quad B+\lambda>0,$$

or

$$\kappa>-K \quad \text{and} \quad \lambda>-B.$$

These conditions will be used to specify the gains in the controller. Usually, additional criteria such as response time and limitation of the available amplifiers will be imposed to help derive specific values for the gain constants κ and λ. Without such criteria, one can arbitrarily select a set of gains such as:

$$\kappa=K \quad \text{and} \quad \lambda=B.$$

These arbitrary selections will be used in the computer simulation in the next section.

To summarize the resulting control design, the servo mechanic system is modeled with the fuzzy rules:

$$R^{(1)}:\ \text{IF } x_1 \in S_n \text{ AND } x_2 \in S_m \text{ AND } u \in S_p \text{ THEN } \dot{x}_1 \in S_m,$$

$$R^{(2)}:\ \text{IF } x_1 \in S_n \text{ AND } x_2 \in S_m \text{ AND } u \in S_p$$
$$\text{THEN } \dot{x}_2 \in S_{-(K/J)n-(B/J)m+(1/J)p}.$$

A fuzzy controller designed just to drive this type of system to stability will have the form:

$R_C^{(1)}$: IF $x_1 \in S_n$ AND $x_2 \in S_m$ THEN $u \in S_{-Kn-Bm}$.

The final closed loop system will have the fuzzy form:

$R_{CL}^{(1)}$: IF $x_1 \in S_n$ AND $x_2 \in S_m$ THEN $\dot{x}_1 \in S_m$,

$R_{CL}^{(2)}$: IF $x_1 \in S_n$ AND $x_2 \in S_m$ THEN $\dot{x}_2 \in S_{an+bm-c(\kappa n+\lambda m)}$.

These two rules describe a set of systems bounded from above by the upper bound

$$\begin{bmatrix}\dot{\zeta}_1\\ \dot{\zeta}_2\end{bmatrix}=\begin{bmatrix}0 & 1\\ -K/J-\kappa/J & -B/J-\lambda/J\end{bmatrix}\begin{bmatrix}\zeta_1\\ \zeta_2\end{bmatrix}-\begin{bmatrix}\delta\\ \delta\end{bmatrix},$$

and from below by the lower bound

$$\begin{bmatrix}\dot{\xi}_1\\ \dot{\xi}_2\end{bmatrix}=\begin{bmatrix}0 & 1\\ -K/J-\kappa/J & -B/J-\lambda/J\end{bmatrix}\begin{bmatrix}\xi_1\\ \xi_2\end{bmatrix}+\begin{bmatrix}\delta\\ \delta\end{bmatrix}.$$

It can be seen that both these bounds asymptotically converge to zero, thus, the systems bounded by them will also converge to zero, or are asymptotically stable.

C. Computer Simulations and Numerical Results

In this section, three simulations are presented to illustrate the following: (i) demonstration of the implementation of a simple fuzzy controller design, (ii) demonstration of the robustness of the fuzzy controller when parameters of a fuzzy model are not accurately reflecting the actual system, and (iii) demonstration of the flexibility of the fuzzy controller when the actual system is contaminated with interference and thus behaves differently from the assumed model.

Example 7.4. The purpose of this example is to provide step-by-step implementation of a fuzzy controller. A dynamic system

$$\frac{d^2}{dt^2}\theta(t) + 5\frac{d}{dt}\theta(t) + 10\,\theta(t) = \tau(t)$$

represents a particular servo mechanic system. This equation of motion will be used to represent a real system that the fuzzy controller will try to control. The fuzzy model used to estimate this system is:

$R^{(1)}$: IF $x_1 \in S_n$ AND $x_2 \in S_m$ AND $u \in S_p$ THEN $\dot{x}_1 \in S_m$,

$R^{(2)}$: IF $x_1 \in S_n$ AND $x_2 \in S_m$ AND $u \in S_p$ THEN $\dot{x}_2 \in S_{-5n-10m+p}$,

where x_1 is the angle displacement θ, x_2 the angular velocity, and u the control torque. The fuzzy controller will take the form:

$R_C^{(1)}$: IF $x_1 \in S_n$ AND $x_2 \in S_m$ THEN $u \in S_{-5n-10m}$,

which yields the closed loop form:

$R_{CL}^{(1)}$: IF $x_1 \in S_n$ AND $x_2 \in S_m$ THEN $\dot{x}_1 \in S_m$,

$R_{CL}^{(2)}$: IF $x_1 \in S_n$ AND $x_2 \in S_m$ THEN $\dot{x}_2 \in S_{-10n-20m}$.

Figures 7.25 through 7.28 show the fuzzy controller being applied to the simulated system described in the equation of motion. Apparently, the system is driven to stability because the two bounds are driven to stability.

Example 7.5. The purpose of this example is to determine if a fuzzy controller design based on a model with its parameter erroneously estimated is still functional. A dynamic system

$$\frac{d^2}{dt^2}\theta(t) + 10\frac{d}{dt}\theta(t) + 15\,\theta(t) = \tau(t)$$

represents a particular servo mechanic system. This equation of motion will be used to represent a real system that the fuzzy controller will try to control. The fuzzy model used to estimate this system is the same as the one used in Example 7.4:

$R^{(1)}$: IF $x_1 \in S_n$ AND $x_2 \in S_m$ AND $u \in S_p$ THEN $\dot{x}_1 \in S_m$,

$R^{(2)}$: IF $x_1 \in S_n$ AND $x_2 \in S_m$ AND $u \in S_p$ THEN $\dot{x}_2 \in S_{-5n-10m+p}$.

The fuzzy corresponding controller will take the form:

$R_C^{(1)}$: IF $x_1 \in S_n$ AND $x_2 \in S_m$ THEN $u \in S_{-5n-10m}$.

Figures 7.29 through 7.32 show the fuzzy controller applied to the system and the corresponding output, along with the upper and lower bounds of the fuzzy closed loop system, converging to stability.

Example 7.6. The purpose of this example is to determine if a fuzzy controller designed based on a model with fixed parameter estimate is still functional when the actual system has time-dependent parameters. A dynamic system

$$\frac{d^2}{dt^2}\theta(t) + 5[1+\sin(2\pi t)]\frac{d}{dt}\theta(t) + 10[1+\cos(2\pi t)]\,\theta(t) = \tau(t)$$

represents a particular servo mechanic system. This equation of motion will be used to represent a real system that the fuzzy controller will try to control. The fuzzy model used to estimate this system is the same as the one used in Example 7.4, with the fuzzy controller:

$R_C^{(1)}$: IF $x_1 \in S_n$ AND $x_2 \in S_m$ THEN $u \in S_{-5n-10m}$.

Figures 7.33 through 7.36 show the fuzzy controller applied to the system and the corresponding output, along with the upper and lower bounds of the fuzzy closed loop system. In this case, it is interesting to note that the actual system converges to stability despite its nonlinear nature (or time-variant nature of its parameters). This example mainly shows the robustness of the fuzzy controller when handling a system different from its original assumption (which was implemented in the modeling process that leads to the design of the controller). Common sense projects that if one keeps increasing the nonlinear variation of the system (from the assumption), the effectiveness of the controller will eventually decrease. However, the robustness of the fuzzy controller provides a comfortable envelope in which one can afford some variation.

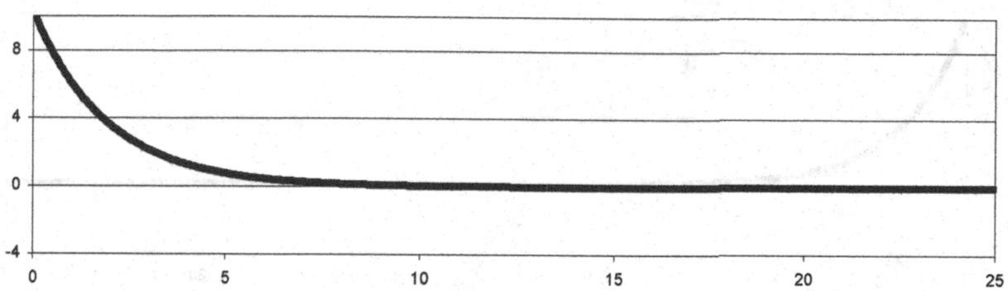

Figure 7.25 Position profile for the servo motor in Example 7.4.

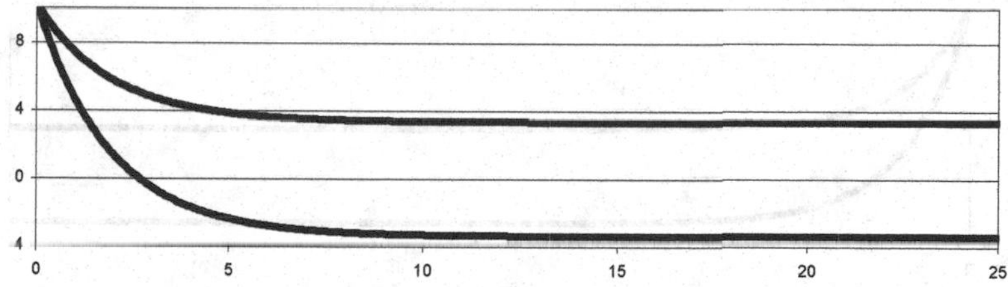

Figure 7.26 Bounds of the positions for the servo motor in Example 7.4.

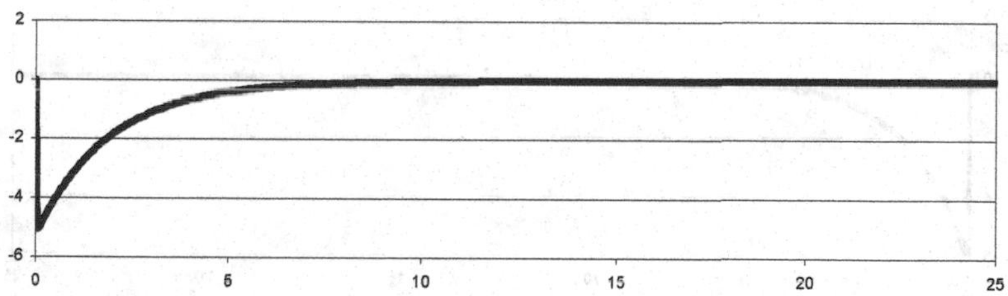

Figure 7.27 Velocity profile for the servo motor in Example 7.4.

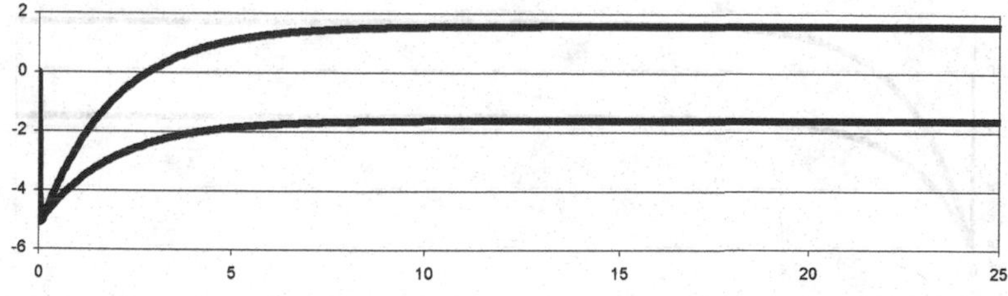

Figure 7.28 Bounds of the velocities for the servo motor in Example 7.4.

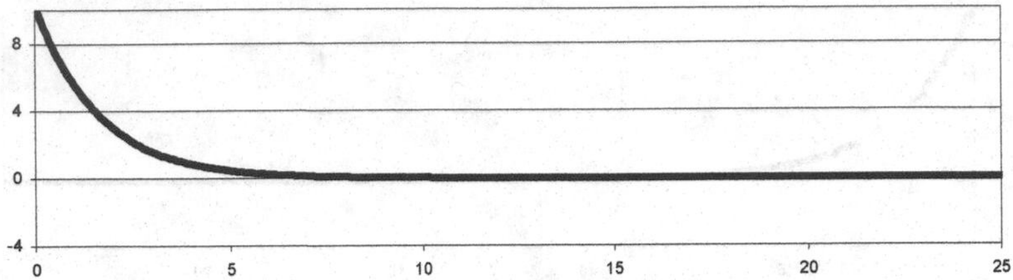

Figure 7.29 Position profile for the servo motor in Example 7.5.

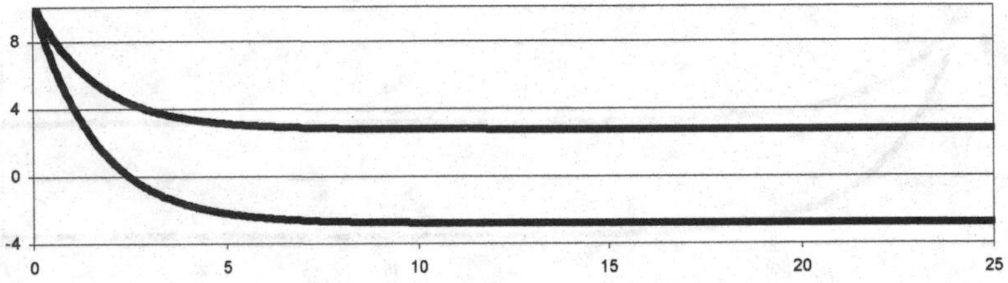

Figure 7.30 Bounds of the positions for the servo motor in Example 7.5.

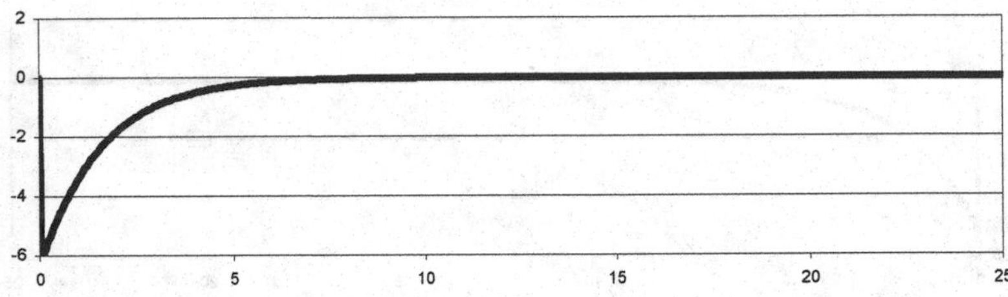

Figure 7.31 Velocity profile for the servo motor in Example 7.5.

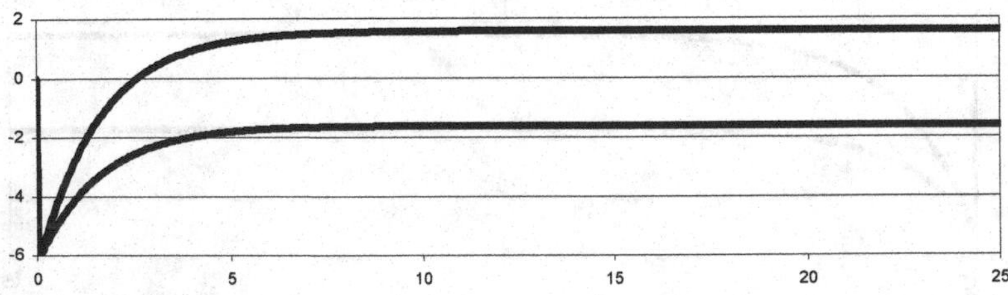

Figure 7.32 Bounds of the velocities for the servo motor in Example 7.5.

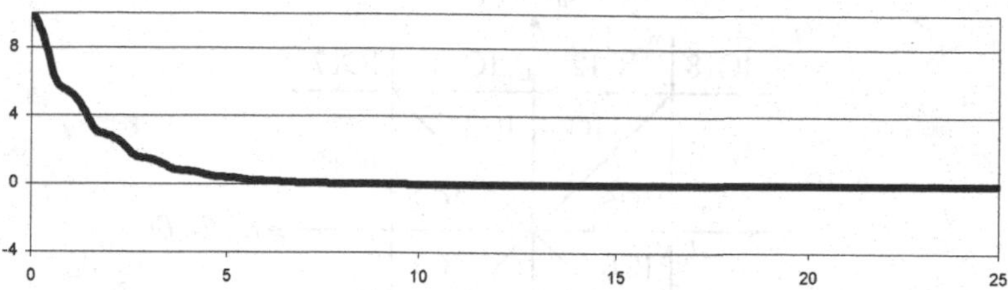

Figure 7.33 Position profile for the servo motor in Example 7.6.

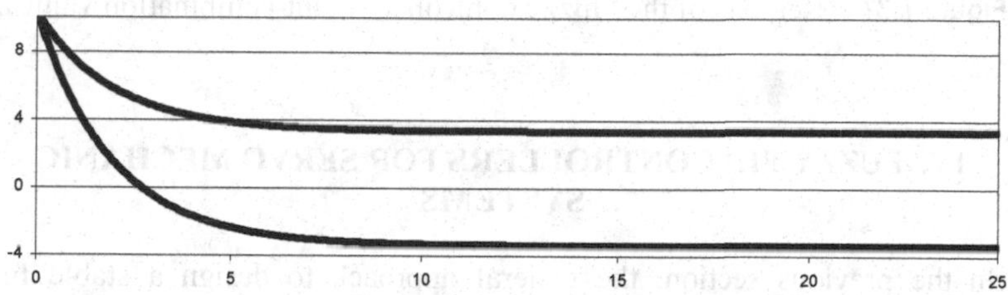

Figure 7.34 Bounds of the positions for the servo motor in Example 7.6.

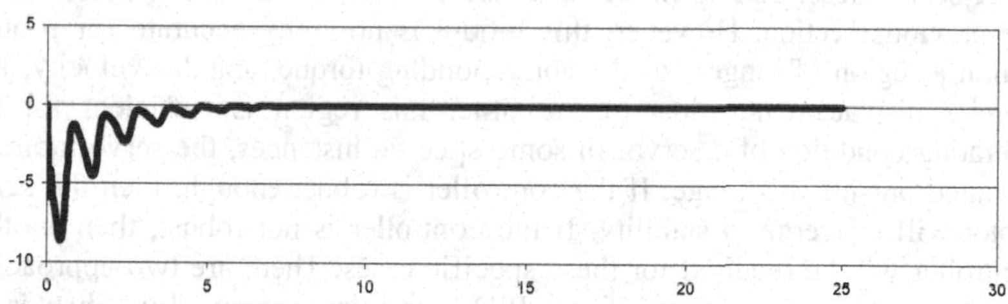

Figure 7.35 Velocity profile for the servo motor in Example 7.6.

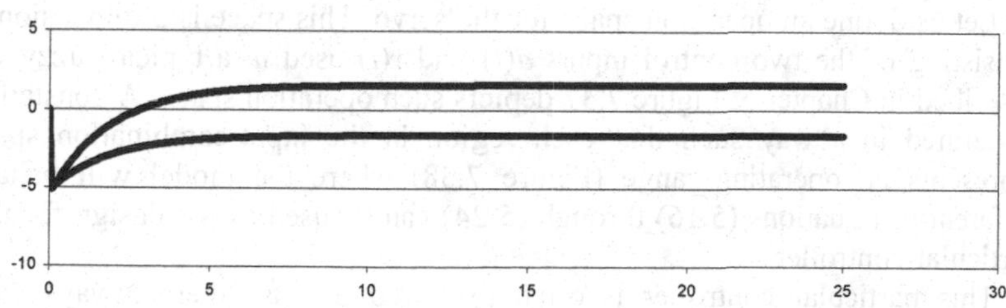

Figure 7.36 Bounds of the velocities for the servo motor in Example 7.6.

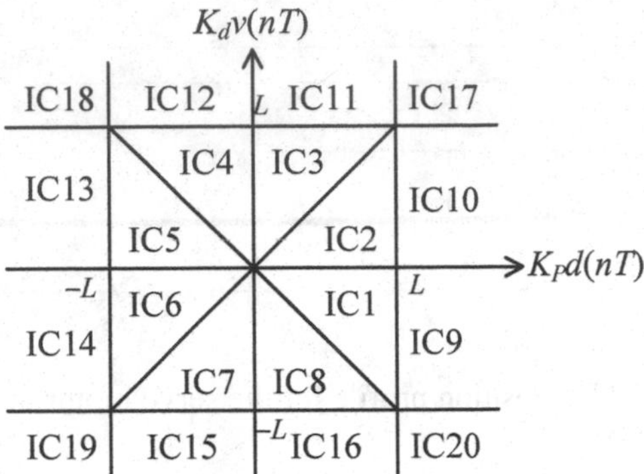

Figure 7.37 Regions of the "fuzzy controller" input-combination values.

IV. FUZZY PID CONTROLLERS FOR SERVO MECHANIC SYSTEMS

In the previous section, the general approach to design a stable fuzzy controller for a servo mechanic system was discussed. This chapter presents more detailed solutions to the servo controller, namely, for the situation when a more complicated model of a servo is needed (or preferred).

A servo motor can be modeled as the IF-THEN fuzzy rules presented in the previous section. However, this model is normally accurate (or robust) within a region of ranges for the corresponding torque, angular velocity, and angular displacement. Most of the time, this region is sufficient for the operating condition of a servo. In some specific instances, the servo motor is operated outside this range. If the controller is robust enough, then the servo motor will converge to stability. If the controller is not robust, then another controller will be required for these specific cases. There are two approaches to this design: (i) designing a fuzzy PID controller (presented in Chapter 5), and (ii) designing an adaptive fuzzy controller (presented in Chapter 6).

A. Fuzzy PID Controller of a Servo Mechanic System

Let us define an operation space for the servo. This space is 2-dimensional, consisting of the two control inputs $d(t)$ and $v(t)$ used in a typical fuzzy PD described in Chapter 5. Figure 7.37 depicts such operation space. A constant L is defined in a way such that each region in the input-combination space represents an operating range (Figure 7.38) where the model will behave differently. Equations (5.16) through (5.24) can be used as the design for this particular controller.

This particular controller is often preferred because there always exist some physical limitations on a mechanical system, i.e., the maximum torque one can provide as input to the system and the maximum achievable angular velocity of the servo motor. By using the fuzzy PID (or PD) controller, one can limit the control signal to some arbitrary constant. Furthermore, this type

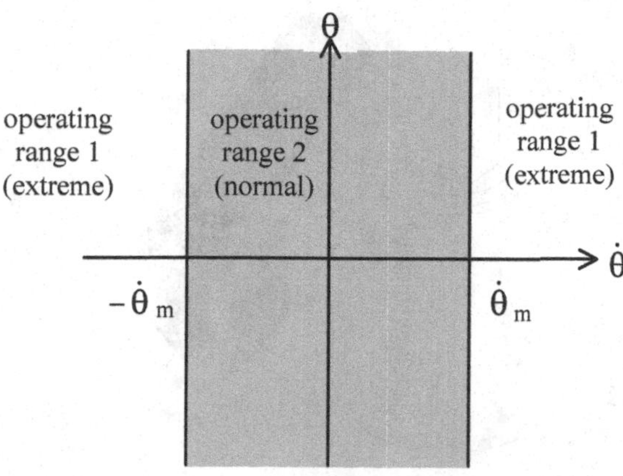

Figure 7.38 Operating ranges for adaptive fuzzy controller.

of controller also exhibits piecewise-linear characteristics that are typical of the servo motor.

B. Adaptive Fuzzy Controller of a Servo Mechanic System

There are two approaches to design an adaptive fuzzy controller for the servo motor: (i) gain scheduling and (ii) fuzzy self-tuning regulator. In the gain scheduling, the fuzzy controller is designed for each operating range, and the observed data are used to determine which operating range the servo is in in order to activate a different gain. In the fuzzy self-tuning regulator, the parameters of the model are adjusted accordingly to the observed data.

The gain scheduling approach basically provides different pre-determined controllers. Whenever observed data signify a particular situation, a corresponding controller is activated.

The fuzzy self-tuning regulator, in theory, will continuously update the parameters of the fuzzy model and the controller is adjusted accordingly. However, due to the nature of the servo motor, this continuous updating requirement can be relaxed: the parameters only need to be updated whenever there is an indication for the need (the system behaves linearly in a piecewise manner). Thus, there will be a scheme, that requires some parameter update, to check to see when the system is moving from one operating range to another.

Figure 7.39 A typical industrial robot manipulator.

V. FUZZY CONTROLLER FOR ROBOTIC MANIPULATOR

Robotic manipulator is a mechanical system consisting of several links connected in a chain configuration, with rotational degree-of-freedoms at each joint (the connection between two links). This system is used in the same function as a human arm and therefore is sometimes referred to as a mechanical arm. The equation of motion for this robotic manipulator system is highly nonlinear. Nonlinear controllers for this type of system are normally difficult to design, too complex for practical implementation, or not robust enough to tolerate any error in the assumptions.

In general practice, the final goal of controlling a manipulator is to put the end-effector (the link furthest from the base) at some specific coordinates. These coordinates are defined in the *xyz*-space (sometimes referred to as the Cartesian space). However, in order to put the end-effector at these coordinates, the joints have to be moved to some angles. A direct transformation exists between these angles and the *xyz* coordinates of the end-effector. This transformation is known as the direct kinematics. Most of the control problems for manipulators are in the joint space, with the desired position of the end-effector translated into a set of desired joint angles. Formulating the problem this way, one avoids adding more complexity to the already complicated problem (due to high nonlinearity in the equation of motion).

For illustrative purposes, a commonly used planar manipulator (as shown in Figure 7.39) is used. This is a manipulator with only one degree of freedom at each joint, and all degree of freedoms move the manipulator in a plane. A typical 2-link planar manipulator with equal-length l and equal-mass m for each link, where the weight is uniformly distributed along the length of each link, follows the equation of motion below:

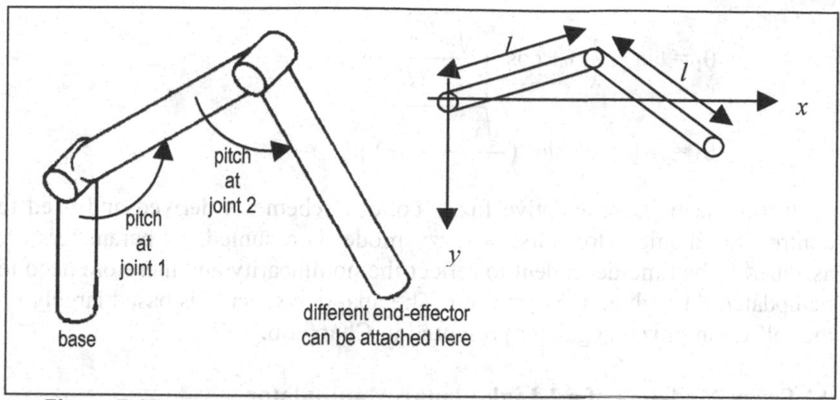

Figure 7.40 A 2-link planar manipulator in 3-D space and in x-y plane.

$$\begin{bmatrix} \tau_1 \\ \tau_2 \end{bmatrix} = \begin{bmatrix} a_{11} & a_{12} \\ a_{21} & a_{22} \end{bmatrix} \begin{bmatrix} \ddot{\theta}_1 \\ \ddot{\theta}_2 \end{bmatrix} + \begin{bmatrix} b_1 \\ b_2 \end{bmatrix},$$

where

$$a_{11} = \frac{3}{2} ml^2 + ml^2\cos(\theta_2), \qquad a_{12} = \frac{1}{4} ml^2 + \frac{1}{2} ml^2\cos(\theta_2),$$

$$a_{21} = \frac{1}{4} ml^2 + \frac{1}{2} ml^2\cos(\theta_2), \qquad a_{22} = \frac{1}{2} ml^2,$$

$$b_1 = -ml^2\, \dot{\theta}_1\, \dot{\theta}_2 \sin(\theta_2) + \frac{3}{2} mlg\cos(\theta_1) + \frac{1}{2} mlg\cos(\theta_1+\theta_2),$$

$$b_2 = \frac{1}{2} mlg\cos(\theta_1+\theta_2).$$

The 2×2 matrix containing a_{11}, a_{12}, a_{21}, and a_{22} represents the moment of inertia of the manipulator. The 2×1 vector matrix containing b_1 and b_2 represents the Coriolis torques. The equation of motion sometimes can be represented simply by $\tau = I\alpha + C$.

It can be seen that under the most favorable conditions (of having equal-length and equal-mass links, of being planar, and of having the mass uniformly distributed along the length) the equation of motion is simplified significantly. However, the final form is still highly nonlinear.

Notice that for the 2-link planar manipulator above, the direct kinematics used to transform the joint angles into the coordinates of the end-effector is given as follows:

$$\begin{bmatrix} x \\ y \end{bmatrix} = l \begin{bmatrix} \cos(\theta_1) \\ \sin(\theta_1) \end{bmatrix} + l \begin{bmatrix} \cos(\theta_1 + \theta_2) \\ \sin(\theta_1 + \theta_2) \end{bmatrix},$$

and its corresponding inverse kinematics formula can be obtained in closed form as:

$$\theta_1 = \tan^{-1}(\frac{y}{x}) \pm \cos^{-1}(\frac{\sqrt{x^2+y^2}}{2l}),$$

$$\theta_2 = \mp[\,\pi - 2\sin^{-1}(\frac{\sqrt{x^2+y^2}}{2l})\,].$$

In this section, an adaptive fuzzy control scheme is derived and used to control this manipulator. First, a fuzzy model is assumed. Its parameters are assumed to be time-dependent to reflect the nonlinearity and therefore need to be updated throughout the operation. The adaptive scheme is based largely on the self-tuning fuzzy regulator presented in Chapter 6.

A. Fuzzy Modeling of a 2-Link Planar Manipulator

A 2-link planar manipulator will be represented by a model with four state variables: x_1 (angle displacement at joint 1), x_2 (angular velocity at joint 1), x_3 (angle displacement at joint 2), and x_4 (angular velocity at joint 2). The two control inputs are u_1 (torque delivered to joint 1 by a servo mechanism), and u_2 (torque delivered to joint 2 by another servo mechanism).

With these notations, the fuzzy model can be set up as a set of fuzzy logic rules as follows:

$R^{(1)}$: IF $x_1 \in S_i$ AND $x_2 \in S_j$ AND $x_3 \in S_n$ AND $x_4 \in S_m$ AND $u_1 \in S_p$ AND $u_2 \in S_q$ THEN $\dot{x}_1 \in S_j$,

$R^{(2)}$: IF $x_1 \in S_i$ AND $x_2 \in S_j$ AND $x_3 \in S_n$ AND $x_4 \in S_m$ AND $u_1 \in S_p$ AND $u_2 \in S_q$ THEN $\dot{x}_2 \in S_{f(i,j,m,n,p,q)}$,

$R^{(3)}$: IF $x_1 \in S_i$ AND $x_2 \in S_j$ AND $x_3 \in S_n$ AND $x_4 \in S_m$ AND $u_1 \in S_p$ AND $u_2 \in S_q$ THEN $\dot{x}_3 \in S_n$,

$R^{(4)}$: IF $x_1 \in S_i$ AND $x_2 \in S_j$ AND $x_3 \in S_n$ AND $x_4 \in S_m$ AND $u_1 \in S_p$ AND $u_2 \in S_q$ THEN $\dot{x}_4 \in S_{g(i,j,m,n,p,q)}$.

The functions $f(\times)$ and $g(\times)$ can be selected to best fit the behavior of the system. In general practice, in order to simplify the problem, $f(\times)$ and $g(\times)$ are often assigned the linear form:

$$f(i,j,m,n,p,q) = a_{11}i + a_{12}j + a_{13}m + a_{14}n + b_{11}p + b_{12}q,$$
$$g(i,j,m,n,p,q) = a_{21}i + a_{22}j + a_{23}m + a_{24}n + b_{21}p + b_{22}q.$$

The constants a_{11}, a_{12}, a_{13}, a_{14}, a_{21}, a_{22}, a_{23}, a_{24}, b_{11}, b_{12}, b_{21}, and b_{22} are assumed to be time dependent, i.e., they will change with time and therefore will need to be updated from time to time. There are two popular techniques of updating these constants: (i) updating according to a nonlinear model (e.g., the equation of motion mentioned earlier), and (ii) updating according to the solution of a curve-fitting problem.

When updating these constants according to a nonlinear equation of motion of the form $\tau = I\alpha + C$ mentioned earlier, the update can be derived based on the equation of motion. This equation of motion can be rewritten in the form:

$$\begin{bmatrix}\ddot{\theta}_1\\ \ddot{\theta}_2\end{bmatrix}=\begin{bmatrix}\alpha_1\\ \alpha_2\end{bmatrix}+\begin{bmatrix}\beta_{11} & \beta_{12}\\ \beta_{21} & \beta_{22}\end{bmatrix}\begin{bmatrix}\tau_1\\ \tau_2\end{bmatrix},$$

where

$$\alpha_1=\frac{1}{2\gamma}[\,\dot{\theta}_1\,\dot{\theta}_2\,\sin(\theta_2)+\frac{3}{2l}g\cos(\theta_1)+\frac{1}{4l}g\cos(\theta_1+\theta_2)-\frac{1}{2l}g\cos(\theta_2)\cos(\theta_1+\theta_2)],$$

$$\alpha_2=\frac{1}{\gamma}\{[\frac{1}{4}+\frac{1}{2}\cos(\theta_2)][\,\dot{\theta}_1\,\dot{\theta}_2\,\sin(\theta_2)-\frac{3}{2l}g\cos(\theta_1)-\frac{1}{2l}g\cos(\theta_1+\theta_2)]+\frac{g}{l}[\frac{3}{2}+\cos(\theta_2)]\cos(\theta_1+\theta_2)\},$$

$$\gamma=\frac{11}{16}+\frac{1}{4}\cos(\theta_2)-\frac{1}{4}\cos^2(\theta_2),$$

$$\beta_{11}=\frac{1}{2ml^2\left[\frac{11}{16}+\frac{1}{4}\cos(\theta_2)-\frac{1}{4}\cos^2(\theta_2)\right]},$$

$$\beta_{12}=\frac{-\frac{1}{4}-\frac{1}{2}\cos(\theta_2)}{ml^2\left[\frac{11}{16}+\frac{1}{4}\cos(\theta_2)-\frac{1}{4}\cos^2(\theta_2)\right]},$$

$$\beta_{21}=\frac{-\frac{1}{4}-\frac{1}{2}\cos(\theta_2)}{ml^2\left[\frac{11}{16}+\frac{1}{4}\cos(\theta_2)-\frac{1}{4}\cos^2(\theta_2)\right]},$$

$$\beta_{22}=\frac{\frac{3}{2}+\cos(\theta_2)}{ml^2\left[\frac{11}{16}+\frac{1}{4}\cos(\theta_2)-\frac{1}{4}\cos^2(\theta_2)\right]}.$$

From this equation one can derive different linearized models that yield the corresponding constants for the fuzzy model. One particular set of constants can be derived from the nonlinear formula above using the popular truncated Taylor series expansion method around a set of fixed point constants:

$$a_{11}=\frac{\partial}{\partial\theta_1}\alpha_1,\qquad a_{21}=\frac{\partial}{\partial\theta_1}\alpha_2,$$

$$a_{12}=\frac{\partial}{\partial\dot{\theta}_1}\alpha_1,\qquad a_{22}=\frac{\partial}{\partial\dot{\theta}_1}\alpha_2,$$

$$a_{13}=\frac{\partial}{\partial\theta_2}\alpha_1,\qquad a_{23}=\frac{\partial}{\partial\theta_2}\alpha_2,$$

$$a_{14} = \frac{\partial}{\partial \dot{\theta}_2} \alpha_1, \qquad a_{24} = \frac{\partial}{\partial \dot{\theta}_2} \alpha_2,$$
$$b_{11} = \beta_{11},$$
$$b_{12} = \beta_{12},$$
$$b_{21} = \beta_{21},$$
$$b_{22} = \beta_{22}.$$

When updating the constants a_{11}, a_{12}, a_{13}, a_{14}, a_{21}, a_{22}, a_{23}, a_{24}, b_{11}, b_{12}, b_{21}, and b_{22} according to the solution of a curve-fitting problem, one takes the observed data $X(t_n) = [x_1(t_n)\ x_2(t_n)\ x_3(t_n)\ x_4(t_n)]^T$ and the calculated data $\dot{X}(t_n) = [\dot{x}_1(t_n)\ \ \dot{x}_2(t_n)\ \ \dot{x}_3(t_n)\ \ \dot{x}_4(t_n)]^T$, and finds a solution that minimizes the squared-norm of the error function:

$$\min \| \dot{X}(t_n) - A(t_n)X(t_n) - B(t_n)\tau(t_n) \|^2.$$

Notice that the optimization problem above yields an infinite number of solutions (there are more unknown variables than constraint equations). In order to derive a unique solution, one often imposes additional constraints, e.g., the smoothness in the constants $A(t_n)$ and $B(t_n)$ when compared to $A(t_{n-1})$ and $B(t_{n-1})$. In this setting, the optimization problem becomes

$$\min \| A(t_n) - A(t_{n-1}) \|^2 + \| B(t_n) - B(t_{n-1}) \|^2$$

such that

$$\dot{X}(t_n) = A(t_n)X(t_n) + B(t_n)\tau(t_n).$$

The optimization problem above now has a unique solution.

There are controllers designed on the basis of selecting the constants a_{ij} and b_{mn} in such a way that the variables appear to be independent of each other. This process is called decoupling the variables, i.e., the model can be separated into two independent models, each of a simpler nature and much easier to handle. However, it has been noted that such decoupled systems are often sensitive to rapid changes and often need continuous update that will consume considerable computing power.

In the case of a 2-link manipulator, each of the two decoupled systems only consists of two variables and one control input, making it much easier to design a controller with closed form solution. One example of such decoupling systems is the selection of the constants a_{ij} and b_{mn} as follows:

$$a_{11} = 0, \qquad a_{21} = 0,$$
$$a_{12} = \frac{\alpha_1}{\dot{\theta}_1}, \qquad a_{22} = 0,$$
$$a_{13} = 0, \qquad a_{23} = 0_2,$$
$$a_{14} = 0, \qquad a_{24} = \frac{\alpha_2}{\dot{\theta}_2},$$
$$b_{11} = \beta_{11}, \qquad b_{21} = 0,$$
$$b_{12} = 0, \qquad b_{22} = \beta_{22}.$$

B. Fuzzy Controller of a 2-Link Planar Manipulator

The fuzzy controller of a 2-link planar manipulator can be designed based on the updating of its fuzzy model. There are two ways to update the parameters of the fuzzy model: (i) using the equation of motion as a guideline to calculate the parameter using the truncated Taylor series expansion, and (ii) using the observed data and the curve-fitting algorithm to estimate the parameters. Once the parameters have been updated, a controller can be systematically determined based on these newly updated parameters.

The decoupling method works in a similar manner: the parameters of each decoupled system are updated either using the equation of motion or the curve-fitting. However, the system is much more sensitive in that it often requires updating at a much faster frequency (i.e., the interval between two operating actions must be sufficiently short) to avoid deviation.

C. Numerical Simulations

In this section, there are three examples illustrating three methods of controlling the 2-link manipulator.

Example 7.7. This example illustrates the adaptive fuzzy control in which the parameters of the fuzzy model are updated according to the equation of motion of the manipulator. At any time interval $[t_n,t_{n+1})$ where $t_{n+1} = t_n + \Delta t$, the parameters a_{ij} and b_{mn} are updated using the truncated Taylor series expansion and the observed angular displacements $\theta_1(t_n)$ and $\theta_2(t_n)$, and the angular velocities $\dot{\theta}_1(t_n)$ and $\dot{\theta}_2(t_n)$. Figures 7.41 through 7.44 plot the angular displacements and angular velocities of both joints 1 and 2, along with the upper and lower limits defined by the fuzzy model.

Example 7.8. This example illustrates the adaptive fuzzy control in which the parameters of the fuzzy model are updated using the curve-fitting method. At any time interval $[t_n,t_{n+1})$ where $t_{n+1} = t_n + \Delta t$, the parameters a_{ij} and b_{mn} are updated by minimizing the deviation from their values in the previous interval $[t_{n-1},t_n)$ and the observed angular displacements $\theta_1(t_n)$ and $\theta_2(t_n)$, and the angular velocities $\dot{\theta}_1(t_n)$ and $\dot{\theta}_2(t_n)$. Figures 7.45 through 7.48 plot the angular displacements and angular velocities of both joints 1 and 2, along with the upper and lower limits defined by the fuzzy model.

Example 7.9. This example illustrates the adaptive fuzzy control in which the parameters of each decoupled fuzzy model are updated according to the equation of motion of the manipulator. At any time interval $[t_n,t_{n+1})$ where $t_{n+1} = t_n + \Delta t$, the parameters a_{ij} and b_{mn} are updated using the truncated Taylor series expansion and the observed angular displacements $\theta_1(t_n)$ and $\theta_2(t_n)$, and the angular velocities $\dot{\theta}_1(t_n)$ and $\dot{\theta}_2(t_n)$. Notice that the duration Δt is much smaller than the values used in the previous two examples. Figures 7.49 through 7.52 plot the angular displacements and angular velocities of both joints 1 and 2, along with the upper and lower limits defined by the fuzzy model.

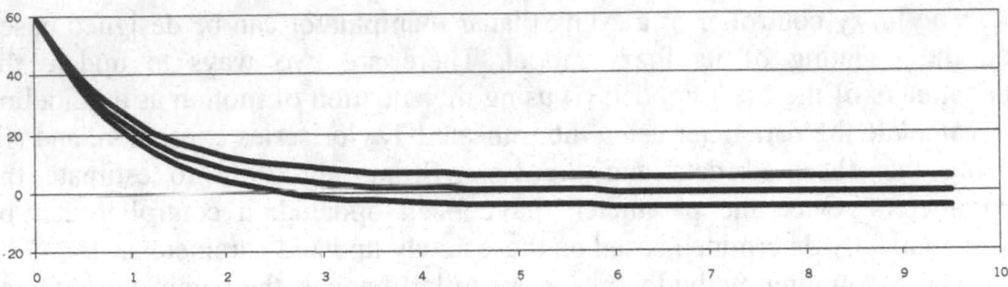

Figure 7.41 Angular position and bounds profiles for joint 1 in Example 7.7.

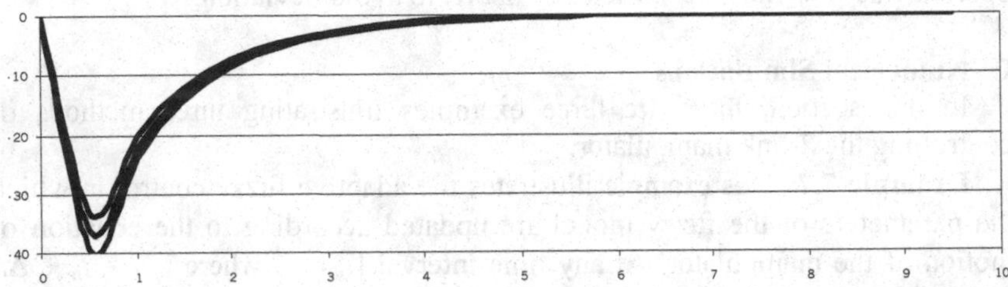

Figure 7.42 Angular velocity and bounds profile for joint 1 in Example 7.7.

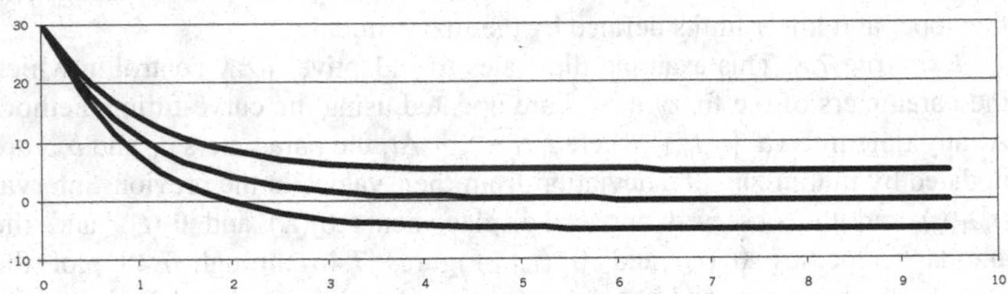

Figure 7.43 Angular position and bounds profiles for joint 2 in Example 7.7.

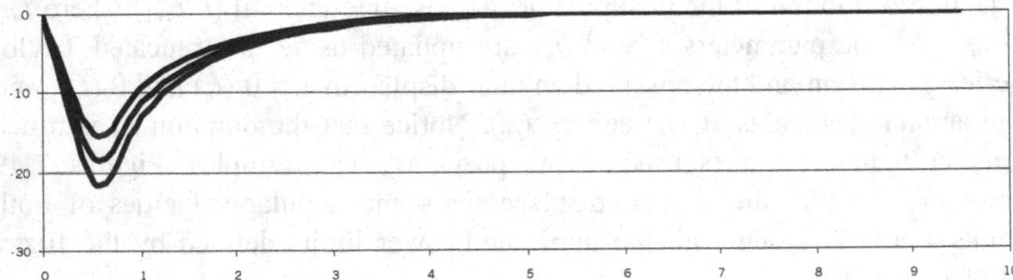

Figure 7.44 Angular velocity and bounds profile for joint 2 in Example 7.7.

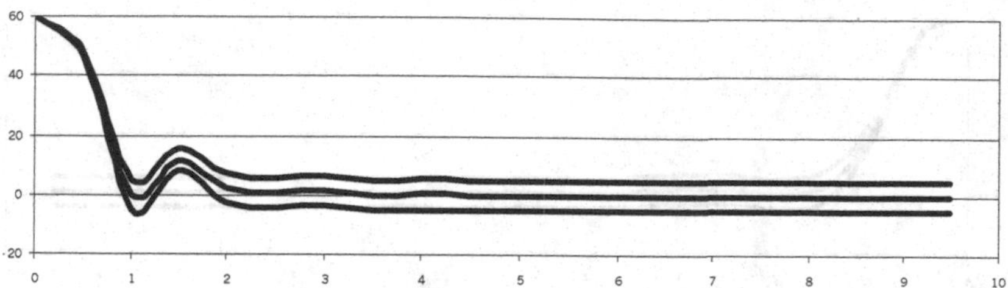

Figure 7.45 Angular position and bounds profiles for joint 1 in Example 7.8.

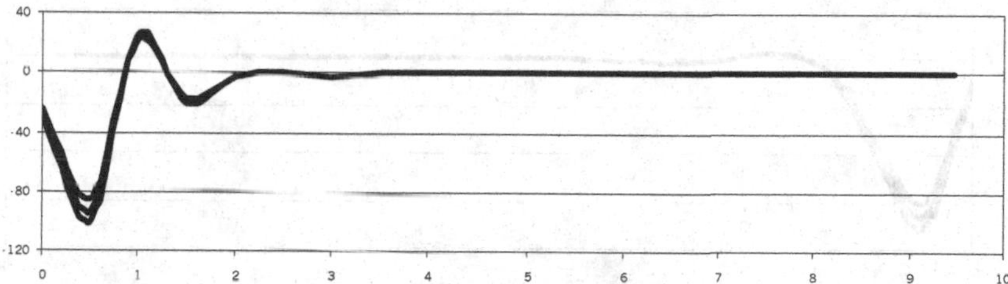

Figure 7.46 Angular velocity and bounds profile for joint 1 in Example 7.8.

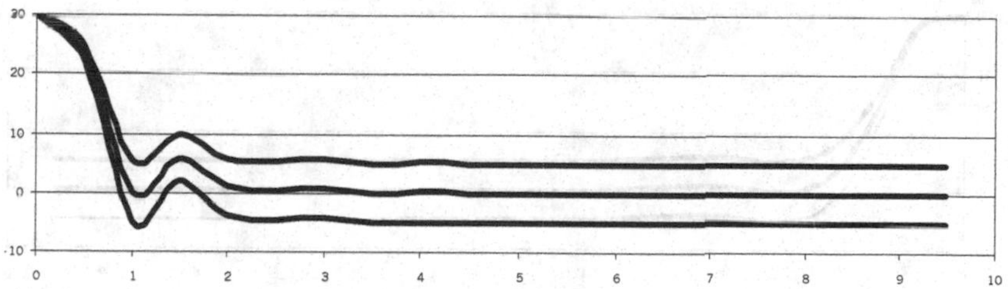

Figure 7.47 Angular position and bounds profiles for joint 2 in Example 7.8.

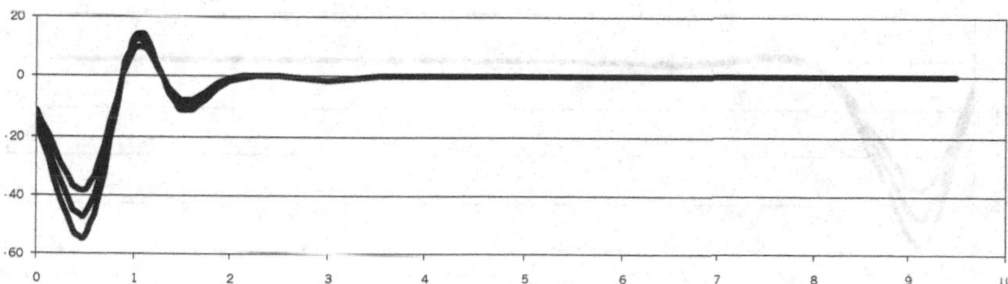

Figure 7.48 Angular velocity and bounds profile for joint 2 in Example 7.8.

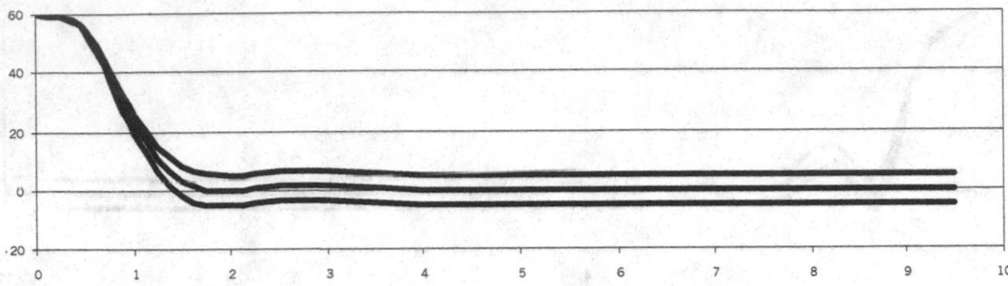

Figure 7.49 Angular position and bounds profiles for joint 1 in Example 7.9.

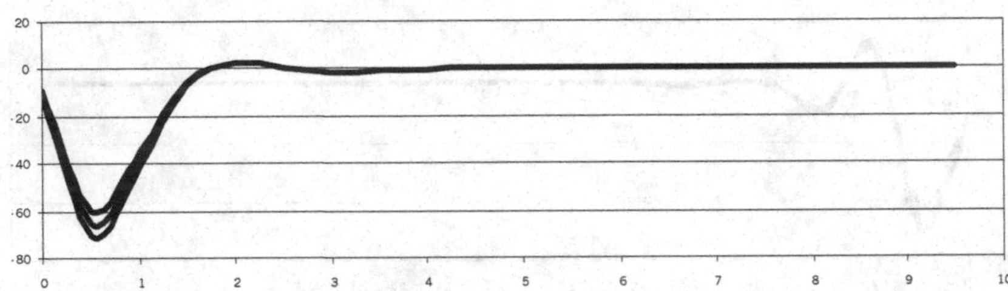

Figure 7.50 Angular velocity and bounds profile for joint 1 in Example 7.9.

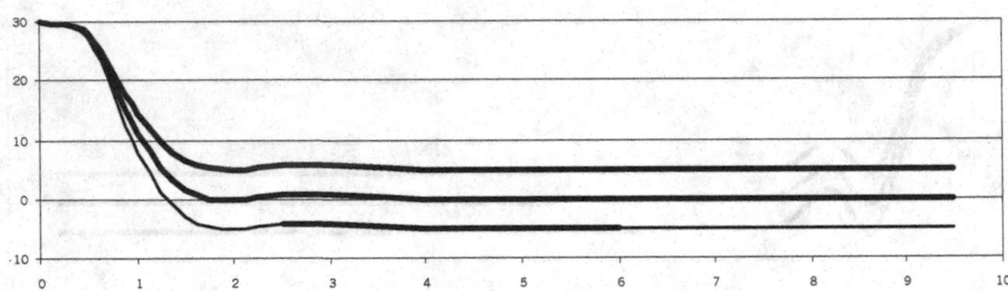

Figure 7.51 Angular position and bounds profiles for joint 2 in Example 7.9.

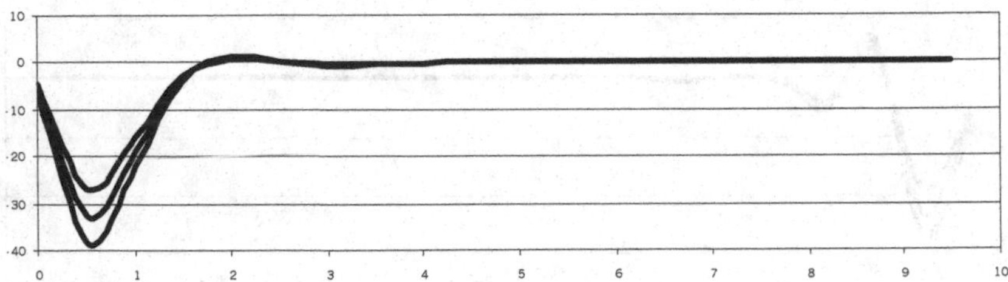

Figure 7.52 Angular velocity and bounds profile for joint 2 in Example 7.9.

PROBLEMS

P7.1 Given the health monitoring diagnostic system described in Section I of this chapter, write a computer program to simulate the following sensor data for the time $t \in [0,300.0]$ for every interval $\Delta t = 1$ second:

(a) $x_1(t) = 98 + 4\sin(\frac{\pi}{80}t)\,e(t)$,

$x_2(t) = 120 + 40\sin(\frac{\pi}{60}t)\,e(t)$,

$x_3(t) = 70 + 60\sin(\frac{\pi}{40}t)\,e(t)$.

(b) Implement the 27 rules mentioned and plot out the membership value for each rule against time t. What diagnostic conclusion can one draw from the plot of these 27 membership values?

(c) Change the data in (a) to include additive random noise of Gaussian distribution. Repeat both steps (a) and (b).

$x_1(t) = 98 + 4\sin(\frac{\pi}{80}t)\,e(t) + n(t)$,

$x_2(t) = 120 + 40\sin(\frac{\pi}{60}t)\,e(t) + n(t)$,

$x_3(t) = 70 + 60\sin(\frac{\pi}{40}t)\,e(t) + n(t)$.

P7.2 Obtain a header format of digital image data stored under the bitmap format (with extension .bmp). Write out the format of this header in the following table:

filetype (2 words)		data (1 word)	data (1 word)
data (1 word)	data (1 word)	data (1 word)	data (1 word)
data (1 word)	data (1 word)	data (1 word)	data (1 word)
data (1 word)	data (1 word)	data (1 word)	data (1 word)
data (1 word)	data (1 word)	data (1 word)	data (1 word)
data (1 word)	data (1 word)	data (1 word)	data (1 word)
data (1 word)	data (1 word)	data (1 word)	data (1 word)
data (1 word)	data (1 word)	data (1 word)	data (1 word)
data (1 word)	data (1 word)	data (1 word)	data (1 word)
data (1 word)	data (1 word)	data (1 word)	data (1 word)
data (1 word)	data (1 word)	data (1 word)	data (1 word)

P7.3 Obtain a digital image stored in the binary bitmap format (whose filename ended with the extension .bmp) described in Problem ***P7.2***. Write a computer program that simulates the off-focused data using the smooth (average) filter. Each off-focused image will be saved to the same binary bitmap format as the original image. Assuming that for each unit away from the correct focal point, the radius of the smooth filter is larger. The smooth filter is given as

$$I_{\text{smooth}}(n,m) = \frac{1}{(2R+1)^2} \sum_{i=n-R}^{n+N} \sum_{j=m-R}^{m+R} I_{\text{original}}(i,j)\,.$$

P7.4 Given a sequence of digital images generated in Problem ***P7.3***, for each digital image, write a computer program to calculate the normalized sharpness index as

$$\sigma = \frac{1}{NM} \sum_{n=1}^{N} \sum_{m=1}^{M} e(n,m)\,.$$

Is there another algorithm to calculate the sharpness index of a digital image? Outline and implement your own algorithm and compare the results to that of the above formula.

P7.5 Implement the fuzzy control for the autofocus camera using the simulation results from Problems ***P7.3*** and ***P7.4***. Set the control step $\Delta\theta$ arbitrarily large and adjust the control steps using various adjustment parameters γ.

REFERENCES

[1] K. R. Ames. *A Relational Approach to the Development of Expert Diagnostic Systems*. Hampton, VA: NASA Langley Research Center (1984).

[2] N. Ayache (editor). *Computer Vision, Virtual Reality, and Robotics in Medicine*. Berlin, Germany: Springer-Verlag (1995).

[3] R. C. Berkan and S. L. Trubatch. *Fuzzy Systems Design Principles: Building Fuzzy IF-THEN Rule Bases*. New York, NY: IEEE Press (1997).

[4] K. S. Fu, R. C. Gonzalez, and C. S. G. Lee. *Robotics: Control, Sensing, Vision, and Intelligence*. New York, NY: McGraw-Hill (1987).

[5] S. E. Gordon. *Expert Systems: Their Impact on Performance and Cognitive Strategies in Diagnostic Inference*. Brooks Air Force Base, TX: Air Force Human Resources Laboratory, Air Force Systems Command Report (1987).

[6] M. Klein and L. B. Methlie. *Expert Systems: a Decision Support Approach: With Applications in Management and Finance*. Reading, MA: Addison-Wesley Publishing Company (1990).

[7] F. A. Lootsma. *Fuzzy Logic for Planning and Decision Making*. Boston, MA: Kluwer Academic Publishers (1997).

[8] R. W. Miller. *Servomechanisms: Devices and Fundamentals*. Reston, VA: Reston Publishing Company (1977)

[9] C. V. Negoita. *Expert Systems and Fuzzy Systems*. Menlo Park, CA: Benjamin/Cummings Publishing Company (1985).

[10] M. J. Patyra and D. M. Mlynek (editors). *Fuzzy Logic: Implementation and Applications*. New York, NY: Wiley (1996).

[11] B. Paul. *Kinematics and Dynamics of Planar Machinery*. Englewood Cliffs, NJ: Prentice-Hall (1979).

[12] I. S. Shaw. *Fuzzy Control of Industrial Systems: Theory and Applications*. Boston, MA: Kluwer Academic Publishers (1998).

[13] M. Sonka, V. Hlavac, and R. Boyle. *Image Processing, Analysis, and Machine Vision*. Pacific Grove, CA: PWS Publishing (1999).

[14] C. Torras (editor). *Computer Vision: Theory and Industrial Application*. Berlin, Germany: Springer-Verlag (1992).

[15] M. Vukobratovic and M. Kircanski. *Kinematics and Trajectory Synthesis of Manipulation Robots*. Berlin, Germany: Springer-Verlag (1986).

Index

A

Adaptive control 239
Applications (of fuzzy control) 271
  Health monitoring 271
  Image sharpness 281
  Robotic manipulator 302
  Servo mechanical system 291
Approximate reasoning 66
α-cut 38

B

Boolean algebra 62

C

Center-of-gravity formula158
Control 103, 109
  Adaptive control 239
  Dual control 257
  Fuzzy control system, 139
    Model-based 170
    Model-free 145
    Set-point tracking 147
    Sub-optimal fuzzy control 258
  Fuzzy controller (FC) 148
    Adaptive fuzzy controller 245
    PID controller 183
  Fuzzy logic controller (FLC) 148
  LMI control toolbox 108
Controllability 129
Characteristic function 3
  Programmable logic control 140

D

Defuzzification 157
  Center-of-gravity formula 158
Dual control 257

E

Extended function 41
Extension principle 40

F

Function
  Characteristic 3, 38
  Extended 41
  Logical 61
  Lyapunov 104
  Membership 6
Fuzzification 148, 150
Fuzzy
  Control system 139
    Adaptive 239
    Model-based 170
    Model-free 145
      Sub-optimal fuzzy control 258
  Controller (FC) 148
    PID controller 183, 192
      IF-THEN rule 76
      Interval partition 116
  Logic 57
  Logic controller (FLC) 148
  Logic rule base 75
  Maximum 50
  Minimum 50
  Number 42
  Relation 69
  Rule base 151
    Self-tuning regulator 252
    Set 5,7
  Subset 7,37
  System modeling 89, 114
    Dynamic102
    Static 90

σ-stability 125

G
Gain scheduling 246
General rule (for fuzzy arithmetic) 43

H
Hansen inverse method 33
Hansen matrix inversion algorithm 34

I
Inclusion monotonic property 22
Inference rule 67
Interval
Arithmetic 9,12
Confidence 9
Equation 16
Evaluation 22
Expression 22
Matrix equation 30
Matrix inversion 30
Matrix operation 25
Rational interval expression 22
Width 20

L
Logic
Multi-valued logic 65
n-valued logic 65
Operations 61
Three-valued logic 65
Two-valued logic 57
Logical formula 62
Logical function 61

M
Measure 3,17
Model reference adaptive fuzzy system 255
Monotonic property 22

N
Nonlinear fuzzy system 133

O
Observability 137

P
Parameter identification 96
Least-squares 98
Programmable logic control (PLC) 140

R
Region of confidence 10
Region of input combination (IC) 195
Representation theorem 40
Resolution principle 40
Rule 59
Rule base 151

S
Set 1
Fuzzy set 7
Fuzzy subset 7
Level set 38
Subset 2
Universal set 1
Set-point tracking 147, 187
Small gain theorem 224
Stability 102, 104, 124, 223
BIBO stability 223
Fuzzy σ-stability 125
Graphical stability analysis 232
PID control system 223
Sub-optimal fuzzy control 258
Sufficient information 244
System identification 243
System parameterization 242

T
Truth table 59

Printed in the United States
by Baker & Taylor Publisher Services